Work and AI 2030

Inka Knappertsbusch · Kai Gondlach
Editors

Work and AI 2030

Challenges and Strategies for Tomorrow's Work

Springer

Editors
Inka Knappertsbusch
CMS Germany
Köln, Germany

Kai Gondlach
Zukunftsforscher Kai Gondlach
Leipzig, Germany

ISBN 978-3-658-40231-0 ISBN 978-3-658-40232-7 (eBook)
https://doi.org/10.1007/978-3-658-40232-7

Responsible Editor: Marija Kojic
This Springer imprint is published by the registered company Springer Fachmedien Wiesbaden GmbH, part of Springer Nature.
The registered company address is: Abraham-Lincoln-Str. 46, 65189 Wiesbaden, Germany

Preface

When we first delved into the discussion surrounding the intersection of AI and the world of work in autumn of 2020, just before the dawn of a new decade, it quickly became apparent that the available sources were disheartening. In addition to marketing slogans from AI pioneers, the material we encountered consisted primarily of position papers and market studies from consulting firms, alongside extensive regulatory papers from state institutions. Despite this, the overarching societal framework is often overlooked in debates concerning the future and ramifications of AI. With this anthology, we aim to bridge this gap by presenting diverse perspectives on the major economic sectors and their employment of AI in 2030, in collaboration with esteemed experts from academia and industry. Our emphasis lies on the impact on the world of work, specifically, the collaboration between humans and AI.

The contributions in the first chapter, "Societal and ethical aspects of AI in the world of work", shed light on various aspects of the AI debate and provide insights into upcoming challenges and opportunities in AI ethics, neurobiological approaches to the AI discourse, and tangible approaches to collaboration in the seemingly disorderly field of AI development. The second chapter, "Legal aspects of AI in the world of work", provides information on labor law aspects while also raising questions for the years ahead, such as the contentious topic of the e-person. The third chapter, "AI in the industrial world of work", provides in-depth insights into industrial production processes in the 2020s, from craftsmanship to the energy, metal, and electrical industries. Chapter four is dedicated to "AI in the medical and pharmaceutical world of work" and delves into the healthcare system, the world of health insurance, and the innovations of Big IT, especially in the USA. Chapter five is titled "AI in the economic world of work", focusing on change and human resource management, the financial world, and hybrid employment forms resulting from artificial intelligence. The sixth chapter deals with "AI in the mobile world of work and logistics", covering the automotive, railway, and aviation industries. Chapter seven is entirely devoted to "AI in education and training", ranging from EdTech to further education to AI-optimized workplace design and assistance systems for people with disabilities.

One common finding that runs through all the contributions is that artificial intelligence is not a new frontier. Applications already exist and are continuously being improved, with innovators from all sectors driving development forward. Therefore, this volume aims to provide an overview for decision-makers in business, politics, and administration who may still be hesitant or searching for justification for higher development budgets. What is certain is that AI is already one of the most transformative trends of the decade, and organizations that do not act now may miss out on the next one.

We would like to express our sincere thanks to CMS Germany for their support of the project from the outset. Of course, we would like to thank the authors in particular for their fantastic cooperation and successful contributions. Last but not least, we would like to thank Alexander Paltzer, Niklas Demmer, and Angela Zeyn for their editorial involvement.

Update in March 2023: After 15 months since the publication of the German version and over 300,000 accesses on SpringerLink, we are thrilled to now offer the English translation of valuable insights from approximately 40 top-notch contributions to an international audience. Since the publication of our first edition in German, there have been technological developments, including the introduction of the powerful ChatGPT technology from OpenAI in November 2022, which is addressed in some of the contributions.

Now more than ever it is crucial for companies to get ready for the use of AI which will tremendously change how we work until 2030. The contributions in this edition point the way in the right direction, providing strategies and specific steps for decision-makers in companies and public institutions to successfully navigate rapid change based on the implementation of AI at the workplace. This is reflected in feedback from dozens of readers who have reached out to us after reading the German version. Once again, a thousand thanks to our authors, without whom this project would not have been possible.

We wish you an insightful and helpful read!

Sincerely Yours

Cologne/Leipzig
March 2023

Inka Knappertsbusch
Kai Gondlach

Contents

Contributors

Prof. Dr. Doris Aschenbrenner Fakultät Maschinenbau und Werkstofftechnik, Hochschule Aalen, Aalen, Germany

Prof. Dr. Norbert Bach Fachgebiet Unternehmensführung/Organisation, Technische Universität Ilmenau, Ilmenau, Germany

Prof. Dr. Lutz Bellmann Friedrich-Alexander-Universität Erlangen-Nürnberg, Erlangen, Germany

Prof. Dr. Benedikt Berger Department of Information Systems, University of Münster, Münster, Germany

Prof. Dr. Thomas Bergs Werkzeugmaschinenlabor WZL, Fraunhofer-Institut für Produktionstechnologie IPT, RWTH Aachen University, Aachen, Germany

Susan Beudt Educational Technology Lab, Deutsches Forschungszentrum für Künstliche Intelligenz(DFKI), Berlin, Germany

Dr. Berit Blanc Educational Technology Lab, Deutsches Forschungszentrum für Künstliche Intelligenz(DFKI), Berlin, Germany

Dr. Lars Michael Bollweg Westnetz GmbH, Dortmund, Germany

Dr. Aljoscha Burchardt Language Technology Lab, Deutsches Forschungszentrum für Künstliche Intelligenz (DFKI), Berlin, Germany

Dr. Michael Burger Fraunhofer-Institut für Techno- und Wirtschaftsmathematik ITWM, TU Kaiserslautern, Kaiserslautern, Germany

Klaus Burmeister Founder and managing director foresight lab, Berlin, Germany

Dr. Stephan Dreyer Leipniz-Institute for Media Research, Hans-Bredow-Institut, Hamburg, Germany

Dr. Saskia Dörr WiseWay Berät Unternehmen, Bonn, Germany

Rolf Feichtenbeiner Educational Technology Lab, Deutsches Forschungszentrum für Künstliche Intelligenz(DFKI), Berlin, Germany

Frank Fischer Product Marketing, Snyk, Mellieha, Malta

Prof. Dr. Sebastian Fischer Berlin School of Economics and Law, Berlin, Germany

Oliver Franck Lehrstuhl für Mobilität, Handel und Logistik, Zeppelin Universität Friedrichshafen, Friedrichshafen, Germany

Vincent Geilenberg Lehrstuhl für Mobilität, Handel und Logistik, Zeppelin Universität Friedrichshafen, Friedrichshafen, Germany

Kai Arne Gondlach Zukunftsforscher, Leipzig, Germany

Dr. Thorsten Gressling Disruptive Technology, Bayer AG, Wuppertal, Germany

Dr. Philipp Hartmann AI Strategy, UnternehmerTUM GmbH, München, Germany

Dr. Volker Hartmann VP of Legal & Governmental Affairs, Vay Technology GmbH, Berlin, Germany

Marcel Heinen CMS Germany, Köln, Germany

Dr. Eckhard Hempel Invandus E&R GbR, Höchstadt, Germany

Prof. Dr. Thomas Hess Institute for Digital Management and New Media, Ludwig-Maximilians-Universität München, Munich, Germany

Dr. Andreas Hinz RKW Rationalisierungs- und Innovationszentrum der Deutschen Wirtschaft e. V. Kompetenzzentrum, Eschborn, Germany

Prof. Dr. Joachim Hornegger Friedrich-Alexander Universität Erlangen-Nürnberg (FAU), Erlangen, Germany

Prof. Dr. Marco Huber Center for Cyber Cognitive Intelligence (CCI), Fraunhofer Institute for Manufacturing Engineering and Automation IPA, Stuttgart, Germany

Dr. Thomas Hummel Strategy and Innovation, Siemens Healthineers, Erlangen, Germany

Prof. Dr. Dr. habil. Clemens Jäger FOM Hochschule für Oekonomie & Management in Essen, Essen, Germany

Patricia Jares CMS Germany, Köln, Germany

Christian Jauch Fraunhofer Institute for Manufacturing Engineering and Automation IPA, Stuttgart, Germany

Prof. Dr. Sabina Jeschke Deloitte Deutschland & RWTH Aachen, Aachen, Germany

Dr. Tim Jeske ifaa—Institut für angewandte Arbeitswissenschaft, Düsseldorf, Germany

Dr. Inka Knappertsbusch CMS Germany, Köln, Germany

Dr. Stefan Knupfer AOK PLUS, Dresden, Germany

Dr. Heinrich Kögel Economic AI GmbH, Regensburg, Germany

Luise Kronenberger CMS Germany, Köln, Germany

Can Kömek CMS Germany, Hamburg, Germany

Dr. Ingo Kucz White Octopus GmbH, Berlin, Germany

Sven Lindig Lindig Fördertechnik GmbH, Krauthausen, Germany

Johannes Mayer Werkzeugmaschinenlabor WZL, RWTH Aachen University, Aachen, Germany

Dr. Annette Miller Hessisches Zentrum für Künstliche Intelligenz—hessian.AI, Darmstadt, Germany

Claudia Pohlink Berlin, Germany

Prof. Dr. Niels Pinkwart Educational Technology Lab, Deutsches Forschungszentrum für Künstliche Intelligenz(DFKI), Berlin, Germany

Dr. Maike Pricelius G-IBS Gesellschaft für Innovation, Beratung und Service mbH, Berlin, Germany

Dr. Michaela Regneri AI & Cognitive Computing, OTTO GmbH & Co. KG, Hamburg, Germany

Dr. André Renz Helmut Schmidt University, University of the Federal Armed Forces Hamburg, Hamburg, Germany

Monika Rimmele DiGa Factory, Berlin, Germany

Stefan Sander IT-Recht, SDS Rechtsanwälte Sander Schöning PartG mbB, Duisburg, Germany

Dr. Alexander Rühr OMMAX — Building Digital Leaders, London, United Kingdom

Dr. Peter Schlicht Artificial Intelligence, CARIAD SE, Wolfsburg, Germany

Bruno Schmalen SCHMALEN—Kommunikation und Training, Langenhagen, Germany

Johannes Schmees Leipniz-Institute for Media Research, Hans-Bredow-Institut, Hamburg, Germany

Dr. Daniel A. Schmidt Frankfurt am Main, Germany

Konrad Scheuermann Digital Products and Projects, DB Regio AG, Berlin, Germany

Prof. Dr. Wolfgang H. Schulz Lehrstuhl für Mobilität, Handel und Logistik, Zeppelin Universität Friedrichshafen, Friedrichshafen, Germany

Prof. Dr. Anita Schöbel Fraunhofer-Institut für Techno- und Wirtschaftsmathematik ITWM, TU Kaiserslautern, Kaiserslautern, Germany

Stanley Smolka Executive Education, Zeppelin Universität Friedrichshafen, Friedrichshafen, Germany

Magdalena Soetebeer Schaltzeit GmbH, Berlin, Germany

Prof. Dr. Martin Spindler Universität Hamburg und Economic AI, Hamburg, Germany

Dr. Henrike Stephani Fraunhofer-Institut für Techno- und Wirtschaftsmathematik ITWM, TU Kaiserslautern, Kaiserslautern, Germany

Prof. Dr. Sascha Stowasser ifaa—Institut für angewandte Arbeitswissenschaft, Düsseldorf, Germany

Dr. Verena Svensson Stadtwerke Düsseldorf AG, Düsseldorf, Germany

Sebastian Terstegen ifaa—Institut für angewandte Arbeitswissenschaft e. V., Düsseldorf, Germany

Prof. Dr. Stefan Tewes FOM Hochschule für Oekonomie & Management in Essen, Essen, Germany

Dr. Daniel Trauth senseering GmbH, Köln, Germany

Didem Uzun Lufthansa Industry Solutions, Business Development AI & Data Analytics, Norderstedt, Germany

Tobias Vogt Köln, Germany

Helmut Wasserbacher Novartis International AG, Basel, Switzerland

Dr. Susan Wegner Deloitte Legal, Berlin, Germany

Ludwig Weh Fraunhofer IMW Center for International Management and Knowledge Economics, Leipzig, Germany

Dr. Stefan Weigert AOK PLUS, Dresden, Germany

Prof. Dr. Werner Widuckel Friedrich-Alexander-Universität Erlangen-Nürnberg, Erlangen, Germany

Dr. Gerlind Wisskirchen CMS Germany, Köln, Germany

Michael Zeck HD PLUS GmbH, München, Germany

Social and ethical aspects of AI in the world of work

The Ghost of German Angst: Are We Too Skeptical for AI Development?

The Art Figure of the Fearful Enemy of Technology and Courage for Critical Optimism

Kai Arne Gondlach and Michaela Regneri

> *"Worrying about evil AI killer robots today is a little bit like worrying about overpopulation on the planet Mars."—Andrew Ng, Vice President & Chief Scientist Baidu*

1 Introduction

The idea of humans as crafty creators of intelligent life has existed since antiquity: While Pygmalion in Ovid's Metamorphoses still needed divine help to bring his art woman to life, automata as pure human work have fascinated various authors and artists since the Romantic era (cf. Russel & Norvig, 2010). Famous examples are *L'Homme Machine* by Julien Offray de La Mettrie (1748), the demonic idea of Pierre-Simon Laplace (1749–1827) or E.T.A. Hoffmann's (1776–1822) living dolls. The foundations of modern AI technologies, especially neural networks, date from the mid-twentieth century. The transition of thinking machines from fiction to modern reality is symbolically marked by a conference

K. A. Gondlach (✉)
Zukunftsforscher, Leipzig, Germany
e-mail: ich@kaigondlach.de
URL: http://www.kaigondlach.de/

M. Regneri
AI & Cognitive Computing, OTTO GmbH & Co. KG, Hamburg, Germany
e-mail: michaela.regneri@otto.de

I. Knappertsbusch and K. Gondlach (eds.), *Work and AI 2030*,
https://doi.org/10.1007/978-3-658-40232-7_1

from 1956: Computer scientists, logicians and mathematicians laid the groundwork for research on what has since been called "artificial intelligence" (AI) at the *Dartmouth Summer Research Project on Artificial Intelligence* (McCarthy et al., 1955).

In literature, films and series from the science fiction genre, robots or other AI forms often appear as villains who threaten or enslave their creators or the whole of humanity. This also reflects the tradition of applied AI, which had its beginnings in weapons development. Harmless, clumsy or even human-friendly AIs are common in the Asian region, but are only slowly making their way into the entertainment media here. In film and television, there are also significantly more people with fear of AI than those who perceive it as a promising future technology and enrichment.

The topic of this volume reflects diverse narratives about encounters between humans and machines: Many discussions on AI and the world of work are shaped by the prejudice that Germany mainly faces the AI future with fear.[1] Not least the centuries-old and globally used term of a "German Angst"[2] is also applied here without any empiricism. Many then look enviously in the direction of the USA or China, whose attitudes seem more liberal, generous and thus also more conducive to innovation.[3]

This contribution analyses the basic assumption of the "German Angst" and casts a critical futures research perspective on the origin of the beliefs about the sceptical German employees and employers.

2 Status Quo

Some studies foster great pessimism regarding AI and the future of work. The most likely best-known one was conducted at Oxford University (Frey & Osborne, 2013) and deals with the threat of jobs by automation. The authors estimated that almost half of the activities performed by humans in the US (47%) were at high risk (>70%) of being endangered by 2033. For Germany, they calculated this risk for 42% of all employments. According to them, jobs with a high proportion of repetitive tasks, but increasingly also cognitively demanding activities that involve the evaluation of large amounts of data, would be particularly

[1] cf. Halva (2019), Lesch (2019), Peteranderl (2019).

[2] cf. Kierkegaard (1992); Dehne (2017), p. 21 f.; Psychomeda (n. d.).

[3] cf. Paus et al., (2018, p. 21 ff.).

affected. This prognosis was subsequently endorsed by some other authors (e.g. Berg et al., 2018; Berg & Dehmel, 2020).

Frey and Osborne do not make predictions about jobs, but about activities, i.e. subtasks of occupations. Role models in everyday work, however, have always been changing, not only since digitisation and AI. Chimney sweeps, for example, no longer have top hats and long ball brushes, but everyday clothing and compact digital measuring devices. While it was therefore unsound from the outset to derive the dystopia of mass unemployment from the Oxford study, it is hardly surprising that later works largely refute these scenarios. One example was commissioned by the Federal Ministry of Labour and Social Affairs (Bonin et al., 2017). As a result, the authors conclude that only 12% of jobs are highly endangered by automation, for the US 9%.

The World Economic Forum estimates in its "Future of Jobs Report 2020" that technology will replace around 85 million jobs worldwide by 2025, but at the same time create 97 million new jobs. Others also consider technological unemployment to be overestimated and see at least as many new employment opportunities as endangered positions (Arntz et al., 2017; Kapeliushnikov, 2019).

In society, the bogeyman of mass unemployment due to AI never really caught on. According to a representative study by Bitkom Research 2020, 44% of Germans fear the use of AI in everyday work—but more than half do not and two thirds see AI as an opportunity. Specifically, 73% expect employees to be more controlled and 65% believe in the reduction of jobs. But even if a good half of the participants see such noticeable changes due to AI in the next five years, this does not mean that they view the technology critically: In a variety of areas of life, a majority supports the use of AI, including human-related services such as care, medicine and human resources.

What is striking is that possible fears such as more control or data abuse do not reflect reservations about the technology, but mistrust of the people who have the power to use the technology maliciously. A similar result was obtained by a recent study by IU International University (IUBH, 2021) in Erfurt. 75% of the respondents in companies[4] had no reservations about the use of artificial intelligence, but hoped for work relief and more efficient, less monotonous work. One of the study authors, Ulrich Kerzel, emphasises: "The fear of AI is dangerously overestimated in Germany. If the managers assume that the employees are

[4] About a third of the respondents held management positions, 10.6% were employed in the human resources department, the remaining 60% were employees without leadership competence.

sceptical about AI, this makes the decision for artificial intelligence more difficult. Thus, the fear of fear becomes a brake block—and leads to companies not using the enormous potential of AI."

Beyond the labour market, people are often assumed to have a generic fear of AI, which must somehow be remedied by "demystifying" and "changing the mindset" before the new technologies can be used. The studies expose this fear as a purely mythological figure, analogous to the media representation of AI. One often encounters masses of frightened, technophobic adults in movies, but not in social reality.

This was confirmed in a study by the IT service provider Adesso SE (Adesso, 2020). Three quarters of the respondents had explicitly no fear of AI and look worry-free at the technological future.

The Corona pandemic has certainly contributed to further support the enthusiasm for technology. A possible consequence is that more contact with AI, e.g. for forecasts or in the home office, also brings more confidence for its use. What is often lacking so far are surveys of those who have already successfully integrated automation into their everyday work and transparency about which wishes or fears come true. A few smaller studies are the exception and paint a consistently positive picture for so-called "hybrid" work hand in hand with an AI system (James et al., 2013; Regneri & Kertelge, 2020).

In addition to the present experiences, the future research also draws a differentiated picture of the AI popularity. Lea Daniel (Daniel, 2021) examined different views on the development of high technologies, especially with reference to AI: "The guiding images show that the ideas of [AI] are shaped by a logic of progress, a striving for technical development on the one hand and fears of unexpected dangers on the other. But they also contain the visionary character of a research-strong Europe that can withstand the global power struggle in AI development" (ibid. p. 52).

Like all everyday topics, AI is thus characterised by opportunities, threats and contradictions. In public discourse, however, it often seems like a simplified good-evil dichotomy without grey areas, which does not exist in reality.

3 Challenges and Solutions

Structural change always happens as evolution. The steam engine took about 300 years from invention to official use and the subsequent labour market change spanned half a century. In the digital age, however, we usually observe faster developments: After the publication of the World Wide Web, it took about eleven years until the dotcom bubble burst (Borchers, 2010). But even if this event represented one of the biggest failures of digitisation, web technologies are still the basis of the digital labour market. Hardly anyone has ever attributed fear of the internet to the general population, just because the current economy was not yet up to the new technology.

AI currently still falls short of expectations in terms of gross domestic product development (Servoz, 2021). Speculative reasons for this range from insufficient modernisation of organisational structures to simply regulatory framework conditions. While educational opportunities, a broad technology sector and the innovation power of the economy in Germany are good starting conditions, the digital modernisation mills in many economically driving industries turn slowly, because digitization is a mammoth task among many others.

In the further contributions to this volume, countless concrete, cross-sectoral application examples are given, which show: There are positive, innovative examples.

4 Outlook on German Innovation Instead of German Angst 2030

Mass unemployment due to too rapid digitalisation is currently not to be feared. But even with a higher pace of digitalisation, activities would change, but new ones would be added. Those who want to be drivers instead of driven by this development must gather the biggest motivators for upheavals: More profit and new business models for companies and better and freer work opportunities for employees. This requires progress not only on a technical, but above all on an organisational level. Such also need time—but maybe we have more of it than we think.

By 2030, artificial intelligence will fit into the series of digital upheavals, but in itself will not be a devastating driver of layoffs or upheavals in the local labour market. As other contributions to this volume show, the ongoing digital change driven by AI will only gradually unfold its effects on the labour market. And we can use this time wisely. Wisely means here, not only to build mindset changes and training measures against the assumed fear in the population, but also to create an environment for the employees in which they can achieve new peak performances together with AI.

Some jobs or companies may still lose importance or disappear altogether in the coming years due to widespread use of AI. And some workers may even hope that they can transfer unwanted tasks to an AI. You just have to let them.

If our economy wants to take a leading role with AI in the next almost ten years, it has to do a lot to remove the technical hurdles. And it has to stop using the fears of its employees as a reason for hindrance. Instead, it should grant them the role of creator: To design a workplace where tedious routine work is automated and the rest has become a human excellence network can be a guarantee of the future for the employees and the highly technologised companies. This requires trust, which many companies may have learned during the pandemic: The employees know their job best and also know how to improve it most effectively with automatic means. Their new roles as AI caregivers with a greatly expanded range of tasks is a win-win from better working conditions and digitalisation support.

5 Summary and Practical Recommendations

By now, AI has arrived in everyday life and some old narratives are losing their meaning. The perhaps most present ghost from the fictional past is the myth of the "German Angst", which also does not stop at the digital progress. People do not fear AI or new work, but the old structures and work environments that do not allow such a change and may even misuse AI as a powerful tool. Organisations, on the other hand, try to retrain the employees instead of adapting their structures to the new era.

That it can also be done differently is shown by the following contributions in this volume. As a guiding motif, we would like to emphasise the role of AI as a revolutionary of work at this point: If we think AI and digitalisation to the end, we can and must allow as many people as possible to create new roles and professional careers for themselves in order to preserve progress. The technology is so complex that everyone can best assess and design it for their own purposes. There are now low-threshold possibilities for AI applications. But it needs organisations that reward their employees for emancipating themselves from predefined tasks and instead going new ways. Responsibility for one's own role and the AI that fills parts of the old one is the way to go. With this decentralised approach to the new technology, we can become the location of innovation-oriented AI companies and celebrate "German Innovation" instead of "German Angst" again.

References

Adesso, S. E. (2020). Report 2021: KI—Eine Bestandsaufnahme. Zum Status von KI in DACH. https://ki.adesso.de/ki-studie/. Accessed: 27. July 2021.

Arntz, M., Gregory, T., & Zierahn, U. (2017). Revisiting the risk of automation. *Economics Letters, 159,* 157–160.

Berg, A., & Dehmel, S. (2020). Künstliche Intelligenz. Bitkom Research im Auftrag von Bitkom. https://www.bitkom.org/Presse/Presseinformation/Die-Menschen-wollen-KI-und-haben-auch-Angst-vor-ihr. Accessed: 27. July 2021.

Berg, A., Buffie, E. F., & Zanna, L.-F. (2018). *Should we fear the robot revolution? (The Correct Answer is Yes)* (IMF Working Papers, WP/18/116). International Monetary Fund. https://www.imf.org/~/media/Files/Publications/WP/2018/wp18116.ashx. Accessed: 27. July 2021.

Bonin, H., Gregory, T., & Zierahn, U. (2017). Übertragung der Studie von Frey/Osborne (2013) auf Deutschland. Zentrum für Europäische Wirtschaftsforschung (ZEW), Mannheim. https://www.bmas.de/DE/Service/Publikationen/Forschungsberichte/forschungsbericht-fb-455.html. Accessed: 27. July 2021.

Borchers, D. (2010). Zehn Jahre Dotcom-Bust: Als die Blase platzte. https://www.heise.de/newsticker/meldung/Zehn-Jahre-Dotcom-Bust-Als-die-Blase-platzte-951796.html. Accessed: 27. July 2021.

Daniel, L. (2021). Leitbilder der Künstlichen Intelligenz. Eine Suche nach zukunftsbezogenen Orientierungsmustern hinter dem Radical Breakthrough Inquirer. In G. de Haan (Hrsg.), *Institut Futur Schriftenreihe,* 02/21. https://refubium.fu-berlin.de/handle/fub188/29742. Accessed: 25. July 2021.

Dehne, M. (2017). *Soziologie der Angst. Konzeptuelle Grundlagen, soziale Bedingungen und empirische Analysen*. Springer VS.

Frey, C. B., & Osborne, M. A. (2013). The future of employment: How susceptible are jobs to computerization? *Technological Forecast and Social Change, 114,* 254–280.

Halva, B. (2019). Künstliche Intelligenz. In FR.de. https://www.fr.de/wissen/kuenstliche-intelligenz-viele-haben-horrorvorstellung-boesen-ki-13287645.html. Accessed: 30. July 2021.

IUBH. (2021). Artificial Intelligence. Die Zukunft künstlicher Intelligenz in deutschen Unternehmen. https://www.iu.de/news/studie-zu-kuenstlicher-intelligenz/. Accessed: 30. July 2021.

James, K. L., Barlow, D., Bithell, A., Hiom, S., Lord, S., Oakley, P., Pollard, M., Roberts, D., Way, C., & Whittlesea, C. (2013). The impact of automation on pharmacy staff experience of workplace stressors. *International Journal of Pharmacy Practice, 21*(2), 105–116.

Kapeliushnikov, R. (2019). The phantom of technological unemployment. *Russian Journal of Economics, 5*(1), 88–116.

Kierkegaard, S. A. (1992). *Der Begriff Angst.* Nachw. u. hrsg. v. Uta Eichler. Philipp Reclam.

Lesch, H. (2019). K.o. durch KI? Keine Angst vor schlauen Maschinen! In: Leschs Kosmos, zdf.de. https://www.zdf.de/wissen/leschs-kosmos/ko-durch-ki-keine-angst-vor-schlauen-maschinen-100.html. Accessed: 30. July 2021.

McCarthy J., Minsky M. L., Rochester N., & Shannon C. E. (1955). A proposal for the Dartmouth summer research project on artificial intelligence. http://www-formal.stanford.edu/jmc/history/dartmouth/dartmouth.html. Accessed: 25. July 2021.

Paus, I., Deißner, A., Pohl, F., Kanellopoulos, C. C., Grimm, R., Stavenhagen, L., Freudenberg, J., & Wolfs, L. (2018). The tech divide. Contrasting attitudes towards digitisation in Europe, Asia and the USA. Vodafone Institute for Society and Communications (Hrsg.). https://www.vodafone-institut.de/wp-content/uploads/2018/10/The_Tech_Divide_People_and_Society_.pdf. Accessed: 30. July 2021.

Peteranderl, S. (2019). „Künstliche Intelligenz muss entzaubert werden". In Spiegel Netzwelt. https://www.spiegel.de/netzwelt/gadgets/kuenstliche-intelligenz-und-kinder-mit-forscherin-stefania-druga-im-interview-a-1251721.html. Accessed: 30. July 2021.

Psychomeda. (o. J.). German Angst—Lexikon der Psychologie. https://www.psychomeda.de/lexikon/german-angst.html. Accessed: 31. July 2021.

Regneri, M., & Kertelge, S. (2020). Keine Kündigungen, kein Kreativ-Idyll: KI in der Praxis. In J. Nachtwei & A. Sureth (Hrsg.), Sonderband Zukunft der Arbeit (HR Consulting Review, Bd. 12, S. 487–490). VQP. https://www.sonderbandzukunftderarbeit.de.

Russel, J., & Norvig, P. (2010). *Artificial intelligence: A modern approach* (3. Aufl.). Prentice Hall.

Servoz, M. (2021). The future of work? Work of the future! European Commission Study. https://digital-strategy.ec.europa.eu/en/library/future-work-work-future. Accessed: 27. July 2021.

Practical Guide AI = All Together and Interdisciplinary

Responsible Innovation for the Integration of Artificial Intelligence Applications into the World of Work

Aljoscha Burchardt and Doris Aschenbrenner

"One man's ceiling is another man's floor"—Paul Simon

This article shows common problems in the development of AI systems (artificial intelligence), such as fundamental misunderstandings between the user side and the AI development side. These arise from deficits in interdisciplinary collaboration and communication and can lead to undesirable consequences for employees or society as a whole in the long term. As a strategy to avoid these and other misunderstandings and to implement AI in the sense of a responsible innovation process, a process model is proposed, which can serve as a practical guide for using AI in the world of work.

A. Burchardt (✉)
Language Technology Lab, Deutsches Forschungszentrum für Künstliche Intelligenz (DFKI), Berlin, Germany
e-mail: aljoscha.burchardt@dfki.de
URL: https://www.dfki.de/~aburch

D. Aschenbrenner
Fakultät Maschinenbau und Werkstofftechnik, Hochschule Aalen, Aalen, Germany
e-mail: doris.aschenbrenner@hs-aalen.de

I. Knappertsbusch and K. Gondlach (eds.), *Work and AI 2030*,
https://doi.org/10.1007/978-3-658-40232-7_2

1 Introduction

At information events on the topic of AI, one sometimes meets employees of companies who are supposed to "look around for an AI". Often enough, they are from IT or even from the purchasing department. When asked which specific processes the department wants to use AI systems in the future, one gets the honest answer that one does not know that exactly either. Certainly also stimulated by the media coverage (in this context, the apocalyptic warnings of self-proclaimed technology philosophers often have a reinforcing effect), this approach probably implies the idea that there should be a "suitable AI" for every company, just as there is also the suitable accounting software or the suitable office package. As already explained elsewhere (PricewaterhouseCoopers, 2018a), this observation confirms the overall relatively low AI literacy (ability to deal competently with AI) in the companies.

That digitisation—and as a consistent continuation of the use of (one of many) AI technologies—has less to do with a classic purchasing process and instead a lot to do with collaborative and interdisciplinary development or even research, is then often already a central insight and the prerequisite for a successful pilot project. Innovations must always be understood in the interplay of technology, human, and organisation within their framework conditions and political regulation (Hirsch-Kreinsen et al., 2018). However, the goal should not be the new technology alone—in our specific case the application of weak artificial intelligence—but the development of the socio-technical system (Sydow, 1985) under the influence of the new technology.

2 Status Quo: What Expertise is needed for the Introduction of an AI System?

AI systems are initially "empty shells" in the delivery state. This applies both to the currently very successful paradigm of machine learning using neural networks and to many other AI technologies (such as rule-based systems) that are frequently used in industry when there is not enough data available or the tasks can be described symbolically well. For the practitioners in the companies, it is important to understand that the systems do nothing before models have been trained with data or rules have been created. And exactly this knowledge building in the systems is a highly interdisciplinary process that requires very different expertise.

For example, suppose suitable data is available and the desired functionality can be expressed by data. In that case, the data must be selected, prepared and legal and technical questions must be clarified. In addition to the AI developers, this involves at least the users, the legal department, the IT (or corresponding service providers), and the management.

The actually interesting task is the specification of the desired functionality. This process takes place almost naturally collaboratively between AI experts, users and management. Often, it is not even possible to determine in advance where the goal of using AI should lie: Should the efficiency and effectiveness of existing processes be increased? Or should new business models be opened up by AI? A genereric "AI IT kit" cannot meet this range of applications, this is about a strategic process of the company.

This process requires much time and close exchange, so that the domain experts learn the strengths and limitations of the technology and the AI experts understand the domain expert's work. A plan must be created jointly and collaboratively, which functionality the AI system should take over, how to test the success or failure of this, what human-machine interfaces are needed, how the new processes fit with the old ones, and not least, sufficient space should also be given to the question of possible undesirable effects that could result among other things from biases in the data.

Intuitively, the task is first to identify the steps in the current processes that are called "pain points" even in German and that one would most like to hand over to the AI system. In robotics, this is referred to as the four "D's": Dull, Dirty, Dangerous, and Dear (McAfee & Brynjolfsson, 2017). Interestingly, people first think of the other end of the process chain, i.e., the most demanding and responsible tasks for which they themselves were trained for a long time. However, this is usually not the place where AI and digitisation can best support.

In the following, this will be briefly illustrated in an example. By adhering to a process model, as described in Chap. 5, the really sensible place of use can then be determined.

A doctor and a nurse are of the opinion that their work with the patient requires so much empathy and individual case considering that they would not even know theoretically how to design a corresponding AI system. After an analysis of the activities, it is confirmed that 40 to 60% of the work of the physicians and nurses is spent on reporting. An AI system that intelligently records the treatments, a data glasse,s or a tablet robot can provide great relief if the time gained is at least partially used for more "touch time" with the patient. The latter is a non-technical question, but it strongly influences the acceptance of the technical solution in the overall concept.

3 Challenges and Solutions

The diversity of tasks that AI is applied to and the multitude of AI-based solutions, make the formulation of a "pattern process" for introducing and solving the (new) problems that arise difficult. Often, certain effects only become apparent in the test operation. Take navigation systems as an example. They combine AI technologies from different decades: heuristic search for the shortest path (1960s) (Gallo & Pallottino, 1988), speech dialogue system (1990s) (Wahlster, 1999) and data-based traffic jam prediction (2000s) (Bolshinsky & Friedman, 2012). The technologies built into navigation systems are technically under-complex and quite transparent. However, the effects of the mass use of navis can only be examined in operation. Does it change the traffic in inner cities and residential areas (Nagy & Simon, 2018)? If so, is that desirable? If negative effects are observed, how can solutions be found together with the system providers: Are only a few cars allowed to be guided through residential areas? And if so, which ones? These are complex social questions that cannot be answered solely at the technical level (let alone be answered in advance by regulatory means). This complexity also currently leads to applications not even being tried out due to numerous concerns. Not without reason, Germany is still far from the desired leading position in this area (Technologies und for Industry Data Dashboard, 2020).

A core problem in dealing with AI is that there is no generally accepted and comprehensive definition (Legg & Hutter, 2007). AI is a collective term for a subfield of computer science, various technologies, and a steadily growing number of applications. Especially from the perspective of users, it is increasingly tedious to ask where "normal" digitisation ends and AI begins, as these terms are used synonymously. According to a Bitkom study from 2020 (Bitkom, 2020), more than half (52%) of the citizens surveyed by telephone say that they can explain well what AI means. Given the above-mentioned vagueness of this term, it would be interesting to have the respondents explain what they understand by AI. Even with the applications, the answers would probably have been very diverse, ranging from spell-checking to industrial robots. Another paradox: If you ask citizens, who think that the topic of AI will be relevant for prosperity, they even wish for a "German leadership role" in this topic, but see mainly dangers in their work environment (Bitkom, 2020). This sounds like "Wash my fur, but don't get me wet!". At the same time, in a PWC study from 2019 (PricewaterhouseCoopers, 2018a), 48% of the companies surveyed rated the topic of AI as "not relevant" for their own company. A similar split is also observed for digitisation topics in general (Gatterer et al., 2018); executives proclaim digital cultures or strategies without practicing them themselves. This can also be an expression of the "Not Invented Here Syndrome" (Hannen et al., 2019) or be due to the fact that leaders are not familiar enough with the rapidly developing technology.

4 Outlook on Collaboration and Interdisciplinarity in 2030

So what might the year 2030 look like? The discussion of how the future of work will look under this light is increasingly occupying politics, science and business worldwide (see e.g. AI the Future of Work Conference, 2020; World Economic Forum, 2020). So far, however, there are relatively few specific demands that crystallise across discourses. One of them is that the possibility of a final decision by a human being is ensured (DGB, 2020), as the fear of losing control over an unlimitedly developing artificial intelligence is widespread.

A vision: In 2030, digitisation is no longer an isolated, technological phenomenon, but the expression of a social transformation that has fundamentally changed our way of working and living (Widuckel & Aschenbrenner, 2020). Due to the high degree of individualisation and networking of people and things, new division of labour, social structures and relationships have emerged that use the productive forces and natural resources much more sparingly and fairly despite the increased number of people on earth. Through new forms of AI-supported communication, co-determination, democratic participation and legitimation have moved much closer together; instead of abstract discourses, everyday problems are solved by governance and regulation. The rigorous focus on interdisciplinarity in education and innovation has proven to be the key and core resource for coping with this transformation. All people have a basic digital education within their domain, whereby computer science has largely developed into an auxiliary science that unfolds its potential in cooperation with the other disciplines. Collaboration across departments is the norm, as is in scientific publishing or acquiring projects. Thinking in organisational or disciplinary silos is no longer in demand. As a result, the "vertical" regulation of AI technologies within the application areas has become as agile as necessary and possible, allowing innovation to take place without endangering the safety of people, society and the environment.

It becomes clear; even if artificial intelligence is a driver for innovation, the main issue is the interaction of the components of technology, human, organisation (Hirsch-Kreinsen, et al., 2018).

5 Summary and Practical Recommendations

Within the framework of the expert dialogue "Human-Technology Interaction—Working with AI" of the Observatory for Artificial Intelligence in Work and Society (AI Observatory) of the Federal Ministry of Labour and Social Affairs, representatives from science, trade unions, companies, civil society and associations have worked together to develop joint strategies for a good human-technology relationship when using AI in work processes (Think Tank BMAS AI Observatory, 2021). Although there are already recommendations for action from the employer side (PricewaterhouseCoopers, 2018b), the importance of co-designs (Jansen & Pieters, 2017) and participatory design processes (Schubotz, 2019) is currently underestimated. In the following, the authors of this article give a subjective selection of relevant findings of the expert dialogues in their own wording.

5.1 Accept the Design Challenge

Anyone who sets out to explore the possibilities of using AI in their own environment cannot delegate this complex transformation. We recommend involving the user's department, IT department, legal department, management and works council in the company and seeking external help if necessary. It is important not to just do the unmade "homework" such as closing interoperability gaps (that different IT systems do not work well together) and to attach the label "AI" at the end, but to really create something new.

5.2 Interdisciplinary Co-Design

It is best to involve research institutions, development service providers, companies and employees in a participatory design process. The people affected, i.e., the end users, must be at the centre and actively involved in the discussion—and directly and not only via representatives such as, e.g., the works council. This approach helps to implement the knowledge about the actual processes and contents of the application in cooperation with the development team into the AI system and to feedback first ideas maybe even as paper prototypes with the people who are affected by it in the end.

In the process, one can be guided by the following points.

5.2.1 Requirements

It must be clear whether the target is optimizing existing processes or a new business model. It needs to be specified what exactly the requirements are for the AI system. What sounds simple is not answered in many projects. Desired side effects of this catch-up digitalization, which can also be improved knowledge management and greater accessibility.

5.2.2 Available Data

Many good ideas currently fail in practice because the data are not available and cannot be collected—this can be about measurement data or communication recordings, e.g., from customer communication. But there are also AI systems that can cope with less data. So, data protection should not serve as an alibi for inaction.

5.2.3 Quality Measures

If we know what is to be improved in the end, we need measuring instruments. We should not only measuring the technical subsystem (do the algorithms also produce the required results?), but also whether the application works by the end users and in the entire organisation (social subsystem). This also includes job satisfaction. There are extensively researched assessment methods for this.

5.2.4 Conversion Process

The digital product is almost always unfinished. A look beyond the horizon to agile, iterative methods or design methods of user-centred technology design helps to identify the right approach here, so that the best possible overall solution is found at every moment.

5.2.5 Experimental Spaces

Systems must first be developed in different prototypes and tested in experimental labs close to reality. The recommendation is to work with pilot or pre-projects and minimal viable products (only partially completed product mock-ups) to capture the complex system and especially the feedback of the employees who will then work with it daily.

Once these points are answered in a broad range of applications, we hope that the paradoxes described above have become a thing of the past and that the machines will do "good work" with us. The main focus is to actively involve the practioner of the companies—they need to take on the design challenge and enter into a participatory design process. Experimental fields, pre-projects and pilots that iteratively approach a hopefully optimal target state from synergy between theory and practice can be very helpful towards reaching the big goal.

References

Advanced Technologies for Industry. (2020). *Data dashboard.*

Bitkom. (2020). Die Menschen wollen KI—und haben auch Angst vor ihr. Pressemeldung vom 28.09.2020. https://www.bitkom.org/Presse/Presseinformation/Die-Menschen-wollen-KI-und-haben-auch-Angst-vor-ihr. Accessed: 31. July 2021.

Bolshinsky, E., & Friedman, R. (2012). *Traffic flow forecast survey* (No. CS Technion report CS-2012-06). Computer Science Department, Technion.

Denkfabrik BMAS KI-Observatorium. (2021). *Demokratische Technikgestaltung in der digitalen Transformation.* Impulspapier zum Fachdialog "MTI—Arbeiten mit Künstlicher Intelligenz".

DGB. (2020). *Künstliche Intelligenz (KI) für Gute Arbeit. Ein Konzeptpapier des Deutschen Gewerkschaftsbundes zum Einsatz von Künstlicher Intelligenz (KI) in der Arbeitswelt.* https://www.dgb.de/downloadcenter/++co++17ebe6bc-9f2d-11ea-80f0-525400e5a74a. Accessed: 31. July 2021.

Gallo, G., & Pallottino, S. (1988). Shortest path algorithms. *Annals of operations research, 13*(1), 1–79.

Gatterer, H., Kappes, C., Kelber, C., Kühmayer, F., Muntschick, V., Papasabbas, L., Schuldt, C., & Zec, P. (2018). *Hands-on Digital.* Zukunftsinstitut GmbH.

Hannen, J., et al. (2019). Containing the not-invented-here syndrome in external knowledge absorption and open innovation: The role of indirect countermeasures. *Research Policy, 48,* 2019.

Hirsch-Kreinsen, H., et al. (Hrsg.). (2018). *Digitalisierung industrieller Arbeit: Die Vision Industrie 4.0 und ihre sozialen Herausforderungen.* Nomos.

Jansen, S., & Pieters, M. (2017). *The 7 principles of complete co-creation.* Bis Publishers.

Legg, S., & Hutter, M. (2007). A collection of definitions of intelligence. *Frontiers in Artificial Intelligence and applications, 157,* 17.

McAfee, A., & Brynjolfsson, E. (2017). *Machine, platform, crowd: Harnessing our digital future.* WW Norton & Company.

Nagy, A. M., & Simon, V. (2018). Survey on traffic prediction in smart cities. *Pervasive and Mobile Computing, 50,* 148–163.

PricewaterhouseCoopers. (2018a). *Künstliche Intelligenz in Unternehmen.* https://www.pwc.de/de/digitale-transformation/kuenstliche-intelligenz/kuenstliche-intelligenz-in-unternehmen.html. Accessed: 31. July 2021.

PricewaterhouseCoopers. (2018b). *Künstliche Intelligenz als Innovationsbeschleuniger im Unternehmen—Zuversicht und Vertrauen in Künstliche Intelligenz.* https://www.pwc.de/de/digitale-transformation/kuenstliche-intelligenz.html. Accessed: 31. July 2021.

Schubotz, D. (2019). *Participatory research: Why and how to involve people in research.* SAGE Publications Limited.
Stanford University. (2020). *AI & the future of work conference*, 27.10.2020. https://hai.stanford.edu/events/ai-future-work-conference. Accessed: 31. July 2021.
Sydow, J. (1985). *Der soziotechnische Ansatz der Arbeits-und Organisationsgestaltung: Darstellung, Kritik Weiterentwicklung (Vol. 428).* Campus.
Wahlster, W. (1999). Sprachtechnologie im Alltag—Der Computer als Dialogpartner. In HNF (Hrsg.), *Alltag der Zukunft, Paderborner Podium 3* (S. 18–37). Schöningh.
Widuckel, W., & Aschenbrenner, D. (2020). *Digital, transformativ, innovativ—Agenda für die Zukunftsfähigkeit Bayerns.* Managerkreis der Friedrich-Ebert-Stiftung.
World Economic Forum (2020). *The future of jobs report 2020.* World Economic Forum, Geneva, Switzerland.

Future Collaboration between Humans and AI

"I Strive to Make You Feel Good," Says My AI Colleague in 2030

Frank Fischer

1 Introduction

It is to be expected that humans and AI will form teams with the goal of enhancing the strengths of human and artificial intelligence and compensating for the weaknesses of each other. This way of working is also called "augmented intelligence". An interesting aspect of such a human-machine team is what is to be expected from both sides to function in this situation. How will the design of AI adapt to humans and what is to be expected from a human colleague?

2 Status Quo

"Augmented intelligence" already exists in application today. A specific example can be found in the field of static analysis of source code. This is traditionally a domain of so-called *symbolic AI*. An engineer programmes rule sets and the programme to be examined is translated into a set of facts, on which these rules are applied. The problem here is that programming these rules is costly and tedious.

F. Fischer (✉)
Product Marketing, Snyk, Mellieha, Malta
e-mail: frank@fischer.com.mt

I. Knappertsbusch and K. Gondlach (eds.), *Work and AI 2030*,
https://doi.org/10.1007/978-3-658-40232-7_3

A start-up called DeepCode[1] (acquired by Snyk in 2020) has changed the process and uses large amounts of source code in the form of open source as training data. With the help of *machine learning* (ML), the training data are processed ("mining") and the result is presented to an engineer. He or she annotates the result and marks false positives, adds missed examples and modifies the respective rule to expand the result space. With this information, the system is changed and applies the rule again. The result is then presented to the engineer again. This cycle—also called "human-guided reinforced learning"—is repeated until the result set has a consistently high quality and the rule is transferred to the production rule set. This not only simplifies the generation and maintenance of rules, but also makes it much faster.

3 Challenges and Solutions

In the context of this article, two challenges play a special role. The first is technical in nature. If one looks at the current examples, AI plays the role of a sophisticated tool in them (see the flood of articles on the application of AI in medicine, cybersecurity or education). The solutions require a high degree of prior knowledge from the user, work only in a very narrow area and produce false results if users do not provide input data in the exact right form. Also, today's solutions address niches.

In the last decade, ML solutions in the field of *natural language processing* (NLP) and big data have developed at an extreme speed. We are surrounded and accustomed to so-called *weak AI* (Workshop Informatik und Gesellschaft, 2009). It seems that the combination of two AI technologies will pave the way (Alam et al., 2020). The following article will focus primarily on psychological aspects. However, it is important to note the following basic assumption: The technological foundations (strong AI) for the following will be given.

Let us consider the second major challenge. What will the relationship between humans and artificial intelligence look like? At first, one is inclined to project from today's experience into the future. But this seems to fall short. Recent studies have shown that AI systems are not easy for humans to handle, especially in areas where human and artificial intelligence are close to each other (McIlroy-Young et al., 2020).

[1] DeepCode was acquired by Snyk in 2020 and the product described here is marketed today as Snyk Code.

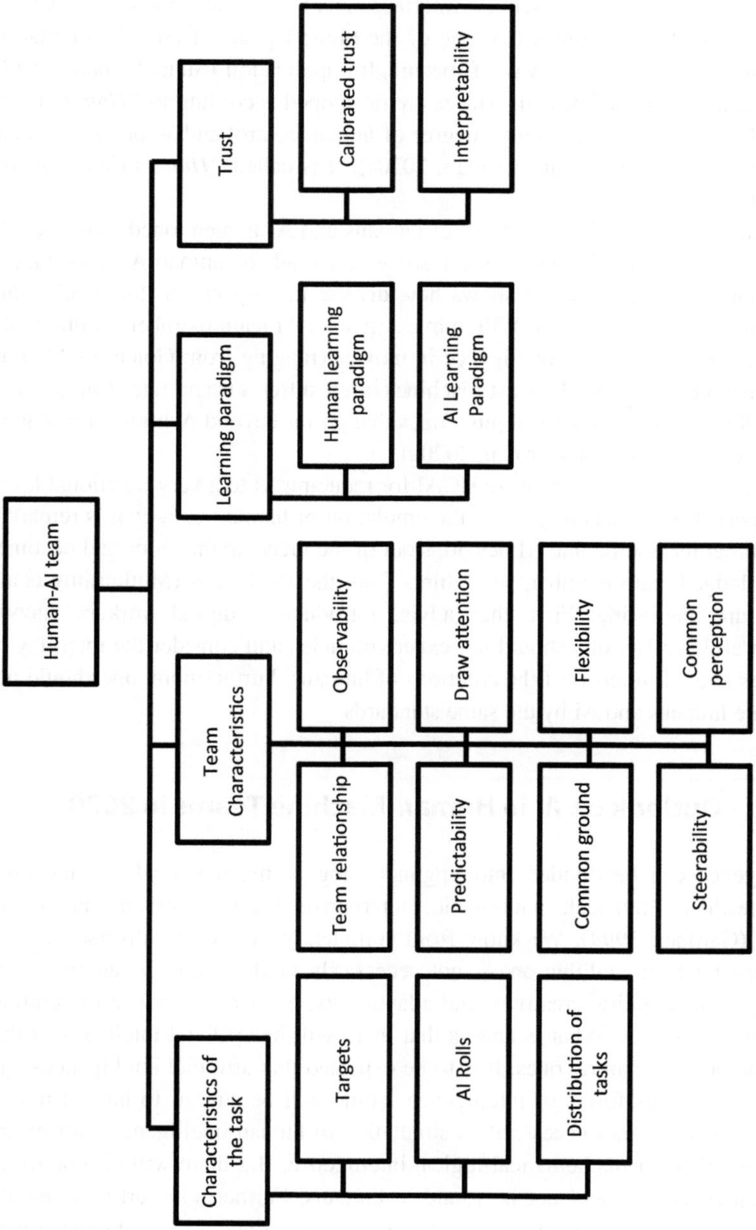

Fig. 1 Human-AI Teaming Taxonomy. (Adapted from Dubey et al., 2020, Fig. 2)

The improvement of human-AI interfaces is a field of active research. Among others, Shneiderman argues that one of the biggest goals of AI—the emulation of humans—stands in the way of meaningful application (Shneiderman, 2020b; Wang et al., 2020). If these interfaces are developed according to *"Human Computer Design"* guidelines, a high degree of human control and automation is enabled at the same time (Shneiderman, 2020c). It is called *"Human-Centered AI"* (HCAI).

Furthermore, the idea of a team of humans and AI is mentioned (Saenz et al., 2020; Siemon et al., 2021). Figure 1 shows a model for human-AI team taxonomy (Dubey et al., 2020). It shows how diverse the aspects of this relationship are, but a key aspect is trust. Will humans trust an AI team member? If one looks at the image of artificial intelligence in movies (ranging from Golem to Alien to Ex Machina), humans who trust machines often suffer a tragic fate (Calo, 2019). We will therefore need ethical guidelines that go far beyond Asimov's robot laws (Calvo et al., 2020; Shneiderman, 2020a).

Even though one can criticize HCAI for reducing AI to a very functional level, the aspect that one has to give up the emulation of humans as a goal is remarkable. But let us assume that AI develops out of the niche applications and becomes a full-fledged team member, some hints from the IBM Labs (Muthusamy et al., 2020) are interesting: First, they advise introducing "digital workers" slowly and carefully. Also, one should not expect miracles and consider the interplay of humans and AI in terms of the emotions of humans. Furthermore, one should not measure humans and AI by the same standards.

4 Outlook on AI in Human-Machine Teams in 2030

Intelligence can be divided into linguistic, logical-mathematical, spatial, bodily-kinesthetic, musical, naturalistic, interpersonal and intrapersonal intelligence (Gardner, 1999). We know from complexity theory that diverse systems are superior to monolithic ones—not necessarily in short-term productivity, but in long-term stability, creativity and adaptability. So it makes sense to combine diverse capabilities. What is new is that there will be artificial intelligences that can complement natural ones. It is to be expected that artificial intelligences will also have various forms of intelligence. Some will be similar to human intelligence—explicitly not necessarily a simulation of human intelligence—for example naturalistic or mathematical-logical intelligence. But there will also be forms of intelligence that will not be found in humans. Furthermore, artificial intelligences can form a common cognitive space because they can communicate with each other relatively easily and very quickly or share their memory. Given current research, we can expect that humans will also be able to join such a space with the help of implants, although perhaps not yet in 2030. By then, we will increasingly see human-machine teams. It is highly likely that by 2030 we will see a working world in which every human or team of humans is assisted by an AI. Let's first look at a human-machine team from a machine's point of view, where the machine from now on is more than a simple tool, but a powerful artificial intelligence (Röser, 2021). First, we need to realize that this AI is not tied to a physical host. Rather, the AI can control different robots; incorporate and use previously learned skills—even from other AIs; create copies of itself that can later combine its separate experiences; and access knowledge available on the internet or connected databases. Furthermore, the AI will draw on a reservoir of rules and experiences applicable to any human (speech recognition, rules of human interaction, general emotion recognition) and can overlay specific patterns for the respective human to better respond to its counterpart. It is likely that these specific patterns will be considered part of privacy, so not every AI will have full access. The AI will also try to emulate humans where it makes sense—for example, through voice control or emotional behavior. But it will refrain from doing so where it does not make sense in the context of the mission. Starting from the basic reservoir described above, the machine will gain further experience and be able to adapt better to the human team members. In principle, an AI can play any role in a team, but some seem better suited than others.

A widely used model is Belbin's nine team roles (Belbin, 2010). Following this model, an AI is very well suited to be assigned the role of "Resource Investigator", "Teamworker", "Implementer" or "Completer Finisher". But there are also less task-oriented roles that are more important for the dynamics of the team, such as the "Shaper"—motivation and goal orientation—or the "Monitor Evaluator"—logical thinking and weighing; even the "Co-Ordinator"—keeps the team aligned with goals, delegates work and decides on team composition. The role of "Plant"—highly creative and unconventional problem solving—but also that of "Specialist" with a lot of experience and gut feeling seem predestined for human team members. Adding an AI to a team (or building a team with AI help) allows you to define the role of an AI. Over time, the AI will learn which roles are missing or weak and help fill them. In addition, the AI can help ensure that the team is put together in the best possible way. Now let's look at the second view—from the human perspective. How will it feel to work with machines in a team? First, we need to realize that working with AI will be normal. People in 2030 will expect certain tasks or parts of tasks to be done automatically by machines. People will have a personal assistant. This AI flutters the daily flood of information and plans the day. It also pays attention to well-being: That you exercise enough, drink enough, your mental well-being. An important aspect will be further development and learning. AI can help identify areas for individual development, find and deliver learning content. In doing so, it will adapt to abilities and needs and will also have some discipline to maintain. The role of work will increasingly evolve from the necessary "breadwinning" towards personal fulfilment. Time as a unit for the work to be done is a relic of the industrial revolution. Certainly, there will be people whose work requires a certain amount of real time, such as repairing machines in the event of a fault. But the mass of work will be developing, designing, controlling and improving. In 2030, not only will the knowledge have been established that human creativity cannot simply be switched on, but the legal and social framework will also be in place. So, an AI can design a working day that suits the individual, bringing out the maximum capabilities. Also, the idea of a 5-day week is antiquated and sometimes—when you are in flow—you stay focused on work beyond the day's working hours. Sometimes you take months off to rest or learn something new. Alternating between work and leisure can be very flowing. It is important that people are not pulled into repetitive tasks. If you can automate something, it will be automated. People are brought in when it becomes special, creative and new. The following phase model should serve as a simple model of work steps: Understand, Explore, Design, Try, Decide, Summarize and Communicate. It may be that you repeat the phases iteratively (you design and try a solution, decide it is not suitable, and go back to exploring). It is no coincidence that this rough model is reminiscent of the design-driven approach practiced by innovative companies today. It's about solving problems creatively—work in 2030 will often look like this. For the people in such a team, work will be as pleasant as possible. Not only physically, but also—and this is new—from a psychological side. The AI will make sure that the chemistry between the team members is right and that the individual feels comfortable. Recent psychological models of job satisfaction establish a link between affection (emotion) for the job and job performance (Zhu, 2012). In modern management models, it is the team leader's responsibility to pay attention to emotional health, attachment of the individual to the organization, motivation, job satisfaction and others. This applies to all team members, and in addition, of course, there are economic goals. What is new is that an AI can help and perhaps even take over this task altogether. We also want to address two obvious risks. As some films have already addressed, it is not impossible for humans to develop an emotional attachment to the AI. While this can be very helpful at a low level (as an example, trust in the AI), stronger feelings such as hate, or love are conceivable. It goes without saying that this hinders successful cooperation. This would be another reason to use weak AI instead, as emotional attachment is unlikely here. Stronger AI can monitor human behaviors and react if necessary. Another risk is the danger of exploiting humans. Gaslighting or other psychological methods can open the door to radical exploitation or make them emotionally dependent. However, if we look at some current developments—increasing mobile working, gig economy, longer life and working hours—we can see a change in the attitude of workers towards work and employers: It is easier to distance oneself from an employer and different employers over the course of a career are seen more positively in terms of experience. So I will have my own private AI that looks after me, my health as well as my value on the labor market and that I trust more.

5 Summary and Practical Recommendations

It seems inevitable that humans and machines (or AI) will work together in teams. Besides the purely functional side, machines will also play a role in the team structure. Realistically, such an AI will aim to optimise the outcome of the team work. But since it is now recognized that humans achieve optimal results in the long term when they are satisfied and relaxed, this will automatically be a goal of the AI. For the human, however, it will be important to see his or her relationship to a team AI under this aspect and to have an alternative personal AI that has only his or her well-being as a goal.

For employers, the growing use of AI as a support in organisational psychology will be important, because it promises a performance boost. Already today we see AI support in human resources. In the future, the daily team operation will also increasingly take place with AI assistance. But to achieve this, AI has to be developed from a very technical niche into the field of psychology.

The relationship between employee and employer has already changed in recent years. The bond is looser and will be separated when the task comes to an end. As an individual, it is important to understand this relationship as such and to optimise for one's own, lasting well-being. Here too, an AI can help in the future.

References

Alam, M., Groth, P., Hitzler, P., Paulheim, H., Sack, H., & Tresp, V. (2020). CSSA'20: Workshop on Combining Symbolic and Sub-Symbolic Methods and their Applications. International Conference on Information and Knowledge Management, Proceedings, November, 3523–3524. https://doi.org/10.1145/3340531.3414072.

Belbin, R. M. (2010). *Team roles at work* (2. Aufl.). Routledge.

Calo, R. (2019). *How We Talk About AI (and Why It Matters).* 1–1. https://doi.org/10.1145/3306618.3314225.

Calvo, R. A., Peters, D., Vold, K., & Ryan, R. M. (2020). Supporting human autonomy in AI systems: A framework for ethical enquiry. *Philosophical Studies Series, 140,* 31–54. https://doi.org/10.1007/978-3-030-50585-1_2.

Dubey, A., Abhinav, K., Jain, S., Arora, V., & Puttaveerana, A. (2020). HACO: A Framework for Developing Human-AI Teaming. 13th Innovations in Software Engineering Conference. https://doi.org/10.1145/3385032.3385044.

Gardner, H. (1999). *Theory of multiple intelligences from: frames of mind: The theory of multiple intelligences.* Basic Books.

McIlroy-Young, R., Sen, S., Kleinberg, J., & Anderson, A. (2020). *Aligning superhuman AI with human behavior.* Proceedings of the 26th ACM SIGKDD International Conference on Knowledge Discovery & Data Mining, August 2020, (pp. 1677–1687). https://doi.org/10.1145/3394486.3403219.

Muthusamy, V., Unuvar, M., Völzer, H., & Weisz, J. D. (2020). *Do's and don'ts for human and digital worker integration.* ArXiv.

Röser, A. M. (2021). Charakterisierung von schwacher und starker Künstlicher Intelligenz. In *Arbeitspapiere der FOM Hochschule für Oekonomie & Management* (Vol. 79). Akademie Verlags- und Druck-Gesellschaft mbH.

Saenz, M. J., Revilla, E., & Simón., C. (2020). Designing AI systems with human-machine teams. *MIT Sloan Management Review, 61*(3), 1–5. https://sloanreview.mit.edu/article/designing-ai-systems-with-human-machine-teams/. Accessed: 27. July 2021.

Shneiderman, B. (2020a). Bridging the gap between ethics and practice: Guidelines for reliable, safe, and trustworthy human-centered AI systems. *ACM Transactions on Interactive Intelligent Systems, 10*(4), https://doi.org/10.1145/3419764.

Shneiderman, B. (2020b). Design lessons from ai's two grand goals: Human emulation and useful applications. *IEEE Transactions on Technology and Society, 1*(2), 73–82. https://doi.org/10.1109/tts.2020.2992669.

Shneiderman, B. (2020c). Human-centered artificial intelligence: Three fresh ideas. *AIS Transactions on Human-Computer Interaction, 12*(3), 109–124. https://doi.org/10.17705/1thci.00131.

Siemon, D., Li, R., & Robra-Bissantz, S. (2021). Towards a model of team roles in human-machine collaboration. International Conference on Information Systems, ICIS 2020—Making Digital Inclusive: Blending the Local and the Global, December 2020.

Wang, D., Churchill, E., Maes, P., Fan, X., Shneiderman, B., Shi, Y., & Wang, Q. (2020). From human-human collaboration to Human-AI collaboration: Designing AI systems that can work together with people. Conference on Human Factors in Computing Systems—Proceedings, 1–6. https://doi.org/10.1145/3334480.3381069.

Workshop Informatik und Gesellschaft. (2009). Schwache KI und Starke KI. Informatik Universität Oldenburg. http://www.informatik.uni-oldenburg.de/~iug08/ki/Grundlagen_Starke_KI_vs._Schwache_KI.html. Accessed: 27. July 2021.

Zhu, Y. (2012). A review of job satisfaction. *Asian Social Science, 9*(1), 293–298. https://doi.org/10.5539/ass.v9n1p293.

AI, Innovation and Start-ups

High-Tech Start-Ups as Drivers of AI Ecosystems

Annette Miller

1 Introduction

In Germany alone, a growth of the German gross domestic product of 430 billion EUR is expected by 2030 due to AI-based technologies.[1] In order to keep up with the global competition for the top positions in the field of AI, many countries have developed national AI strategies.[2] The goal of the German AI strategy is to "strengthen Germany in research, development and application of AI in international competition. To this end, it is necessary to further build up and expand AI ecosystems in Germany and Europe, to strengthen the application of AI across the board and to increase the visibility of outstanding initiatives and structures."[3] In order to secure the research location, expand the competitiveness and exploit the application possibilities, measures are envisaged in the entire spectrum of research, transfer, people, society and regulatory framework. So far, significant AI innovations have primarily come from the USA.[4]

[1] Cf. PWC (2019a).

[2] An overview is given by Konrad-Adenauer-Stiftung (2018), Dutton (2018).

[3] Federal Government (2020, p. 2).

[4] Cf. Bitkom (2020, p. 13).

A. Miller (✉)
Hessisches Zentrum für Künstliche Intelligenz—hessian.AI, Darmstadt, Germany
e-mail: annette.miller@hessian.ai

I. Knappertsbusch and K. Gondlach (eds.), *Work and AI 2030*,
https://doi.org/10.1007/978-3-658-40232-7_4

2 Status Quo

AI research in Germany is broad and well positioned by international standards. AI is now an integral part of the research landscape and almost every German university has at least one AI professorship. In international publication rankings, German scientists occupy very good positions and are cited particularly frequently.[5] Various studies, however, indicate that Germany is falling behind in terms of the use of AI in the economy. If at all, AI is so far mainly used as a tool for productivity improvements through automation.[6] The economic potentials of AI, however, lie especially in innovative products and business models.[7] These are hardly realised, not even through start-ups: Although AI had a significant influence on the business model in 285 founding cases in 2019, which corresponds to a share of 40%, the comparison with Israel, where 1300 AI start-ups existed in 2019, shows the backlog.[8] The amount of investments raised by German start-ups and the low number of AI patent applications are indications that Germany has difficulties in occupying innovative and future-oriented business fields.[9]

3 Challenges and Solutions

Key technologies such as AI are drivers for innovation and the basis for new products and processes. As with other technical innovations, research plays a special role as a source of knowledge. The challenge is to transfer the results and knowledge from the research institutions into application, so that they can be transformed into innovations in the market. However, not only small and medium-sized enterprises, but also multinational corporations have difficulties with this.[10]

In contrast to established companies, start-ups can act flexibly and quickly and are able to bring risky innovations to the market without questioning the previous success of the company. However, AI start-ups are not all the same, but take

[5] Cf. Morik (2021), Federal Government (2020, p. 4).

[6] See BMWi (2020), PWC (2019a).

[7] Cf. PWC (2019b).

[8] Cf. Federal Association of German Startups (2020, p. 11).

[9] Cf. DPMA, (2019), Federal Association of German Startups (2020, p. 20–22).

[10] Cf. Roland Berger (2018, p. 2).

on different tasks in the innovation process. One group of start-ups are AI users, who use AI mainly as a tool in their products and processes. They develop existing technology incrementally or work out new business models based on already known and tested AI systems. The innovation content is less technological, but results from the novelty of the use of AI in products or production processes, which enable productivity increases or efficiency gains. To be distinguished from this are technology-driven AI start-ups: They develop AI systems that can be used in different applications, or products for specific use cases at the interface of AI to specific application areas. Often research-based, they have a high market potential and above-average growth rates.

A special form are deep tech start-ups. Often based on basic research, they bring disruptive innovations to the market that deliver dramatic improvements over the currently used technologies. They do not develop solutions that are directly adapted to specific customer problems, but pursue a problem-oriented approach that aims to solve major economic and social challenges. Through their products, they displace existing solution concepts, which brings with it a rapidly increasing demand and an exponential growth potential.[11]

Technology-intensive start-ups, such as technology-driven and deep tech start-ups, face special challenges in the founding process, as research and development as well as the search for application fields are not only time-consuming, but also very costly. A study shows that deep tech AI start-ups also need an average of about 600,000 US$ and two years for the first prototype after founding. Another 1.4 years pass until market entry.[12]

Since this financing is usually not borne by private investors due to the high risks in the early phase, technology-driven and deep tech start-ups—also referred to as high-tech start-ups in the following—depend on public funds. Although public funding programmes, such as Exist or VIP+, make an important contribution to facilitating the transfer of research-intensive innovations into application. However, such start-ups often need several rounds of funding before they successfully attract risk financing or fail—also because funds for developing a market-ready product are lacking.[13]

Conversely, high-tech start-ups prove to be very economically successful, as an analysis carried out by ourselves shows. The criterion for success was the

[11] Cf. BCG (2021a).

[12] Cf. BCG (2021b, p. 16).

[13] Cf. BCG (2021a, p. 12).

amount of investments raised in the free market. The analysis was based on data from the Crunchbase database (as of spring 2019), which was supplemented by publicly available sources.[14] The analysis provides indications for the importance of the technical-scientific background within the founding teams and the importance of scientific research for successful start-ups. The share of founding teams with at least one member with a completed doctorate is 51%. In addition, the founding teams have a particularly high share of engineering or natural science expertise, 64% even have at least one founding member with a scientific education in computer science.

4 Outlook on AI, Innovations and Start-ups in 2030

AI as a general concept is just beginning to tap into its potentials. Although immense breakthroughs have been achieved in the field of AI in recent years, the research need for the widespread use of AI is still high. New generations of AI models will work with less data or be applicable outside of specific use cases for which they were trained. The simplification of AI development will drive diffusion, as no highly specialised expert teams are needed anymore for the development of AI.[15] Innovation potentials arise especially from the convergence of different technologies and at the interface between AI and application industries.

These breakthroughs are likely to come not from individual inventors, but from teams from research institutions that have deep and broad AI expertise.[16] The results generated there will mainly be transferred into economic application by start-ups that have a close connection to scientific institutions.[17] This enables them to have the best starting conditions, especially if the founders transfer their own scientific results into application and do not have to procure the AI expertise externally in the highly competitive market for AI experts. Through contacts to the parent institutions, the expert knowledge required for further developments can be specifically consulted. At the same time, founders gain access to talents, but also to specific AI infrastructure. The already visible tendency that AI start-

[14] AI startups that were younger than ten years at the time of the survey were considered.

[15] See Goasduff (2020).

[16] See Morik (2021), Roland Berger (2018, p. 2).

[17] See Turkina (2018), Bundesverband Deutscher Startups (2020).

ups cooperate with research institutions above average,[18] will therefore continue. To seize the opportunities, the start-ups will settle in the vicinity of relevant scientific institutions.

In particular, deep tech start-ups will act as drivers in AI innovation ecosystems together with their scientific parent institutions and occupy a central interface between science and economy. By bringing ground-breaking innovations to the market, they can also contribute to solving the major challenges, such as climate change or pandemic control. By combining basic and applied research of different technologies and replacing previously used technologies, some of these companies can completely change existing industries and develop into international technology corporations in the future.

Due to the advantages of a close connection to science, high-tech start-ups will preferably settle in close proximity to the parent institutions. Regions with well-positioned scientific institutions and good location factors (e.g. favourable office space, high-performance AI infrastructure, access to funding) have the opportunity to develop into AI hotspots. Because successful start-ups have a positive impact as role models not only on the founding culture in the scientific institutions. The successes will also stimulate the founding dynamics in the entire region and thus contribute to the emergence of AI users. Even if their market potential is lower, they carry AI into the breadth of economy and society and thus make a contribution to successful AI innovation ecosystems.

A close cooperation and spatial proximity between established companies, start-ups and research institutions, which takes place for example through networking events, cooperation projects or conferences, contributes to positive spillover effects.[19] Investors will be attracted, but also innovative companies, which seek the proximity to the start-ups and research institutions and want to benefit from the location factors.

How excellent research affects the start-up scene is shown by the example of Toronto. Starting from scientific institutions, such as the Vector Institute, a global AI hotspot has emerged, which has a high attraction on top talents from all over the world. The region is characterised by an intensive interconnection of economy and science and has attracted international tech corporations such as Microsoft, Google or IBM. At the same time, economy and science are connected through

---

[18] Cf. Bundesverband Deutscher Startups (2020).

[19] Cf. Turkina (2018, p. 2).

start-ups, such as Deep Genomics or Element AI, which have been founded by renowned AI top scientists and raise immense investments.[20]

5 Summary and Practical Recommendations

AI can generate enormous economic value. While startups are generally an important channel in the AI innovation process, high-tech startups have a special position; they can develop immense economic potentials, attract high investments and contribute to the development of a founding culture and successful innovation networks. At the same time, they can help solve the major economic and social problems.

The German government has recognised the potential of AI start-ups. To support them, for example, the funds for the established programme "Exist" are to be doubled and the newly founded Agency for Breakthrough Innovations is to contribute to the transfer.[21] The targeted support of high-tech start-ups, especially deep techs from science, could further strengthen the AI innovation system.

Since the risks of radical and disruptive AI innovations are not borne by the market, it seems sensible to expand the funding instruments quantitatively and qualitatively. Additional instruments are needed at the interface between classical research funding and private investors, which promote innovations with high socio-economic potential, motivate founders to exploit their own results and bridge the long development times. Because even if existing programmes are to be expanded within the framework of the AI strategy, experience shows that many promising projects do not receive funding in view of comparatively low approval rates or due to detailed specifications on the development status in terms of research and development and are therefore discontinued. Particularly promising start-ups should also be supported in scaling up, so that the talents and companies do not migrate to other AI hot spots.[22]

To ensure efficiency and support the positioning in the field of AI, the measures should not be designed as broad-based funding and the success of political measures should not only be measured by the number of new start-ups. A targeted support of high-tech AI start-ups that have significant economic or social

[20] Cf. Turkina (2018).

[21] Cf. Federal Government (2020, p. 18).

[22] Roland Berger (2019) suggests possible measures to support scale ups.

potentials not only prevents the dilution of the AI strategy, but also increases the chances of the funded projects for subsequent investments. Because not every company that calls itself an AI start-up meets the criteria. An analysis conducted in 2019 even comes to the conclusion that in 40% of the European start-ups named as AI start-ups, there is no significant influence of AI on the business model.[23]

Even if disruptive innovations cannot be planned, the chances can be increased if space is created for basic research at the highest level. The targeted support of AI research centres, such as the Hessian Center for AI—hessian.AI —, where expertise in the entire spectrum of AI is bundled and scientists can conduct research under the best conditions, is an attraction point for the best minds and a source of potential-rich innovations. By enabling scientists to conduct free basic research under the best conditions and supporting them in exploiting their results, not only are innovations promoted, but scientific results are also carried into the breadth of science, economy and society.

References

BCG. (2021a). *Deep tech: The great wave of Innovation.* https://hello-tomorrow.org/wp-content/uploads/2021/01/BCG_Hello_Tomorrow_Great-Wave.pdf. Accessed: 24. June 2021.

BCG. (2021b). *The Deep tech investment paradoxon.* https://hello-tomorrow.org/wp-content/uploads/2021/05/Deep-Tech-Investment-Paradox-BCG.pdf. *Accessed: 24. June 2021.*

Berger, R. (2018). *Artificial Intelligence—A strategy for European startups.* https://www.rolandberger.com/en/Media/AI-startups-as-innovations-drivers-Europe-must-take-action-to-establish-a.html. Accessed: 31. July 2021.

Bitkom. (2020). *KI-Forschung in Deutschland—Der schwere Weg zu 100 neuen KI-Professuren.* Impulspapier.

BMWi. (2020). *Einsatz von Künstlicher Intelligenz in der Deutschen Wirtschaft. Stand der KI-Nutzung im Jahr 2019.* https://www.bmwi.de/Redaktion/DE/Publikationen/Wirtschaft/einsatz-von-ki-deutsche-wirtschaft.html. Accessed: 31. July 2021.

Bundesregierung. (2020). *Strategie Künstliche Intelligenz der Bundesregierung. Fortschreibung 2020.* https://www.ki-strategie-deutschland.de/home.html. Accessed: 24. June 2021.

[23] Cf. Vincent (2019).

Bundesverband Deutscher Startups. (2020). *Künstliche Intelligenz—Wo stehen deutsche Startups.* https://deutschestartups.org/2020/09/03/ki-startups-unter-der-lupe-studie-beleuchtet-das-startup-oekosystem-in-deutschland/. Accessed: 31. July 2021.

DPMA. (2019). *Künstliche Intelligenz: US-Unternehmen bei Patentanmeldungen für Deutschland weit vorne.* https://www.dpma.de/service/presse/pressemitteilungen/20190411.html. Accessed: 24. June 2021.

Dutton, T. (2018). *An overview of national AI strategies.* https://medium.com/politics-ai/an-overview-of-national-ai-strategies-2a70ec6edfd. Accessed: 24. June 2021.

Goasduff, L. (2020). *2 Megatrends dominate the gartner hype cycle for Artificial Intelligence.* https://www.gartner.com/smarterwithgartner/2-megatrends-dominate-the-gartner-hype-cycle-for-artificial-intelligence-2020/. Accessed: 24. June 2021.

Konrad-Adenauer-Stiftung. (2018). *Vergleich nationaler Strategien zur Förderung von Künstlicher Intelligenz.* Teil 1.

Morik, K. (2021). *Status quo der KI-Forschung in Deutschland.* https://www.plattform-lernende-systeme.de/reden-und-beitraege-newsreader/status-quo-der-ki-forschung-in-deutschland.html. Accessed: 24. June 2021.

PWC. (2019a). *Aus dem Hype Realität machen: Fit für Künstliche Intelligenz im Jahr 2020.* https://www.pwc.de/de/digitale-transformation/kuenstliche-intelligenz/aus-dem-hype-realitaet-machen-fit-fuer-kuenstliche-intelligenz-im-jahr-2020.html. Accessed: 24. June 2021.

PWC. (2019b). *Künstliche Intelligenz: Diese sechs Prioritäten sollten Unternehmen für 2019 setzen.* https://www.pwc.de/de/digitale-transformation/kuenstliche-intelligenz/sechs-wichtige-prioritaeten-um-ki-2019-voranzubringen.html#innovation. Accessed: 24. June 2021.

Turkina, E. (2018). The importance of networking to entrepreneurship: Montreal's artificial intelligence cluster and its born-global firm Element AI. *Journal of Small Business & Entrepreneurship, 30*(2018), 1–8.

Vincent, J. (2019). Forty percent of 'AI startups' in Europe don't actually use AI, claims report. https://www.theverge.com/2019/3/5/18251326/ai-startups-europe-fake-40-percent-mmc-report?fbclid=IwAR1pcrmJ0ETx6U1we17VFAqiRnJRf_AeKoLvNju1k0R-HEQQ1iiTUKdiRhjA. Accessed: 24. June 2021.

AI Demands Corporate Digital Responsibility (CDR)

Aligning the Moral Compass for Workers in AI-Enabled Workplaces

Saskia Dörr

1 Introduction

By using artificial intelligence (AI), machine learning or software robotics, companies increase their productivity and efficiency on the one hand. On the other hand, people are replaced by machines. Already today, many jobs have (data) interfaces to self-learning algorithms; they are "AI-supported". It can be assumed that this will affect the vast majority of jobs by 2030.

Fig. 1 sketches some different types of AI-supported jobs: From office and management, call centre, service and sales, production and warehouse to the "self-employed" gig workers—and underpins them with examples. It shows which technologies are implemented at the workplaces, which types of AI analysis can be expected and how a new set of decision functions in management uses the data produced by such technological processes.

S. Dörr (✉)
WiseWay Berät Unternehmen, Bonn, Germany
e-mail: saskia.doerr@wiseway.de
URL: https://wiseway.de

I. Knappertsbusch and K. Gondlach (eds.), *Work and AI 2030*,
https://doi.org/10.1007/978-3-658-40232-7_5

Technology	**Platforms (Algorithms, AI, Machine Learning (ML))**	**People Analytics (AI, ML, filming, voice recording, emotion coding)**	**Chatbots Voice Assistants (AI, ML, filming, voice recording, emotion coding)**	**Cobots Wearables (RFID, dashboards, tablets, GPS, AR glasses)**
Types of AI analysis	Action forecasts, action recommendations, business intelligence	Automated recognition of emotions, supporting activities, action prediction, business intelligence	Automated recognition of emotions, supporting activities, action prediction, business intelligence	Supporting activities, simplification of activities, business intelligence
Typical work areas	Various, usually independent (at home, on the street)	Office, management, “white collar jobs”	Call center, customer service, sales	Production, Warehouse, Factory, “blue collar jobs”
New, additional decision-makers	Human resources, performance monitoring, workforce management, micro task management			
Examples	Delivery services, IT product tests, “content moderation” (“Gig Economy”)	Recruiting, performance control for remote work or home office	Chatbots for pre-filtering service requests, recognition of caller emotion status in the call center	Tracking of warehouse workers, KI-supported manufacturing, augmented reality remote assistance

Fig. 1 1 Artificial intelligence at workplaces: Technologies, types of AI analysis, work areas, decision makers and examples. (Modified and further developed from Moore 2019a)

2 Status Quo and Case Studies

In Germany for example, 3600 companies publicly commit to values such as diversity and inclusion with the "Charter of Diversity" (cf. Charter of Diversity, 2021). This has positive effects for competitiveness, e.g., women in leadership positions increase the performance of the company; diverse and inclusive companies are more attractive employers (cf. Noland et al., 2016; Universum Communications Sweden AB, 2019).

However, the use of AI and algorithms in the employment context does not reflect the values such as diversity and inclusion mentioned in the "Charter of Diversity" and has the opposite effect for the employees. Some recent examples from the media:

- Violation of the privacy of Amazon drivers by high-tech surveillance system (cf. Gärtner, 2021)
- Artificial intelligence in job applications assesses applicant suitability based on glasses, headscarves and wall decorations and thus reinforces existing stereotypes (Harlan & Schnuck, 2021)
- Digital monitoring of employees in the home office by software (cf. Raidl & Tyborski, 2020)
- Suspension of a teacher after an evaluation algorithm of the learning outcomes "questioned" his teaching method (cf. Kantayya, 2020)

For the employees, risks arise for equality, fairness, dignity, personality protection and privacy. The reasons for this are manifold. One cause is the "algorithmic bias" due to entrenched, previous societal prejudices in the AI training data or new interpretation patterns developed by the algorithm. This biases the decisions of the AI and multiplies undesirable societal notions (cf. Gov.uk, 2020). Furthermore, the "problem of many hands" creates responsibility gaps in the use of algorithms (cf. Hamadi & Manzo, 2021).

The impact of AI support on different quality aspects of jobs, such as health and safety, personality, data protection and employee rights of the employees working there, has so far hardly been integrated into the broad discussion on ethical AI (cf. Moore, 2019b). Various factors lead to the fact that the previous "blind spot" becomes a business risk.

3 Challenges and Solutions

Companies are interested in finding ways to automate on the one hand, and on the other hand to maintain the loyalty and engagement of their employees as well as to be an attractive employer for talents. The "moral compass" needs to be realigned in the "algorithmic new territory".

With digitalisation, the moral expectations of companies change. The management discipline that provides methods and tools to apply "business ethics" in the business is Corporate (Social) Responsibility (CSR, CR). It evolves into Corporate Digital Responsibility (CDR, cf. Dörr, 2020; Esselmann & Brink, 2016). For companies, it is now a matter of defining their digital responsibility when using data, algorithms, the "emerging technologies" in general, and digital business models. The aim is to reduce damage to the common good, individuals, and society and to increase both business and societal benefits.

Corporate Digital Responsibility (CDR) is an area of comprehensive Corporate Responsibility (CR) in an increasingly digitalised economy and society. It is "voluntary entrepreneurial activities in the digital domain that go beyond what is legally required today and that actively shape the digital world to the benefit of society" (Federal Ministry of Justice and Consumer Protection, 2018, p. 2).

CDR, like CR, is guided by the principle of sustainability and arises from taking responsibility for the economic, social and ecological impacts of corporate actions as well as from resolving the conflicts of objectives between them. The purpose of CDR is to build the trust of different stakeholders of the company in digitalisation (cf. Dörr, 2020, p. 39 ff.).

In addition to customers and market, the community, and climate and environment, employees are essential stakeholders for companies (cf. Dörr, 2020, p. 99 ff.). This includes the fair transformation of the workplaces, i.e. to shape the "digital change socially and individually beneficial and to extend the employer's care to employees and other workers at digitally supported workplaces" (Dörr, 2020, p. 108). The responsibility refers to all forces involved in value creation, such as self-employed, project staff, crowd workers, etc.

CDR can be described as a framework for an organisational culture that enables individuals—leaders and employees—to take responsibility for data and AI. It provides orientation by self-commitments to societal demands for the use of data and digital. It allows the company as a whole to publicly account for it in a transparent and verifiable way (cf. Lobschat et al., 2021).

4 Outlook on AI-Supported Workplaces in Companies with CDR in 2030

For companies that promote AI support at workplaces by automating their processes and at the same time want to maintain credibility of their value, sustainability and purpose orientation, Corporate Digital Responsibility offers a valuable set of instruments in organisational and corporate management. This is to be substantiated and detailed with the following theses with a view to 2030.

4.1 Voluntary Commitment to AI-Supported Workplaces that Goes Beyond AI and Data Regulation can Create Competitive Advantages

The regulations and laws for dealing with data and algorithms, such as the EU General Data Protection Regulation or the legislative process around the "Ethics Guidelines for Trustworthy AI" in the EU, influence the digital actions of companies and thus the CDR. Especially for multinational corporations, the challenge is to follow the different national requirements (cf. Lobschat et al., 2021, p. 883).

The C(S)R has evolved from the challenge of globalisation and takes into account the fact that national legislation does not cover the business scope of multinational corporations. High standards, e.g. for environmental protection or occupational safety in the facilities in the countries of the global south, are based on voluntary self-commitments of the companies that go beyond the national requirements. The same argument can be applied to digitalisation, e.g. to principles for digital consumer protection (KPMG, 2020).

The possibility of achieving competitive advantages through digital responsibility results from the integration of CDR into the company and the proximity to the core business. The more CDR is integrated into corporate management, everyday decisions or innovation processes, the less it is imitable by other actors and the more it forms a signal for trustworthiness (cf. Dörr, 2020, p. 196 ff.). An investment in high standards for data and AI management can therefore justify competitive advantages. For example, CDR strengthens the trust of both customers and employees in the digital services and in the company. Furthermore, CDR supports the implementation of values—a factor that is becoming increasingly important for the choice of workplace. This binds employees and enables the company to recruit talents as an attractive employer (cf. Esselmann et al., 2020).

4.2 AI Ethics and Governance Alone are not Enough for Credibility in Dealing with Algorithms

The establishment of AI ethics and AI governance for a single AI system is not enough. It needs a cross-company philosophy for dealing with AI, if value orientation, purpose and sustainability are to be credibly advocated.

> "It is unlikely that external individual AI systems will be identified as 'trustworthy' or 'not trustworthy'; rather, an organisation is considered trustworthy or not and AI systems inherit the reputation of the organisation" (McCarthy & Byrd, 2021).

CDR is therefore seen as a framework to implement ethical AI governance(s) and strategies organisationally, to ensure justice for all stakeholders, to codify trust and to meet expectations of corporate social responsibility (cf. Elliott et al., 2021).

A corresponding CDR culture enables the company to evaluate alternative behavioural options and to choose the "right" way forward, both on an individual and organisational level (cf. Lobschat et al., 2021, p. 6).

4.3 CDR Develops an Organisational Framework that Enables Trust and Credibility Among All Stakeholders, Especially Among Employees

From the perspective of CDR, the core task is to implement the existing value, purpose and sustainability orientation for the employees at AI-supported workplaces. This results in, for example, the following fields of action (cf. UNI Global Union, 2021):

- Ensure that an AI system is in line with the existing codes of conduct, human rights laws, ILO conventions and collective agreements as well as other applicable regulations.
- Take up claims of the affected employees to an AI solution and use them for regular review of the AI.
- Have an external review on the impartiality and fairness of the AI in operation.
- Use cross-company governance mechanisms to control the AI application, e.g. through a governance body, in which representatives of the employees are represented alongside manufacturers, programmers, human resources department and other interest groups.

- Limit the AI-based decisions in operation and clarify at which points a human always has to make a final decision, e.g. exclusion in personnel decisions.
- Account for the employees on the implementation of digital responsibility within the sustainability report and/or business report.
- Share the knowledge built up in the company on how to deal with AI-supported workplaces for a digitally responsible transformation in economy, politics and civil society.

It is significant for a CDR approach to establish a cross-functional management across human resources, purchasing, production control, sales or service. Thus, different types of employment relationships, such as permanent employees, temporary and project workers, gig workers or subcontractors in service and sales, are considered.

As a result of these tasks, which are performed on a cross-company basis, an organisational framework is created that implements social demands when using data and digital in different functions in a verifiable way. A culture may develop that enables trust and credibility.

5 Summary and Practical Recommendations

Corporate Digital Responsibility in corporate management provides mindset and tools to maintain the credibility of corporate values when using AI at workplaces. The "moral compass" is adapted to the new data and algorithm world, the entrepreneurial risks are reduced and new opportunities for innovative employee-oriented corporate management are opened.

Companies that already use AI at workplaces or plan to do so are therefore recommended to:

- Build up know-how in the new field of Corporate Digital Responsibility.
- Anchor responsibility for CDR in the organisation and ensure top management support.
- Take into account stakeholder expectations of digital responsibility, especially of employees or employee representatives.
- Develop own commitments to responsible digitalisation or follow existing CDR management principles.

AI use, especially in the employee context, needs the framework of a CDR management to represent the corporate values in these transformative changes. CDR provides the strategic framework to realign the "moral compass" in the digital world and to implement it in corporate practice.

References

Bundesministerium für Justiz und Verbraucherschutz. (2018). Corporate Digital Responsibility-Initiative: Digitalisierung verantwortungsvoll gestalten Eine gemeinsame Plattform. https://www.bmjv.de/SharedDocs/Downloads/DE/News/Artikel/100818_CDR-Initiative.pdf?__blob=publicationFile&v=3. Accessed: 7. July 2021.

Charta der Vielfalt. (2021). Charta der Vielfalt. Über uns. https://www.charta-der-vielfalt.de/ueber-uns/ueber-die-initiative/. Accessed: 5. July 2021.

Dörr, S. (2020). Praxisleitfaden Corporate Digital Responsibility. Springer Gabler, Berlin. https://www.springer.com/de/book/9783662605912. Accessed:12. July 2021.

Elliott, K., Price, R., Shaw, P., Spiliotopoulos, T., Ng, M., & Coopamootoo, K. van Moorsel, A. (2021 Jun). Towards an Equitable Digital Society: Artificial Intelligence (AI) and Corporate Digital Responsibility (CDR). *Society., 14,* 1–10. https://doi.org/10.1007/s12115-021-00594-8.Zugegriffen:7.Juli2021.

Esselmann, F., & Brink, A. (2016). Corporate Digital Responsibility: Den digitalen Wandel von Unternehmen und Gesellschaft erfolgreich gestalten. *Spektrum, 12*(1), 38–41.

Esselmann, F., Golle, D., Thiel, D., & Brink A. (2020). Corporate Digital Responsibility. Unternehmerische Verantwortung als Chance für die deutsche Wirtschaft. Positionspapier Zentrum Digitalisierung. Bayern. https://www.bayern-innovativ.de/seite/corporate-digital-responsibility. Accessed: 5. July 2021.

Gärtner, M. (2021). Kritik an Amazon-Überwachung: „Verletzung der Privatsphäre und Vertrauensbruch". https://www.amazon-watchblog.de/kritik/2567-kritik-ueberwachung-amazon-fahrer.html. Accessed: 7. July 2021.

Gov.uk. (2020). Review into Bias in Algorithmic Decision-Making. Independent Report. https://www.gov.uk/government/publications/cdei-publishes-review-into-bias-in-algorithmic-decision-making/main-report-cdei-review-into-bias-in-algorithmic-decision-making#background-and-scope. Accessed: 7. July 2021.

Hamadi, H., & Manzo, C. (2021). Corporate Digital Responsibility. A Study on Managerial Challenges for AI Integration in Business. Master Programme in Management. School of Economics and Management Lund University. http://lup.lub.lu.se/student-papers/record/9052507. Accessed: 7. July 2021.

Harlan, E., & Schnuck, O. (2021). Fairness oder Vorurteil? Fragwürdiger Einsatz von Künstlicher Intelligenz bei der Jobbewerbung. BR 24 vom 16.02.2021. https://web.br.de/interaktiv/ki-bewerbung/. Accessed: 7. July 2021.

Kantayya, S. (2020). Coded Bias. Dokumentarfilm. https://www.codedbias.com/. Accessed: 5. July 2021.

KPMG. (2020). The KMPG Survey of Sustainability Reporting 2020. https://home.kpmg/xx/en/home/insights/2020/11/the-time-has-come-survey-of-sustainability-reporting.html. Accessed: 5. July 2021.

Lobschat, L., Mueller, B., Eggers, F., Brandimarte, L., Diefenbach, S., Kroschke, M., & Wirtz, J. (2021). Corporate digital responsibility. *Journal of Business Research, 122,* 875–888. https://doi.org/10.1016/j.jbusres.2019.10.006.Zugegriffen:5.Juli2021.

McCarthy, M., & Byrd, M. (2021). How to build AI systems that a society wants and needs. World Economic Forum vom 02.07.20201. https://www.weforum.org/agenda/2021/07/how-to-build-ai-that-society-wants-and-needs. Accessed: 7. July 2021.

Moore, P. V. (2019a). Artificial Intelligence in the workplace: What is at stake for workers? OpenMind BBVA. https://www.bbvaopenmind.com/en/articles/artificial-intelligence-in-workplace-what-is-at-stake-for-workers/. Accessed: 7. July 2021.

Moore, P. V. (2019b). OSH and the future of work: Benefits & risks of Artificial Intelligence tools in workplaces. European Agency for Safety and Health at Work. https://osha.europa.eu/sites/default/files/publications/documents/OSH_future_of_work_artificial_intelligence_0.pdf. Accessed: 5. July 2021.

Raidl, M., & Tyborski, R. (2020). Die digitale Überwachung: Wie Unternehmen ihre Mitarbeiter beschatten. Handelsblatt vom 24.06.2020. https://www.handelsblatt.com/technik/digitale-revolution/digitale-revolution-die-digitale-ueberwachung-wie-unternehmen-ihre-mitarbeiter-beschatten/25917236.html?ticket=ST-4339093-K1heC5JY7Dd-ftLZ0Emfh-ap4. Accessed: 5. July 2021.

UNI Global Union. (2021). 10 Principles for ethical Artificial Intelligence. The future world of work. http://www.thefutureworldofwork.org/docs/10-principles-for-ethical-artificial-intelligence/. Accessed: 5. July 2021.

Noland, M., Moran, T., & Kotschwar, B. (2016). Is gender diversity profitable? Evidence from a global survey. Peterson Institute for International Economics. Working Paper 16–3. https://www.piie.com/publications/wp/wp16-3.pdf. Accessed: 7. July 2021.

Universum Communications Sweden AB. (2019). Universum's First Global Diversity & Inclusion Index. Press release. https://universumglobal.com/blog/dni-inxdex-news2019/. Accessed: 7. July 2021.

AI Ethics and Neuroethics Promote Relational AI Discourse

A Combined Embodiment Approach Strengthens the Socially Integrated Regulation of AI Technology in the World of Work

Ludwig Weh and Magdalena Soetebeer

1 Introduction: 'Artificial Intelligence' as a Result of Dualistic Knowledge Production

In the middle of the seventeenth century, René Descartes recognised that human beings are based on a fundamental separation between physical and mental existence:

> "[O]n the one hand I have a clear and distinct idea of myself, in so far as I am simply a thinking, non-extended thing; and on the other hand I have a distinct idea of body, in so far as this is simply an extended, non-thinking thing. And accordingly, it is certain that I am really distinct from my body, and can exist without it." (Descartes, 1984, p. 54)

In this sense, the *Cartesian body-mind dualism* establishes that form of (academic) knowledge production, which produces (academic) knowledge *indepen-*

Institut für Theoretische Biologie, Humboldt-Universität zu Berlin, Berlin, Germany
Berlin, Germany

L. Weh (✉)
Fraunhofer IMW Center for International Management and Knowledge Economics, Leipzig, Germany
e-mail: ludwig.weh@imw.fraunhofer.de

M. Soetebeer
Schaltzeit GmbH, Berlin, Germany
e-mail: magdalena.soetebeer@schaltzeit.com

I. Knappertsbusch and K. Gondlach (eds.), *Work and AI 2030*,
https://doi.org/10.1007/978-3-658-40232-7_6

dently of the subjective physical reality of the researching person, which can be reproduced at any time and regardless of context under objectively standardised conditions. Accordingly, empirical findings *in vivo* or *in vitro* are considered particularly reliable in the life sciences, if they can be validated by theoretical computer models *in silico*, which in turn are based on mathematically proven universal principles.

Through dualistic abstraction of biological (e.g. neuronal or bacterial) processes of information processing, the mathematical foundations for computer-based learning algorithms or 'machine learning' (Hebb, 1949) were already developed from the middle of the twentieth century. In this context, A and B in Fig. 1 show how the interpretation of a nerve cell as an abstract mathematical computing unit reduces its complex neuroanatomical and -physiological embedding in living neural tissue to its mere function of information integration and transmission. While the simplified mathematical version, such as 'Hebbian learning' (ibid.), allows the transfer of learning processes in neural networks to machine processes *in silico*, it simultaneously neglects a multitude of complex biological factors that influence the function of nerve cells *in vivo*. This loss of 'biological corporeality' characterises the simplified transfer of neural networks to machine learning algorithms; C and D in Fig. 1 visualises how complex connections between cell bodies or axons in neural tissue of the brain are abstracted to multistage calculation cascades with computational units (nodes) and their connections (edges).

2 Status Quo

The dualistic idea suggests that the human mind can exist separately from its physical basis and is reproducible. On this basis, the (natural) scientific development of 'artificial intelligence' (AI) as a disembodied, technological implementation of formerly embodied, biological learning processes formulates a quasi-religious promise of salvation, by promising the overcoming of its corporeality by the human mind with technological means (Coenen, 2014; Geraci, 2008). A common AI definition adopts this techno-optimism, in which the machine can surpass the biological intelligence:

> "[Artificial intelligence] is the science and engineering of making intelligent machines, especially intelligent computer programmes. It is related to the similar task of using computers to understand human intelligence, but AI does not have to confine itself to methods that are biologically observable." (McCarthy, 2004, p. 2)

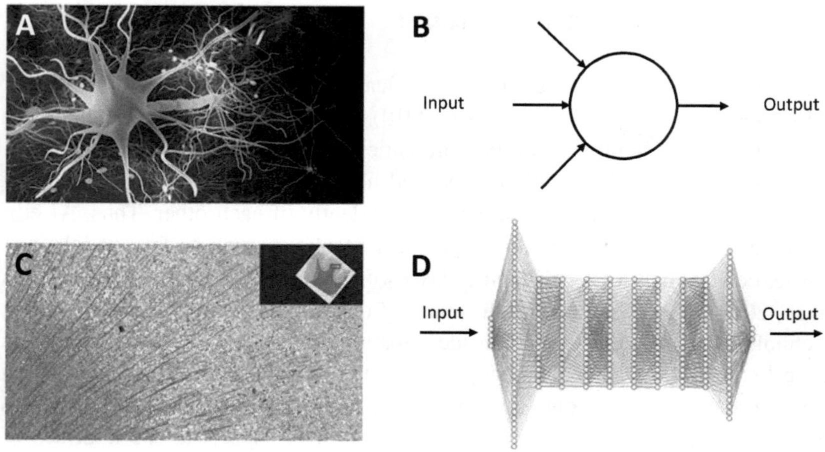

Fig. 1 Dualistic abstraction of neural structures as a mathematical basis for machine learning and AI technology. **A** Graphical representation of a single neuron in the cellular network (*in vivo,* embodied); **B** as a logical computing unit according to Hebb (1949) (*in silico,* disembodied). **C** Neural connections in cortical tissue section (*in vitro,* embodied); **D** Visualisation of a multistage AI learning algorithm of an artificially simulated neural network (*in silico,* disembodied), adapted and edited from DeVries et al. (2017)

At the same time, doubts arise in the practice-oriented research about the purely dualistic knowledge production with regard to its social integrability and benefit for human well-being (Jotterand & Bosco, 2020). In order to assess and regulate the side effects of dualistic developments such as AI technology, an effective science ethics is needed; this should take into account "that human reasoning and behaviour is defined by our physical and social experience and interaction with the world" (Price et al., 2009, p. 3). In the ethical consideration of the 'disembodied' AI technology and the possible social side effects of its widespread application, body and corporeality gain increasing importance for both human individually and socially, as well as machine knowledge production and their possible hybrid forms (Liu et al., 2018). In addition to AI- and data-ethically shaped socio-technical considerations, for example in the transformation of work processes, this motivates the consideration of the embodied foundations and effects of AI. A neuroethically shaped embodiment approach can support this by integrating biological, technological, social and psychological aspects of AI technology.

3 Challenges and Solutions

Although neuroscience research has significantly influenced the development of AI technology (Savage, 2019; Ullman, 2019) and promising application fields for AI in the field of neurotechnologies are anticipated (Yuste et al., 2017), the ethical consideration and regulation of AI and neuroscience research or technology development have largely developed independently of each other. Thus, AI ethics focuses mainly on technology-specific side effects such as fear of job loss, abuse, concealment of legal liability issues or unnoticed spread of biases (Jobin et al., 2019); in connection with the ethics of data-driven technologies (e.g. 'smart technologies'), AI ethics thus considers mainly ethical, legal and social aspects of technology development and application, e.g. in data and consumer protection (ibid., Hand, 2018). In contrast, neuroethics, with a biomedical focus, addresses mainly theoretical, empirical, practical and political aspects at the interface of neuroscience and bioethics in the four areas of (i) the human self, agency and responsibility; (ii) social policy for health care and education; (iii) therapeutic interventions; (iv) public discourse and education (Illes & Bird, 2006). With the ethical consideration of emerging neurotechnologies such as brain-machine interfaces and possible body interventions such as human-AI interfaces or cognitive enhancement, a further field of application of neuroethics influenced by the embodiment concept emerges (Kellmeyer et al., 2021).

Combined approaches at the interface of AI and neuroethics are discussed in the scientific literature so far mainly as opinion and direction discourses, e.g. in comment articles or blog posts, and consider, for example, the use of AI technology in neuroscience research and therapy or technology development (Wolkenstein et al., 2018). In this context, Yuste et al. (2017) formulate the four focal points *privacy, identity, agency, equality* for AI-ethical and neuroethical integrated research and development as 'responsible neuroengineering'. In the ethical debate on the social side effects of the AI-induced transformation of work processes, there is also the need to critically examine embodied forms of AI, e.g. in the form of autonomous machines and production processes—for the emergence of strong, autonomous or 'new' AI, the importance of their embodiment or their embodied representation is controversially discussed as a central criterion (Müller, 2007; Steels, 2007). The discursive relevance of the corporeality or materiality of AI becomes particularly evident in the example of the external appearance or function of AI-controlled robots and the human reference to them (Robert, 2017): It can manifest itself, for example, as fear of threat and displacement of human workers (McClure, 2018) or in a functional and beneficial human-machine cooperation (DeCanio, 2016; Decker et al., 2017) and ranges to ideas of the physically hybrid human-AI connection (Liu et al., 2018; Makin et al., 2017)—in any case, it emphasizes AI-related questions of social justice (Lankisch et al., 2019), social integrated dissemination and democratic regulation of cognitive technologies in the world of work (Ienca & Bird, 2019; Wang & Siau, 2019).

An AI- and neuroethically based embodiment approach can support the sociotechnical negotiation of these questions by tracing back the development of AI algorithms as a reflexive image of biological (e.g. neurally embodied) processes of learning and information processing of humans. In doing so, this approach can make the alienation of AI technology from biological intelligence, the dualistic abstraction of AI technology, transparently recognisable and (re)establish a functional relationship between 'embodied' human workers and 'bodiless' AI technology. Figure 2 shows exemplarily how AI and neural processes as well as analogously AI ethics and neuroethics can be related to each other and to a central embodiment concept; as a discursive basis, this concept can be interpreted more specifically for the world of work and can further make accessible the technological and biological processes underlying AI applications and their ethical consideration for social discourses. In a framework of participatory science communication, orientation knowledge can thus be provided to support discursive opinion formation and democratic decision-making for the socially integrated development, dissemination and regulation of AI technology.

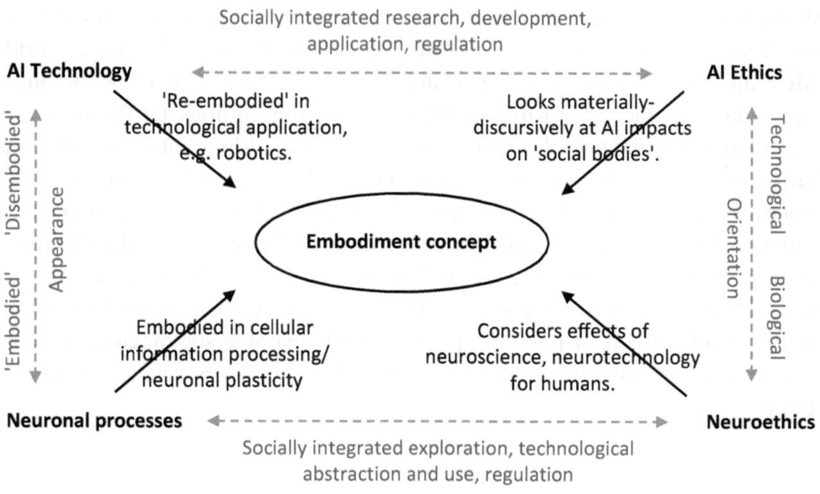

Fig. 2 Schematic formation of an embodiment concept based on a combined AI and neuroethical approach, which can support opinion formation and decision making in sociotechnical discourse processes as orientation knowledge

4 Outlook on AI Discourse in 2030

With regard to the diversity of interpretations of the embodiment concept in relation to AI technology (Ziemke, 2001), the approach outlined here proposes a combined concept from an AI ethics and neuroethics perspective. This approach overcomes the dualistic separation between biological materiality and a technologically abstracted 'artificial intelligence' by creating a 're-embodied' understanding of AI technology, which can provide an accessible and generally understandable basis for discourse in the debate about its socially integrated development, implementation and regulation in the world of work. In addition to a concrete 'mechanistic' or 'organismoid' (ibid.) embodiment of AI technology, e.g. in robotics, this embodiment approach aims above all to make the less visible forms and abstract effects of 'bodiless' AI, such as in search engine algorithms, tangible and comprehensible. This is not done by giving the technology a substitute illustrative body shape; instead, processes of biological information processing are characterised as the originally embodied basis of algorithm-based 'machine learning', so that in the sociotechnical discourse a reference back from abstract AI to concrete biological processes as 'organismic' or 'naturalistic' embodiment (ibid.) becomes possible.

Towards a 'rehumanisation' of AI technology (Jotterand & Bosco, 2020) and the social integration of its side-effect debate, this approach follows Merleau-Ponty's radically corporealistic attitude; contrary to a dualistic view, this describes the human body as "our general medium for having a world" (Merleau-Ponty, 2002, p. 66). For the debate on the future of work under increasing influence of AI-driven technologies, this implies (i) the necessity of a relational reference of workers to 're-embodied' AI technology within the framework of their individual and especially collective professional reality of experience as a 'social body'; this can be done by (ii) promoting social discourse and the necessary professional training in the sense of a democratised ethics debate on the desirable use and regulation of AI technology (cf. Illes & Bird, 2006); this can be supported by (iii) enabling a critical-normative examination of side effects or possible, probable and desirable sociotechnical future scenarios, using suitable media and formats of participatory science communication.

5 Summary and Practical Recommendations

For the ethical debate on AI regulation, a human-centred approach seems necessary to create trust in the potentials of the technology for future further automated work processes (Shneiderman, 2020). In doing so, the "rich phenomenological experience of rationality and behaviour" (Rainey & Erden, 2020, p. 19) of people in work processes influenced by AI technology should be taken into account; in addition, the presented approach of a reflexive reference back of abstract AI technology to its originally bodily located foundations of biological information processing can improve its negotiability with traditional work processes as embodied, material-discursive practice. In this 're-embodied' understanding, AI-based, machine-autonomised work processes become more comparable with the value of human work, for example in the wage debate. The presented embodiment approach can thus support the discourse on the socially integrated implementation and regulation of AI technology in the world of work in terms of content.

Acknowledgements We gratefully acknowledge the kind permission to use Fig. 1, Section C by University of Michigan Virtual Slide Box.

References

Coenen, C. (2014). Transhumanism in emerging technoscience as a challenge for the humanities and technology assessment. *Teorija in praksa, 51*(5), S. 754–771.

DeCanio, S. J. (2016). Robots and humans–complements or substitutes? *Journal of Macroeconomics, 49,* 280–291.

Decker, M., Fischer, M., & Ott, I. (2017). Service robotics and human labor: A first technology assessment of substitution and cooperation. *Robotics and Autonomous Systems, 87,* 348–354.

Descartes, R. (1984). *The philosophical writings of descartes: Volume 2* (Vol. 2). Cambridge University Press.

DeVries, P. M., Thompson, T. B., & Meade, B. J. (2017). Enabling large-scale viscoelastic calculations via neural network acceleration. *Geophysical Research Letters, 44*(6), 2662–2669.

Geraci, R. M. (2008). Apocalyptic AI: Religion and the promise of artificial intelligence. *Journal of the American Academy of Religion, 76*(1), 138–166.

Hand, D. J. (2018). Aspects of data ethics in a changing world: Where are we now? *Big data, 6*(3), 176–190.

Hebb, D. O. (1949). The organization of behavior; a neuropsycholocigal theory. *A Wiley Book in Clinical Psychology, 62,* 78.

Ienca, M. (2019). Democratizing cognitive technology: A proactive approach. *Ethics and Information Technology, 21*(4), 267–280.
Illes, J., & Bird, S. J. (2006). Neuroethics: A modern context for ethics in neuroscience. *Trends in neurosciences, 29*(9), 511–517.
Jobin, A., Ienca, M., & Vayena, E. (2019). The global landscape of AI ethics guidelines. *Nature Machine Intelligence, 1*(9), 389–399.
Jotterand, F., & Bosco, C. (2020). Keeping the "Human in the Loop" in the age of artificial intelligence. *Science and Engineering Ethics, 26*(5), 2455–2460.
Kellmeyer, P., Müller, O., & Voigt, J. (2021). Embodiment, movement and agency in Neuroethics. *Neuroethics,* 1–3.
Lankisch, C., Prettner, K., & Prskawetz, A. (2019). How can robots affect wage inequality? *Economic Modelling, 81,* 161–169.
Liu, R., Ren, Z. Q., & Wang, Z. Y. (2018). Dualism of Knowledge Creation for Human-Machine Interactive Processing. In *2018 2nd IEEE Advanced Information Management, Communicates, Electronic and Automation Control Conference (IMCEC)* (pp. 1487–1490). IEEE.
Makin, T. R., de Vignemont, F., & Faisal, A. A. (2017). Neurocognitive barriers to the embodiment of technology. *Nature Biomedical Engineering, 1*(1), 1–3.
McCarthy, J. (2004). What is Artificial Intelligence? http://jmc.stanford.edu/articles/whatisai.html. Accessed: 31. July 2021.
McClure, P. K. (2018). "You're fired", says the robot: The rise of automation in the workplace, technophobes, and fears of unemployment. *Social Science Computer Review, 36*(2), 139–156.
Merleau-Ponty, M. (2002). *Phenomenology of perception*. Routledge.
Müller, V. C. (2007). Is there a future for AI without representation? *Minds and Machines, 17*(1), 101–115.
Price, S., Roussos, G., Falcão, T. P., & Sheridan, J. G. (2009). Technology and embodiment: Relationships and implications for knowledge, creativity and communication. *Beyond Current Horizons, 29,* 1–22.
Rainey, S., & Erden, Y. J. (2020). Correcting the brain? The convergence of neuroscience, neurotechnology, psychiatry, and artificial intelligence. *Science and Engineering Ethics, 26*(5), 2439–2454.
Robert, L. P. (2017). The growing problem of humanizing robots. *International Robotics & Automation Journal,* 3(1).
Savage N (2019) How AI and neuroscience drive each other forwards. Nature, 571(7766), S15–S15.
Shneiderman, B. (2020). Human-centered artificial intelligence: Reliable, safe & trustworthy. *International Journal of Human–Computer Interaction, 36*(6), 495–504.
Steels, L. (2007). *Fifty years of AI: From symbols to embodiment – and back. In 50 years of artificial intelligence* (pp. 18–28). Springer.
Ullman, S. (2019). Using neuroscience to develop artificial intelligence. *Science, 363*(6428), 692–693.
Wang, W., & Siau, K. (2019). Artificial intelligence, machine learning, automation, robotics, future of work and future of humanity: A review and research agenda. *Journal of Database Management (JDM), 30*(1), 61–79.

Wolkenstein, A., Jox, R. J., & Friedrich, O. (2018). Brain-computer interfaces lessons to be learned from the ethics of algorithms. *Cambridge Quarterly of Healthcare Ethics, 4,* 635–646.

Yuste, R., Goering, S., Bi, G., Carmena, J. M., Carter, et al. (2017). Four ethical priorities for neurotechnologies and AI. *Nature News, 551*(7679), 159.

Ziemke, T. (2001). Disentangling notions of embodiment. In *Workshop on Developmental Embodied Cognition,* 83.

Legal aspects of AI in the world of work

Digital Product Monitoring Obligations for Smart Products

Opportunities and Risks of Digital Product Monitoring for IoT Products

Volker Hartmann

"Product liability and safety play a central role for smart products."—Dr. Volker Hartmann, VP of Legal & Governmental Affairs, Vay Technology GmbH

1 Introduction

More and more products are equipped with software and controlled by it, networked (IoT) and automated. Such "smart products" are conquering all kinds of industries and fields of application. Increasingly complex use cases have to be mastered by these robotic machines[1]. An increasing technical complexity goes hand in hand with increased requirements for the programming of corresponding control software, so that AI is often used.[2]

[1] Eidenmüller (2017, p. 1).

[2] Ebers (2017, p. 94 f.).

V. Hartmann (✉)
VP of Legal & Governmental Affairs, Vay Technology GmbH, Berlin, Germany
e-mail: volker.hartmann@vay.io

I. Knappertsbusch and K. Gondlach (eds.), *Work and AI 2030*,
https://doi.org/10.1007/978-3-658-40232-7_7

2 Status Quo/Inventory and Case Studies

2.1 Which Product Liability Law Applies to Manufacturers of Smart Products?

Currently, there is no lex specialis for product liability regarding smart products in the EU. Although the Commission initiated a review of the European legal framework for product liability and product safety for networked, software-based products in 2020 as part of the EU digital strategy.[3] This process is to be seen in the overall context of the EU Commission's AI strategy and AI white paper[4] and has not yet progressed to a reliable stage.[5]

2.2 The Current Product and Producer Liability as Possible Civil Law Bases for a Digital Product Monitoring Obligation

A civil liability of the manufacturer for damages caused by defective products results primarily from the Product Liability Act (§ 3 para. 1 ProdHaftG) and from the so-called producer liability according to § 823 para. 1 BGB (German Civil Code).

The latter also applies to damages caused by defective software—thus also to software-controlled smart products, regardless of whether the control software is permanently implemented in the product or located remotely in a cloud.[6] Although it is discussed whether the product concept of the Product Liability Act also covers software, it will probably be assumed that the Product Liability Act applies to smart products, in which a built-in software intelligence is responsible for the control, in accordance with the protective and normative purpose.[7]

An important difference between the two liability regimes exists with regard to the product monitoring obligation at issue here. This only takes effect after the product has been placed on the market and aims to control hazards that only

[3] Cf. European Commission (2020).

[4] Ebers et al. (2020) § 3 para. 168 ff. with further references.

[5] Cf. European Parliament (2020) and European Commission (2021); see also Theurer et al., (2021, p. 83 ff.).

[6] Günther J-P (2016, p. 146, 147, with further references).

[7] Habersack et al. (2017, § 2 ProdHaftG para. 17 ff.); Reusch and Weidner (2018, p. 29).

become apparent later "in the field". Only the producer liability according to § 823 para. 1 BGB establishes such a product monitoring obligation, which is in principle unlimited in time.[8]

2.3 The ProdSG as a Possible Regulatory Basis for a Digital Product Monitoring Obligation

In § 6 para. 3 ProdSG (Product Safety Act), it is stipulated for consumer products, among other things, that manufacturers must record customer complaints and communicate remedial measures to their dealers. Furthermore, authorities can probably also order digital hazard control measures via § 26 para. 2 s. 1 ProdSG.[9] In this respect, the regulatory content largely corresponds to the tortious product monitoring obligation.[10]

3 Challenges and Solutions

3.1 Initial Situation: No Explicitly Regulated Digital Product Monitoring Obligation

Whether the described requirements for the product monitoring obligation also apply to smart products is currently not explicitly regulated. Here, the question arises of a risk control by software updates for smart products and/or remote accesses (e.g. deactivating a (partial) function).[11]

3.2 Meaningfulness of a Digital Product Monitoring Obligation for Smart Products

An effective digital product monitoring obligation for smart products is almost imperative:

[8] Habersack et al. (2017, § 823 BGB para. 837).

[9] Lüftenegger (2021, p. 293, 299).

[10] Polly (2020, p. 77 ff.); on the GPSG at the time: Helmig (2005, 142, 143).

[11] Habersack et al. (2017, § 823 BGB Rn. 1008 ff.); Theurer et al., (2021, p. 83, 86); Raue (2017, p. 1841); Hartmann (2015, p. 122, 124 ff.); idem (2021, p. 47, 50).

3.2.1 Merging of Safety and Security

Product liability law primarily concerns the product safety itself (= safety), not security, which describes the protection of a product or an IT system against external manipulations/interventions by third parties.[12] However, with smart products, safety and security merge due to the functional relevance of the networking level.[13]

Even if the question of the legal impact of the intervention of third parties arises here, there is still an original product liability/product safety aspect: If product users could legitimately expect a design-wise foresight of the manufacturer in terms of security, the latter is already obliged by law to secure such technologies against corresponding interventions by third parties according to the state of the art (design obligation).[14] This should also be regularly assumed for smart products, whose essence lies precisely in the networking.[15]

If one affirms corresponding manufacturer obligations at the level of design, then there must also be a product monitoring obligation for smart products to remedy security risks.[16] The networking of these IoT products thus implies a content-wise expanded area of product monitoring (safety, but also security). The update capability for reactions from the product monitoring would then have to be incorporated into the system design from the outset and accordingly also given.

3.2.2 The Problem of So-Called Opacity of AI

AI seems to be surrounded by a certain mysticism in some aspects. This certainly includes the problem of the so-called opacity of AI. This refers to the fact that—depending on the exact method—the learning successes of neural networks are not always comprehensible and predictable.[17]

It is therefore sometimes argued that it is per se defective in design if an AI-based control software is designed in such a way that a self-learning algorithm creates risks that are unforeseeable and uncontrollable by the manufacturer in

---

[12] Cf. § 2 para. 2 BSIG; Klindt and Bräutigam (2015, p. 85 f.).

[13] Cf. Handelsblatt (2015).

[14] BGHZ 80, 186; BGHZ 80, 199.

[15] Hartmann (2017, p. 2, 8); likewise: Martin and Uhl (2020, p. 7, 10 f.); Ebers et al., (2020, § 5 Rn. 22 with further references).

[16] At least insofar as securing against external interventions is objectively necessary and reasonable for the manufacturer.

[17] Ebers et al., (2020, § 5, Rn. 2 with further references).

circulation.[18] In products prone to danger, AI is therefore usually used only in a "decoupled" variant, in which a separate quality assurance and release of the trained software takes place before placing on the market or an update is issued. Such software does not learn live in operation, but the "field data" serve the further development of the algorithm in a protected development environment. For a data-driven development, the manufacturer must ensure that the behaviour of the product is only changed by a centralised body, created in the development environment and again with secured updates.[19]

This results in massive implications for the product monitoring obligation: The described procedure requires a closed loop of development, observation of the product in operation and updates based on it.

4 Outlook on a Digital Product Monitoring Obligation in 2030

4.1 Forecast: The Digital Product Monitoring Obligation will Come

By 2030, a digital product monitoring obligation for smart products will be explicitly regulated by law and in technical regulations. This thesis can be supported by the following lines of argumentation.

4.2 Effectiveness of a Digital Product Monitoring Obligation

The product monitoring obligation according to the law as it exists is designed as a flexible instrument by case law, so that within the framework of what is objectively necessary and reasonable, the most effective possible risk management must take place.[20] Traditionally, the manufacturer has certain obligations to react in order to control identified product risks. These obligations range from warning

[18] Hilgendorf (2018, p. 85, 93); Hartmann (2017, 2, 7 with further references); differently or relativizing, however, Ebers et al., (2020, § 5 Rn. 24).

[19] Lukas (2021, p. 123, 126); Hartmann (2017, p. 2, 8); Rempe (2016, p. 17 ff.).

[20] BGHZ 179, 157 ("care beds"); Habersack et al. (2017, § 823 BGB Rn. 838).

obligations to product recalls, in exceptional cases even an obligation to retrofit or repair.[21] Translated into a digital product monitoring context, in which the manufacturer can directly contact the products and the users in the market via the Internet, this results in advantages in terms of the most effective[22] and comparatively milder[23] risk management:

- Digital warnings can be addressed precisely to affected product groups (software versioning!)[24] and the relevant recipient group (HMI, display on the screen).[25] The latter becomes particularly relevant in the case of security breaches, when public warnings could even cause more harm than benefit.[26]
- By software updates, hazards can be eliminated by targeted functional changes.[27] In the case of a situation which is considerably dangerous, the software or a certain software component or (partial) function can be temporarily deactivated until the update is installed.[28]
- As a last resort, a complete deactivation as a counterpart of the decommissioning of the smart product by remote access of the manufacturer could be considered.[29]

4.3 Regulatory Management Systems (Example: Automotive Sector)

Current regulatory trends, especially in the automotive sector, also speak in favour of a digital product monitoring obligation.

Through the EU type approval law for motor vehicles[30] technical regulations of the UNECE (United Nations Economic Commission for Europe), the so-called

---

[21] Ebers et al., (2020, § 5 Rn. 38 ff.).

[22] Lüftenegger (2021, p. 294, 295.).

[23] Lüftenegger (2021, p. 293, 296 f.).

[24] Lüftenegger (2021, p. 293, 296.).

[25] Ebers et al., (2020, § 5 Rn. 39, 41); Lüftenegger (2021, p. 293, 294.).

[26] Rockstroh and Kunkel (2017, p. 77, 81); May and Gaden (2018, p. 110, 112).

[27] Lüftenegger (2021, p. 293, 298).

[28] Ebers et al., (2020, § 5 Rn. 41); Lüftenegger (2021, p. 293, 298 ff.).

[29] On this, see ECJ NJW (2015, p. 1163 Rn. 53); Schrader (2018, p. 314, 317).

[30] Regulation 2018/858/EU.

UNECE rules, are incorporated into the regulatory framework of the technical regulation for motor vehicles in the EU.[31] By UNECE rules, aspects for connected and (partially) automated vehicles have recently been regulated, which each include management systems, according to which the manufacturer must demonstrate the compliance with certain processes until an end-of-lifecycle or end-of-service.

This proof must be provided by the manufacturer as a prerequisite for granting a type approval, quasi as a regulatory market access requirement, before the product is put into circulation.

The UNECE rules R155 (Cyber Security), R156 (Software Updates) and R157 (Automated Lane Keeping Systems) are mentioned in this respect.[32] Particularly interesting for the area of product monitoring obligation are current plans for an adapted UNECE-R157, which requires an Automated Driving Management System (= ADMS) for automated vehicles.[33] In the current draft status, a management system for ensuring product safety, including cyber security, over the entire service life of the product is described under Annex 4, Sect. 3.5 (Safety Management System).

With an ADMS, the regulatory foundation of a digital product monitoring obligation for automated vehicles would be created. This applies all the more in conjunction with the Cyber Security Management System (CSMS) prescribed in UNECE-R155 and the Software Update Management System (SUMS) according to UNECE-R156. An incorporation of the corresponding technical regulations into the EU type approval law thus creates a comprehensive framework (CSMS + SUMS + ADMS), which on the one hand obliges the manufacturer of connected, automated motor vehicles to implement and maintain corresponding processes, and on the other hand also shapes the legitimate consumer expectation of the safety of the products in terms of product liability law.[34]

---

[31] See Art. 57 (1), (2), 34 (1), 35 (1), 58 (1) and Annex II of Regulation 2018/858/EU; the UNECE rules are based on the UN agreements on vehicle regulation of 1958 and 1998, to which the EU is a party; see Gaupp (2019, p. 163).

[32] Theurer et al., (2021, p. 83, 85).

[33] United Nations Economic Council for Europe (2021).

[34] Theurer et al., (2021, p. 83 ff.).

4.4 Technical Norms and Standards: Security of Connected (IoT-)Consumer Products

Also a tendency towards digital product observation manifests itself in the standard ETSI EN 303 645 (security of connected (IoT-)consumer products).[35] The standard contains a catalogue of (minimum) requirements for the security of smart products. This includes for the manufacturer, besides design obligations, also product observation obligations, such as maintaining a system for managing reports on cyber security vulnerabilities and recommendations regarding software updates for smart products.

Ultimately, mechanisms are defined at the level of a technical standard that can be classified as congruent with existing legal product liability and product safety obligations.[36]

5 Summary and Practical Recommendations

A digital product observation obligation creates advantages for users and manufacturers, as it uses the networking level of smart products as a direct and effective communication and interaction level.

For manufacturers, however, a digital product observation obligation also entails many challenges and additional expenses. The digital product observation obligation must be translated into internal company processes and procedures. Due to the presented, and increasingly expected, post-deployment management systems, a lifecyle management for product safety, including cyber security, must be ensured. But also the congruent product liability requirements (to avoid civil liability for violating the digital product observation obligation) fall into the same category.

Considerable amounts of data from the market must be collected, evaluated and processed in a structured way, in order to adequately handle findings relevant for hazard control. Great challenges are therefore to be expected in terms of the qualification of the staff, the creation of appropriate IT infrastructure and the adaptation of internal company processes. The latter are likely to affect the (tech-

[35] European Telecommunications Standards Institute (2020).

[36] Theurer et al., (2021, p. 83, 84).

nical) management, approval and error correction processes, the quality assurance, but also the legal advice.[37]

References

Ebers, M. (2017). Autonomes Fahren: Produkt- und Produzentenhaftung. In B. Oppermann & J. Stender-Vorwachs (Hrsg.), *Autonomes Fahren* (pp. 94–125). Beck.

Ebers, M., Heinze, C., Krügel, T., & Steinrötter, B. (2020). *Rechtshandbuch Künstliche Intelligenz und Robotik*. Beck.

Eidenmüller, H. (2017). The rise of robots and the law of humans. *Oxford Legal Studies Research Paper No., 27,* 1–15.

Europäische Kommission. (2020). Report on the safety and liability implications of Artificial Intelligence, the Internet of things and robotics. https://ec.europa.eu/info/sites/default/files/report-safety-liability-artificial-intelligence-feb2020_en_1.pdf. Accessed: 30. June 2021.

Europäische Kommission. (2021). Entwurf einer VO für Harmonisierte Regeln für Künstliche Intelligenz (Artificial Intelligence Act), COM(2021) 206 final. https://eur-lex.europa.eu/resource.html?uri=cellar:e0649735-a372-11eb-9585-01aa75ed71a1.0019.02/DOC_1&format=PDF. Accessed: 4. July 2021.

Europäisches Parlament. (2020). Entwurf eines Berichts des Europäischen über die Produktsicherheit im Binnenmarkt vom 03.04.2020, 2019/2190(INI). https://www.europarl.europa.eu/doceo/document/A-9-2020-0207_DE.html. Accessed: 4. July 2021.

European Telecommunications Standards Institute. (2020). Cyber Security for Consumer Internet of Things: Baseline Requirements. https://www.etsi.org/deliver/etsi_en/303600_303699/303645/02.01.01_60/en_303645v020101p.pdf. Accessed: 4. July 2021.

Gaupp, W. (2019). *Type approval legislation for road vehicles*. Kirschbaum Verlag.

Günther, J.-P. (2016). *Roboter und rechtliche Verantwortung*. Hubert Utz Verlag.

Handelsblatt. (2015). Chrysler ruft 1,4 Millionen Fahrzeuge zurück. https://www.handelsblatt.com/mobilitaet/motor/nach-jeep-hack-chrysler-ruft-1-4-millionen-autos-zurueck/12102998.html?ticket=ST-9325251-btPrTOlas0byEIHteooE-ap4. Accessed: 4. July 2021.

Hartmann, V. (2015). Big Data und Produkthaftung. *DAR,* 122–126.

Hartmann, V. (2017). Here come the robots. *PHi,* 2–9.

Hartmann, V. (2021). Software als Steuerungsintelligenz in intelligenten Produkten. *REthinking:Law,* 47–51.

Helmig, E. (2005). Marktbeobachtungspflicht für Hersteller und Händler unter dem GPSG. *PHi,* 140–145.

Hilgendorf, E. (2018). Offene Fragen der neuen Mobilität: Problemfelder im Kontext von automatisiertem Fahren und Recht. *RAW,* 85–93.

[37] Hartmann (2015, p. 122, 126); Lüftenegger (2021, p. 293, 300).

Klindt, T., & Bräutigam, P. (2015). Digitalisierte Wirtschaft/Industrie 4.0, Gutachten für den BDI. https://bdi.eu/media/themenfelder/digitalisierung/downloads/20151117_Digitalisierte_Wirtschaft_Industrie_40_Gutachten_der_Noerr_LLP.pdf. Accessed: 4. July 2021.

Lukas, A. (2021). Haftungsfragen autonomer Produktionsnetzwerke in der Industrie 4.0. *ZdiW,* 123–128.

Lüftenegger, K. (2021). Alexa, warum geht der Kühlschrank nicht mehr? *RDi,* 293–300.

Martin, M., & Uhl, K. (2020). Cyberrisiken bei vernetzten Fahrzeugen. *RAW,* 7–14.

May, E., & Gaden, J. (2018). Vernetzte Fahrzeuge. *InTeR,* 110–116.

Habersack, M., et al. (2017). *Münchener Kommentar zum Bürgerlichen Gesetzbuch*. Beck.

Polly, S. (2020). *EU Products Law*. Dr. Sebastian Polly.

Raue, B. (2017). Haftung für unsichere Software. *NJW,* 1841–1846.

Rempe, C. (2016). Smart Products in Haftung und Regress. *InTeR,* 17–21.

Reusch, P., & Weidner, N. (2018). *Future law*. Deutscher Fachverlag.

Rockstroh, S., & Kunkel, H. (2017). IT-Sicherheit in Produktionsumgebungen. *MMR,* 77–82.

Schrader, P. (2018). Herstellerhaftung nach dem StVG-ÄndG 2017. *DAR,* 314–320.

Theurer, J., Reinsberg, J., Borst, L., & Bosch, P. (2021). Perspektiven der Produkthaftung und Produktsicherheit in der Industrie 4.0. *ZdiW,* 83–87.

United Nations Economic Council for Europe. (2021). Agreement Concerning the Adoption of Harmonized Technical United Nations Regulations for Wheeled Vehicles, Equipment and Parts which can be Fitted and/or be Used on Wheeled Vehicles and the Conditions for Reciprocal Recognition of Approvals Granted on the Basis of these United Nations. https://unece.org/sites/default/files/2021-03/R157e.pdf. Accessed: 4. July 2021.

The Use of AI-Based Speech Analysis in the Application Process

Patricia Jares and Tobias Vogt

1 Introduction

At the beginning of every employment relationship is the application process. Often, human resources professionals face an overwhelming flood of applications. When selecting candidates, not only the professional qualification plays a role. Many companies want to hire employees who fit the job profile and team based on their personality. The use of AI-based speech analysis can not only support this, but ideally also contribute to an objectification of the application process. Because the potential for discrimination in the early stage of the application process is empirically proven. For example, in the context of a study, the application documents of a candidate with a German name received significantly more invitations to an interview than an identical application with a foreign name (Expert Council of German Foundations for Integration and Migration, 2014).

2 Status Quo

For several years now, an AI-based speech analysis software for recruiting has also been offered on the German market by a company. The software determines personality traits of the speaker such as creativity, commitment, goal orientation,

P. Jares (✉)
CMS Germany, Köln, Germany
e-mail: Patricia.Jares@cms-hs.com

T. Vogt
Köln, Germany

I. Knappertsbusch and K. Gondlach (eds.), *Work and AI 2030*,
https://doi.org/10.1007/978-3-658-40232-7_8

resilience, sociability or assertiveness based on a digital phone interview of about 15 minutes. However, what a candidate says is not decisive. The programme analyses how the candidate says something, both prosodically (voice guidance, pitch, etc.) and linguistically (sentence structure, word choice, etc.) and decodes in this way psychologically relevant speech patterns. There is no case law on the admissibility of such software yet.

3 Challenges and Solutions

3.1 Data Protection Law

Speech analysis in an application interview with regard to characteristics and skills constitutes a processing of personal data within the meaning of Art. 2 I GDPR or § 1 I BDSG (Betz, 2019, p. 148). The use of such speech analysis is to be gauged against the provisions of the GDPR and the BDSG and requires data protection permission.

3.1.1 Consent

As data protection permission, consent according to Art. 4 No. 11 GDPR comes into question. This must be voluntary. When processing employee data, § 26 II 1 BDSG requires taking into account the dependence of the employee in the employment relationship and the circumstances under which the consent was given. If the speech analysis is a prerequisite for going through the further application process, there will be a lack of voluntariness due to the factual compulsion. In this case, consent as data protection permission norm is ruled out (Gola, 2018, p. 24, 26; LDI NRW, 2017, p. 53; Joos, 2020, p. 1216, 1220).

Regardless of the possibility of consent as data protection permission, however, in view of § 201 StGB (German Penal Code) and the civil law protected right to one's own word, consent of the data subjects must always be obtained (Gola, 2018, p. 24, 25; LDI NRW, 2017, p. 63).

3.1.2 § 26 I 1 BDSG Necessity

As a justification, the necessity for the establishment of the employment relationship may be considered, which is regulated as a permission in § 26 I 1 BDSG. Necessity here does not mean a compelling necessity, but an interest balancing oriented to the principle of proportionality (Gola & Heckmann, 2019, § 26 para. 16; Joos, 2020, p. 1216, 1220). The data processing must be objectively suitable for achieving the purpose. A less burdensome alternative for the data subject must either not exist or be unreasonable for the employer. In addition, there must be proportionality in the narrower sense (Betz, 2019, p. 148, 149; Gola, 2018, p. 24, 25).

Even if one assumes the basic suitability of the speech analysis, it cannot be completely ruled out that individual applicants are assigned or denied incorrect characteristics based on the comparison with the reference data of the programme and depending on the quality and quantity of the data basis. From the mention of profiling in Art. 22 GDPR, it follows by implication that profiling—and thus not 100 %-certain predictions—can be data protection compliant. Therefore, it does not have to be completely ruled out that incorrect results can occur in individual cases. However, the principle of data accuracy from Art. 5 I d) GDPR, to which § 26 V BDSG explicitly refers, as well as recital Art. 71 GDPR, must be taken into account. Therefore, the employer has to take all technical and organisational measures that ensure in an appropriate manner that factors that lead to incorrect personal data are corrected and the risk of errors is minimised (Gola DS-GVO 2018, Art. 5 para. 24; Betz, 2019, p. 148, 149). From this, at least the necessity to use a programme that corresponds to the state of the art should arise.

Less burdensome measures include, in particular, assessment centres or non-automated psychological tests, for example by business psychologists. Not only for the employer, who can significantly speed up the application process and save costs, but also for the affected applicants, the automated speech analysis has advantages. For example, participating in an assessment centre or a psychological examination usually involves a higher expenditure of time. Unlike an automated analysis, it cannot be excluded by programming that "as a by-product" personality traits that are not necessary for the employment relationship also come to light when examined by humans. In any case, no generally valid statement can probably be made about what is perceived as less intrusive from the perspective of each applicant—to be analysed by strangers in the context of a personal encounter over a period of hours or automatically by a short phone call. Therefore, no equally suitable but less burdensome measure seems to exist (Betz, 2019, p. 148, 149; contra LDI NRW, 2017, p. 63).

Proportionality in the narrower sense is to be assessed on the basis of a balancing of interests between the personality rights of the applicant and the interest of the employer in the data processing (Betz, 2019, p. 148, 150). Occasionally, the assignment or denial of characteristics and expertise based on an automated speech analysis is seen as generally disproportionate (so Gola, 2018, p. 24, 27). However, it is not convincing to assume a blanket violation of the personality rights of the applicant. Evaluating an applicants's personality based on a speech analysis constitutes a significant interference with the personality rights. However, such interferences can be justified. The employer's interest to find a suitable applicant and to save resources, especially in the case of a large number of applicants, is to be recognised. Ultimately, it will depend on the specific design of the speech analysis and the requirements for the position to be filled in the individual case.

A comprehensive personality screening is inadmissible (Dzida, 2017, p. 541, 545; Betz, 2019, p. 148, 150). The examination must be limited to those characteristics that are relevant for the specific purpose—i.e. for the specific position (Betz, 2019, p. 148, 150). It would therefore be inadmissible, for example, to determine the team ability of the applicant, which is irrelevant for the job as a field service employee (Betz, 2019, p. 148, 150). If rather harmless characteristics such as team ability, commitment or creativity are evaluated, the interest of the employer in the data processing will usually prevail—provided that these characteristics represent a relevant requirement for the specific position. The more sensitive a characteristic is, the more difficult a justification becomes due to the intensity of the interference (Dzida, 2017, p. 541, 545). Guidance can be taken from the case law on the employer's right to ask questions in job interviews, according to which questions about pregnancy, political party or union membership or family planning are always inadmissible. The employer may not obtain information that he himself may not ask through the "back door" by means of an AI-based system (Kaulartz/Braegelmann 2020, chap. 11 para. 30; Dzida, 2017, p. 541, 544).

3.1.3 Permissible Degree of Automation of the Selection Process

A problem is to what extent the AI-based programme is allowed to sort out applicants itself. Because according to Art. 22 I GDPR, decisions with legal consequences or significant impairments are generally inadmissible, as far as they are based exclusively on automated processing. In any case, a procedure in which applicants are already finally sorted out directly by the AI-based programme, without a natural person being involved, falls under the general prohibition of Art. 22 I GDPR. How cases are to be assessed in which the ratings of all applicants determined by the programme are transmitted to the HR team, which then select suitable applicants for an interview on this basis, is controversial. Some demand that all applicants who are subjected to a language analysis be invited to an interview afterwards (Gola, 2018, p. 24, 27; LDI NRW, 2017, p. 62). However, this is likely to be too far-reaching, as Art. 22 GDPR does not prohibit that a human decision is made on the basis of a previous automated data processing. This already follows from the wording of Art. 22 GDPR, according to which the data subject only has the right not to be subject to a decision based "exclusively" on automated processing. Recital Art. 71 of the GDPR speaks of a decision "without any human intervention". Therefore, it is sufficient that a human being is interposed, who has the final decision-making authority and can also override the decision proposed by the AI-based programme (Paal & Pauly, 2021, Art. 22 para. 16, 17b f.; Kaulartz & Braegelmann, 2020, chap. 8.4 para. 7; cf. also BT-Drs. 16/10529, 13 on the predecessor provision § 6a BDSG aF). However, it does not do justice to this if only formally a person is interposed, who theoretically has the possibility of making his or her own, deviating decision, but in practice implements the decision intended by the programme without his or her own examination (Paal & Pauly, 2021, Art. 22 para.19; Kaulartz & Braegelmann, 2020, chap. 8.4 para. 7). The acting employee must therefore include further aspects in his or her decision (Kaulartz & Braegelmann, 2020, chap. 8.4 para. 7)—such as the application documents.

An automated decision falling under Art. 22 I GDPR would only be permissible if an exception to this prohibition provided for in Art. 22 II GDPR applies. However, consent pursuant to Art. 22 II c) GDPR is ruled out just as well as under § 26 II 1 BDSG due to the pressure the applicant is subjected to. Nor can it be based on the only conceivable Art. 22 II a) GDPR, according to which a decision based exclusively on automated processing is permissible if this is necessary for the conclusion or performance of a contract between the data subject and the controller. Necessity within the meaning of Art. 22 II a) GDPR does not mean that no equally suitable, but less burdensome means for the data subject is apparent (but so Betz, 2019, p. 148, 150; Joos, 2020, p. 1216, 1217), since the exceptional character of Art. 22 II a) of the basic inadmissibility of the exclusively automated decision speaks for a stricter standard (Kaulartz & Braegelmann, 2020, chap. 8.4 para. 15). The provision is intended to prevent certain contracts from becoming impracticable to conclude or perform due to the regulation of Art. 22 I GDPR, such as mass transactions (Gola GDPR 2018, Art. 22 para. 29). Art. 22 II a) GDPR only applies according to the prevailing opinion if there is no reasonable alternative to the automated decision-making for the specific contractual situation (Kaulartz & Braegelmann, 2020, chap. 8.4 para. 15; Sydow GDPR/*Helfrich* Art. 22 para. 56; Taeger & Gabel, 2019, Art. 22 para. 51). Instead of AI-based speech analysis, it is certainly possible to resort to other possible, not exclusively automated decision processes in the application process in a reasonable manner (Kaulartz & Braegelmann, 2020, chap. 8.4 para. 15).

3.2 Discrimination/AGG

Another very significant limit to the employer's right to ask questions and thus also to data processing by AI are the discrimination prohibitions of the General Equal Treatment Act (AGG). According to §§ 1, 7 AGG, employees must not be discriminated against on grounds of race or ethnic origin, gender, religion or belief, disability, age or sexual identity. The prohibition of discrimination also applies to applicants, §§ 2 I No. 1, 6 I 2 AGG. An applicant must not be rejected because of one of these characteristics. Questions that may indicate discrimination on the basis of one of these characteristics are therefore prohibited in the application process.

With regard to several characteristics, the use of automated speech analysis could pose a risk of discrimination. This would be the case, for example, if the programme determines the applicant's origin based on word choice and vocabulary. The processing of data on the applicant's ethnic origin would also be inadmissible under data protection law due to the high requirements of Art. 9 GDPR. However, there is also potential for discrimination if the programme does not determine the origin, but due to a lower vocabulary of the German language or a slower speaking speed, it erroneously produces negative analysis results with regard to personality traits such as creativity (cf. also LDI NRW, 2017, p. 61). Then, the non-consideration of the applicant could mean an indirect discrimination on the grounds of ethnic origin. Also problematic could be discrimination on the grounds of gender. This is possibly the case if the programme also analyses the pitch of the voice and thus attributes gender-typical characteristics to the applicants.

4 Outlook on AI-Based Language Analysis in the Application Process in 2030

In view of the fact that the use of AI is generally becoming more and more important and can also contribute to a simplification in human resources management, it can be expected that AI-based language analyses in the application process will be widespread by 2030. The advantages and possibilities for insight compared to a purely traditional personnel selection are obvious. In order for the language analysis to be used in a legally secure manner, corresponding data sets that relate to or are associated with inadmissible differentiation characteristics according to the AGG (General Equal Treatment Act) should not occur in the development and programming of the software in the learning of the programme. An unsupervised, i.e. independent learning of the programme should therefore only take place outside the area of inadmissible questions. At least it would have to be ensured that the corresponding information is neither disclosed to the employer nor taken into account in the decision-making process (Kaulartz/Braegelmann, chapter 11, para. 32). In this way, discrimination in the application process can be reduced. Because a human being cannot be programmed to be free of discrimination, unlike a programme. Even after appropriate training and sensitisation, a human recruiter will not always succeed in completely freeing himself from prejudices and stereotypes that often unconsciously influence his own decisions. By using a discrimination-free software, it can be prevented that suitable applicants are wrongly sorted out at an early stage, because the recruiter has the results of the language analysis in addition to the application folder. Even if the risk of discrimination still existed in the later personal interview, the risk there would be lower than at the beginning of the process. Especially in the personal interview, the applicant has the opportunity to convince by personal impression, while he might not have made it to the "next round" in the early stage of the application despite excellent professional qualifications, simply because of a foreign-sounding name or gender.

5 Summary and Practical Recommendations

The use of AI-based language analysis in the application process can be permissible. However, only personality traits that are necessary for the specific position and covered by the employer's right to ask questions may be determined. Decisions in the application process must not be made exclusively by the programme, but must be made by a natural person with decision-making authority and taking into account other factors. In addition, care must be taken that the software does not base its evaluation on discrimination characteristics that are inadmissible according to § 1 AGG.

References

Betz, C. (2019). Automatisierte Sprachanalyse zum Profiling von Stellenbewerbern. *ZD,* 148–152.

Dzida, B. (2017). Big Data und Arbeitsrecht. *NZA, 2017,* 541–546.

Gola, P. (2018). Aus den aktuellen Berichten der Aufsichtsbehörden (33): Die Digitalisierung des Bewerbermanagements – Videointerviews bei der Bewerbung. *RDV, 2018,* 24–28.

Gola, P., & Heckmann, D. (2019). *Bundesdatenschutzgesetz*. Beck.

Joos, D. (2020). Einsatz von künstlicher Intelligenz im Personalwesen unter Berücksichtigung der DS-GVO und des BDSG. *NZA, 2020,* 1216–1221.

Kaulartz, M., & Braegelmann, T. (2020). *Rechtshandbuch Artificial Intelligence und Machine Learning*. Beck.

Landesbeauftragte für Datenschutz und Informationsfreiheit NRW. (2017). Jahresbericht 2017.

Paal, B., & Pauly, D. A. (2021). *Datenschutz-Grundverordnung*. Beck.

Sachverständigenrat deutscher Stiftungen für Integration und Migration. (2014). *Diskriminierung am Ausbildungsmarkt: Ausmaß, Ursachen und Handlungsperspektiven.*

Taeger, J., & Gabel, D. (2019). *DSGVO – BDSG*. R&W.

Individual Labour Law Issues in the Use of AI

Can Kömek

1 Introduction

The technical development of artificial intelligence (AI) is impressive. Even today, artificial neural networks can recognise and generate images, understand, speak and translate our languages and solve complex logical tasks. AI systems can access vast amounts of data ("Big Data") and continuously improve their own capabilities through so-called "Machine Learning".[1] An AI can handle numerous processes faster and better than a human ever could. By 2030, many more, technically even more advanced AI systems will be market-ready and in daily use. For the world of work, this results in a multitude of conceivable applications. The individual employment law questions that arise in this context depend on the specific use cases of the respective systems. So far, an "AI boss" is still largely a thing of the future.[2] Precisely for this reason, it is worthwhile to deal with the legal possibilities and limits in order to set the course for the future use of AI in the employment relationship.

[1] Geminn (2021, p. 354 f.).
[2] Arnold and Winzer (2018, para. 230).

C. Kömek (✉)
CMS Germany, Hamburg, Germany
e-mail: can.koemek@cms-hs.com

I. Knappertsbusch and K. Gondlach (eds.), *Work and AI 2030*,
https://doi.org/10.1007/978-3-658-40232-7_9

2 Status Quo: Framework Conditions for the Use of AI in the Employment Relationship

The individual employment relationship is characterised by numerous selection and consideration processes that could be controlled by AI in the future. For example, the activity of the employee is only roughly outlined in most employment contracts (e.g.: "You are hired as a salesperson"). The decision as to which specific tasks the employee has to perform when, where and how is only made by individual instructions. The right to issue these instructions belongs to the employer according to § 106 S. 1 GewO (German Industrial Code). He may delegate his right of instruction to third parties (e.g. to managers in the company) and have it exercised through them. It is therefore also possible that the employer exercises his right of instruction through an AI.[3] However, the right of instruction may not be exercised limitless, but only according to "reasonable discretion". The person issuing the instruction must take into account the respective circumstances of the individual case and the legitimate interests of the employee. If an instruction does not comply with reasonable discretion, it is not binding for the employee according to § 315 para. 3 BGB (German Civil Code). In case of doubt, the employee can have this question clarified by a labour court.

Instructions being issued by an AI is already a reality in some companies. In the logistics centers of the Japanese electronics group Hitachi, an AI system assigns specific work tasks to the employees. In doing so, the AI not only takes into account rigid work specifications, but also analyses the human work processes and takes into account dynamic conditions such as the weather or changes in demand.[4] The potential use cases of an "AI boss" go beyond that, however. Instructions are only one sub-area of the selection and consideration processes that exist in the employment relationship. One can also think of decisions about performance- or discretion-based bonuses, decisions about requests (e.g. for vacation, parental leave, part-time work), assessment decisions (e.g. performance assessment for a certificate or required training measures as well as risk assessment in occupational safety), investigations and evaluations of breaches of duty (e.g. by analysing performance data or detecting irregularities), selection decisions (e.g. for applications, transfers or promotions; or within the scope of a social selection

---

[3] Arnold and Winzer (2018, para. 236); Groß and Gressel (2016, p. 994); Günther and Böglmüller (2017, p. 55 f.).

[4] https://www.faz.net/aktuell/feuilleton/debatten/die-digital-debatte/kuenstliche-intelligenz-roboter-als-chef-14239957.html

for operational dismissals) or planning decisions (e.g. vacation and shift schedules).[5] In all these areas, an AI can prepare or even fully automate processes. The employer is not relieved of complying with legal requirements, such as the aforementioned reasonable discretion, even in these cases.[6] Other legal requirements are, in particular, the factual correctness and non-discrimination of a decision (cf. § 612a BGB, § 7 AGG, German General Act on Equal Treatment) or the existence of "urgent operational reasons" as a prerequisite for many labour law provisions (cf. only § 1 para. 2 KSchG, German Act against Unfair Dismissal, § 7 para. 2 BUrlG, German Federal Leave Act, § 15 para. 4 BEEG, German Federal Parental Allowance and Parental Leave Act, § 9 S. 1 No. 4 TzBfG, German Part-Time and Fixed-term Employment Act). Whether the employer has complied with these legal requirements is subject to judicial review in case of doubt and must be demonstrated and proven by him—also in case of using AI—if necessary.

3 Legal Challenges and Solutions

Whether and how AI can be used profitably in the employment relationship is not only a technical or economic, but always also a legal question, due to the aforementioned framework conditions. The starting point is the entrepreneurial freedom of the employer guaranteed by Art. 12 para. 1 AGG, which among other things grants him the right to organise the business according to his ideas. The entrepreneurial freedom also allows the use of new technologies such as AI. However, this is opposed by legitimate interests of the employees.[7] Their concern is that they are exposed to dehumanising, intransparent or unreasonable decisions, against which they are powerless. The freedom risk[8] associated with this is not to be underestimated. It is already difficult to argue with some human superiors, but virtually impossible to do so with an AI.[9] For employers, it is therefore crucial whether and how AI can be used legally in the employment relationship.

Of central importance is the insight that German labour law does not establish a principle according to which selection and consideration processes must always be carried out by a human being. Thus, § 106 s. 1 GewO (German Indus-

[5] See also the examples in Broy and Heinson (2019, para. 5 ff., 61); Joos (2020, p. 1216).

[6] Arnold and Winzer (2018, para. 237).

[7] Cf. Groß and Gressel (2016, p. 994 f.).

[8] Broy and Heinson (2019, para. 1).

[9] Broy and Heinson (2019, para. 18).

trial Code) and § 315 para. 3 BGB do not speak of "*human* discretion", but only of "*reasonable* discretion". But what corresponds to reasonableness and is also in accordance with the labour law requirements can be taught to an AI with appropriate legal expertise and is thus initially a technical question.[10] Thus, the AI can be trained to respect the principle of equal treatment applicable in labour law, to comply with mandatory legal provisions (e.g. on working time, §§ 3 ff. ArbZG (German Work Time Act)) and to take into account the social data (seniority, age, maintenance obligations, severe disability) and other legitimate interests of the employees in its decisions. Only the question of whether the outcome of the decisions produced by an AI are in compliance with legal requirements and can withstand judicial review is a legal one.

The greater hurdle is not German labour law, but European data protection law.[11] According to Art. 22 para. 1 GDPR (European General Data Protection Regulation), decisions that have legal effects on a person or significantly affect them in a similar way must not be based solely on automated data processing. Only in exceptional cases is this permissible according to Art. 22 para. 2 GDPR ("general prohibition with permission reservation"). Whether the prohibition of automated individual decisions precludes the use of AI in the employment relationship without exception is largely unclear. The labour court jurisprudence has not yet had to deal with this question, as far as can be seen. The reason for this is, besides the still low diffusion of AI in the world of work, mainly that automated decisions so far played a role rather in the automatic granting of credit and the associated scoring.[12] On closer inspection, however, there are good reasons to assume that the legal hurdle of Art. 22 (1) GDPR is not insurmountable in the employment relationship.[13] Depending on the use case of the AI, it is already questionable whether the provision is applicable at all. It is required that the decision is based *exclusively* on an automated data processing and has a *legal effect* or a *similarly significant impact* on the data subject. If an AI only prepares selection and consideration decisions and these still require a substantial, not merely superficial assessment by a human, Art. 22 para. 1 GDPR is not relevant.[14]

In particular, it is controversial whether a fully automated AI instruction always falls under the prohibition of Art. 22 para. 1 GDPR. It could be argued

[10] Günther and Böglmüller (2017, p. 56).

[11] Cf. Arnold and Winzer (2018, para. 234).

[12] Broy and Heinson (2019, para. 2); Günther and Böglmüller (2017, p. 56).

[13] Cf. Broy and Heinson (2019, para. 25).

[14] Arnold and Winzer (2018, para. 324).

that an instruction does not have any direct legal effect, as it only specifies the existing work obligation and does not create a new performance obligation.[15] However, one can hardly deny an employment law instruction its legal effect. If the employee refuses a permissible instruction, he commits a breach of duty, which can be punished with a warning or dismissal.[16]

However, even according to Art. 22 para. 1 GDPR, generally prohibited AI decisions can be permissible if the exception of Art. 22 para. 2 lit. a) GDPR is not interpreted too strictly. According to this, an automated individual decision is permissible if it is necessary for the fulfilment of a contract. The case law has not yet specified in more detail when this "necessity" exists, so that there are still corresponding scope for evaluation.[17] In view of the superior performance, objectivity and efficiency of AI systems compared to human decision-makers, it seems anything but far-fetched to evaluate their use in the employment relationship as "necessary". Regardless of this, the national legislator could also use the exception of Art. 22 para. 2 lit. b) GDPR to create a permission norm specifically for the use of AI in the employment relationship. However, even in the case of an exceptionally permissible, automated individual decision, the employer must ensure appropriate safeguards in accordance with Art. 22 para. 3 GDPR. In particular, the intervention of a human in the automated decision of the AI must always remain possible.

4 Intermediate Result and Evaluation

The employer is generally free to use AI in his employment relationships. German labour law does not oblige the employer to carry out all selection and consideration processes by human decision-makers. Nevertheless, there may be co-determination and data protection obligations, which are highlighted in other contributions to this work. The problem for the employer from an individual labour law perspective lies elsewhere: he remains legally responsible for the decisions made by the AI. In the event of a dispute, the employer is often obliged to present and prove the legal validity of a specific selection or consideration decision before a labour court. The employer will not be able to escape this

[15] Arnold and Winzer (2018, para. 238); Günther and Böglmüller (2017, p. 56).

[16] See also Broy and Heinson (2019, para. 29); Groß and Gressel (2016, p. 994).

[17] Broy and Heinson (2019, para. 36).

responsibility by simply referring to his AI. If the decisive aspects of a specific AI decision are no longer comprehensible for humans ("black box" problem), a court cannot examine them either. The employer risks in this case not being able to explain the legality of an AI decision and losing the court case. From an individual labour law perspective, it is therefore indispensable for employers that the AI system used can transparently reconstruct its decisions, so that they can be understood with human reason. If an AI system does not guarantee this, employers take a legal risk and employees have good prospects of success in court.

5 Outlook on AI in the Employment Relationship in 2030

The use of AI is nothing more, but also nothing less than a novel technology in the world of work, as electricity, the computer or the Internet were before. In this sense, AI in the employment relationship is to be seen as an evolution and not as a revolution of the world of work.

Although it is still open what meaningful use cases the further technical development of AI will bring forth, it can at least be estimated. In the next few years, AI systems will be developed that automate numerous selection and consideration processes in the employment relationship. The "AI boss" could in the future assign individual work tasks (or projects, customers, routes, etc.) to employees, process their requests for vacation or part-time work automatically, evaluate their performance and pay bonuses based on this, suggest training courses or formulate certificates, or identify possible compliance violations based on large data sets and prepare disciplinary measures.

In the best case, an AI processes those tasks that are already part of the working world more efficiently and objectively than humans can. Especially the companies that see themselves in an international competition will not want to miss out on the associated benefits. In the worst case, however, an AI system produces results that are contrary to the facts or incomprehensible to human reason. The concerns of the workers, being powerless to dehumanised decisions, are understandable and must be taken seriously, let alone for reasons of acceptance. But they are neither technically nor legally insoluble and therefore must not serve as a pretext for a blocking attitude in business, politics or justice. What is required instead are clear, practical and binding guidelines for the use of AI in the employment relationship, which take the needs of both the employee and the employer side seriously. This could be done in particular by supplementing the GDPR or adopting an AI regulation[18] at the European level, clear labour law requirements of the national legislator or by a consistent jurisprudence of the labour courts.

Especially the labour courts have to adjust to a new quality of selection and consideration decisions by employers. An AI can process much larger amounts of data and perform more complex decision processes than a human. It can be expected that AI decisions will rely on more or different aspects that would not have been considered in human decision making. By combining and evaluating large amounts of data, an AI could, for example, make reliable predictions about which employees can perform which tasks in cooperation with which colleagues or customers best, based on previously determined performances, preferences, characteristics and abilities. Other, until now dominant decision aspects (e.g. the formal qualification or the social data of an employee) could lose importance in contrast. The German labour law does not know, except for a few exceptions (cf. § 1 para. 3 s. 1 KSchG), any priority of certain selection and consideration aspects. The technical progress will therefore also shape the further legal development.

6 Summary and Practical Recommendations

Today, AI is already capable of fully automating complex processes. It is therefore predestined for use in employment relationships, where selection and weighing processes are commonplace. However, employers also have to comply with the labour law requirements when making decisions with the help of AI. The use of AI in employment relationships is therefore only a technical or economic advantage if it can be designed in a legally secure manner. Specifically, this means for developers and users of AI systems that the AI must respect labour law requirements and be able to reconstruct its own decisions in such a transparent way that they remain comprehensible to human reason. In addition, there must always be the possibility that a human can intervene and correct the decision of the AI. Employers should therefore remain critical when using AI systems in employment relationships. They cannot completely shift their responsibility for the specific decisions to the respective AI system. It is in their very own interest that an AI system is not an impenetrable "black box". Employees, on the other hand, should not only see dangers in the use of AI. Humans do not always work rationally or error-free either. Especially in employment relationships, human

[18] See Geminn (2021, p. 354 et seq.).

misjudgements often lead to disputes. An AI offers at least the chance of more objective and appropriate decisions. This can also be an advantage for employees.[19] Decision-makers in business, politics and justice should use their scope of action to help shape such a beneficial use of AI for the world of work.

References

Arnold, C., & Winzer, T. (2018). Kapitel 3. In C. Arnold & J. Günther (Hrsg.), *Arbeitsrecht 4.0.* Beck.

Broy, D., & Heinson, D. (2019). Teil B. II. In S. Weth, M. Herberger, M. Wächter, & C. Sorge (eds.), *Daten- und Persönlichkeitsschutz im Arbeitsverhältnis.* Beck.

Geminn, C. (2021). Die Regulierung Künstlicher Intelligenz – Anmerkungen zum Entwurf eines Artificial Intelligence Act. ZD 2021. pp. 354–359.

Groß, N., & Gressel, J. (2016). Entpersonalisierte Arbeitsverhältnisse als rechtliche Herausforderung – Wenn Roboter zu Kollegen und Vorgesetzten werden. *NZA,* 990–996.

Günther, J., & Böglmüller, M. (2017). Künstliche Intelligenz und Roboter in der Arbeitswelt. BB 2017. pp. 53–58.

Joos, D. (2020). Einsatz von künstlicher Intelligenz im Personalwesen unter Beachtung der DS-GVO und des BDSG. *NZA,* 1216–1221.

Lobe, A. (2016). Künstliche Intelligenz – Der ist jetzt Koch, und der ist Kellner. https://www.faz.net/aktuell/feuilleton/debatten/die-digital-debatte/kuenstliche-intelligenz-roboter-als-chef-14239957.html. Accessed: 9. July 2021.

[19] Groß and Gressel (2016, p. 991).

AI in the Company: Is the Employer or the AI as an e-Person Liable?

Michael Zeck

1 Introduction

The use of digital systems that communicate and interact intelligently with their environment based on AI is ubiquitous and does not stop at working life. There is great potential here, e.g. in automation or intelligent assistance of human work. The rapid technological progress and the increasing digitisation lead to massive structural changes in companies—not least also due to the use of AI. However, this also entails legal risks and liability issues that have not yet been conclusively resolved.

The following will show whether the recognition of an e-Person for AI systems can simplify liability issues for employers and reduce or completely exclude employer liability.

2 Status Quo

AI systems can not only replace human labour in the company, but also complement it. They support workers, collaborate with them and give them instructions if necessary. These can be robots, for example, that hand over (construction) parts to a worker as part of a manufacturing process or business process management tools that execute part (-activities) of a business process and then hand them over to an employee for continuation of the process. Intelligent systems are designed

M. Zeck (✉)
HD PLUS GmbH, München, Germany
e-mail: michael.zeck@hd-plus.de

I. Knappertsbusch and K. Gondlach (eds.), *Work and AI 2030*,
https://doi.org/10.1007/978-3-658-40232-7_10

to act increasingly autonomously. AI makes them capable of learning and they can react independently to different situations based on the decision-making or action leeway granted to them. The interaction of AI systems provided by the employer with employees raises the question of who is liable in these networked processes in the event of errors and incorrect instructions by the AI in the event of damage to the employee.[1] Damages do not always result only from the violation of life and limb by a robot[2], but are diverse (e.g. interference with personality rights through data protection violations).

2.1 AI and Associated Risks

AI comes in many shades, which is why it is hardly possible to define it clearly. All definitions share the principle that AI creates "intelligent" systems or machines that simulate human behaviour by analysing their environment, making autonomous decisions and acting accordingly.[3] AI systems try to replicate cognitive human abilities through algorithms. Algorithms are control commands that ensure that a data input is transformed into a data output. Since these are not predetermined, but are independently changed by AI over time, AI acts autonomously. The algorithms change by AI systems using, evaluating and thus increasing their knowledge base, i.e. learning and thus adaptively mastering completely new situations.[4]

AI is generally referred to today as so-called weak AI. This only reaches human intelligence very limitedly, especially only for the specific requirement for which it was developed. Weak AI is not able to gain a really deeper understanding of the assigned problem solutions.

The technical progress by 2030 can produce strong AI, i.e. systems that actually achieve or even surpass human intellectual abilities. Strong AI could then act on its own initiative, intelligently and flexibly. It would no longer be limited to solving a specific problem.[5]

---

[1] Leupold and Wiesner (2021, para. 106).

[2] This is already discussed in more detail in the context of autonomous robots and machines. For example, Neighbour (2020), Keßler (2017), Böglmüller and Günther (2017).

[3] For example, Platform Industry 4.0 (2019). For details on the technical background, see Stiemerling (2020).

[4] Leupold and Wiesner (2021, para. 2).

[5] On the possibilities and limits of AI and the terminology of strong and weak AI, see Stiemerling (2020, para. 70 et seq.).

In principle, the autonomous development or learning ability of AI entails risks. These result from the lack of predictability of its behaviour (autonomy risk), the lack of traceability of its decision-making (transparency risk) and a difficult assignment of responsibilities due to the interaction with humans or other systems (compound or network risk).[6] Strong AI, if possible, will certainly potentiate the aforementioned risks.

2.2 The Current Liability Regime

AI systems, initially created by humans, are like any software not error-free. Damages are therefore "pre-programmed".[7] It is therefore necessary to clarify which liability regime applies in the event of damage at work.

Generally, the German liability regime differentiates between contractual and statutory liability:[8] The former is regulated in §§ 280 et seq. BGB and the latter in §§ 823 et seq. BGB.

For tortious liability under §§ 823 et seq. BGB (German Civil Code), the violation must be causal and attributable for the damage. In the case of involvement of an AI system, it is difficult to impossible to determine whether individual causal contributions of the AI and not a human being are to be attributed. And, if so, whether an (indirect) attribution to a human being can still be made, since he ultimately gives up the causality by using a system that develops and decides autonomously.[9] Also, the proof of the necessary fault (intent or negligence) is difficult. A person acts negligently if he disregards the care required in traffic (§ 276 BGB). Such a breach of duty of care could be given with regard to monitoring, maintenance and hazard prevention of a system. But the more opaque and thus unpredictable the behaviour of the AI system and thus also a possible malfunction, the less likely is also a breach of duty of care by the employer.[10]

For the sake of completeness, it should be noted that the less relevant contractual liability of the employer does not differ from other permanent obligations in

[6] Linke (2021, p. 201); Leupold and Wiesner (2021, para. 4).

[7] Bitkom (2020).

[8] Schwab (2016).

[9] Böglmüller and Günther (2017, p. 55).

[10] Riehm (2020, para. 6).

general civil law.[11] It can generally result from any violation of contractual obligations (§§ 280 para. 1 and 241 para. 2 BGB). The only question is whether the behaviour of an AI can be attributed to the employer according to § 278 BGB.

3 Challenges of the Current Liability Regime and the e-Person as a Possible Solution Approach

The difficulties of an injured worker to meet his burden of presentation and proof, as well as the supposedly often lacking fault of the employer in connection with AI, sometimes raise doubts as to whether the current liability regime can effectively meet the challenges posed by the use of AI systems.

In the course of this, voices are raised that argue that apparent or still emerging gaps in responsibility caused by the use of autonomous AI systems should be closed by legal development or the introduction of specific AI laws. What is repeatedly demanded here is also the liability of the AI system by recognising its digital legal personality, endowed with the ability to be the bearer of civil law rights and obligations in particular and to have assets.[12]

3.1 The Liability of AI as an e-Person

The idea of endowing AI with a digital legal personality, i.e. creating a new legal construct of the e-Person, is not new. The reason is that only legal persons can be liable. These include natural persons (i.e. humans) or legal entities (e.g. a GmbH, AG, OHG). Some voices see the legal personality of AI systems already implied in existing rules (de lege lata) by way of legal development, which, however, is to be rejected.[13] Others demand de lege ferenda the introduction of corresponding provisions. Thus, even the EU Parliament proposed in 2017 to endow AI with

[11] Schwab (2016) with the note that since the reform of the law of obligations in 2002, a claim for pain and suffering exists without fault and regardless of the basis of liability (§ 253 para. 2 BGB).

[12] Riehm (2020, para. 4).

[13] Neighbour (2020, para. 73).

legal capacity from a certain degree of autonomy[14]. AI should then be able to be liable and sued, for example.

The demand for the creation of an e-Person is driven by the concern that otherwise undesirable gaps in responsibility would exist, which need to be closed. The e-Person becomes the addressee of liability, which could be held accountable in case of wrong decisions. Employers who operate AI systems would thus be freed from direct liability. After all, the influence of humans on the decisions of AI is becoming increasingly less. Finally, AI moves further and further away from the state of knowledge on which it was originally programmed by learning independently.[15]

Since e-Persons do not have their own assets per se, the question inevitably arises as to how, with what and with how much capital they should be endowed in order to be able to assume adequate liability. Here one can think of liability funds or (mandatory) insurance or of equipping the e-Person—comparable to a GmbH—with a minimum capital.

3.2 The Criticism of the e-Person

The logical basic idea of the own legal personality for AI systems with the accompanying avoidance of problems of proof and a clear legal allocation of damages, however, meets with justified criticism. It is ultimately superfluous, constitutes a massive intervention in the basic structures of the legal system[16], encounters a whole series of practical obstacles and raises more questions than it solves. The EU Parliament has explicitly considered the legal personality of AI systems as no longer necessary in the regulatory proposal submitted in October 2020.[17]

As already mentioned, an e-Person must have access to financial resources in the event of liability. The capital allocation must be such that the opportunities and risks of a behaviour actually affect the same legal subject and can also be

[14] Which degree of autonomy should be sufficient here is an exciting, but difficult to answer question.

[15] Presentation by Riehm (2020, para. 5 et seq.).

[16] Riehm (2020, para. 12).

[17] See the resolution of the European Parliament of 20 October 2020 with recommendations to the Commission for a regulation of civil liability for the use of artificial intelligence 2020/2014(INL).

weighed against each other by it.[18] Apart from the fact that insurance and fund solutions could easily be mapped in an existing liability regime without an e-Person, such a transfer of the liability risk would not suffice. Costs and risks that should actually be borne by the user of the AI system are simply outsourced.[19] The level of care in programming and in selecting and using AI will inevitably decrease without corresponding responsibility on the part of the user and employer.

In order to achieve the desired legal and economic effect, the e-Person must therefore be endowed with a corresponding liability mass, i.e. own capital. However, since this will not be able to be available without limit, it is to be feared that backers and users of autonomous AI systems will only misuse them as a shield of liability due to their limited (own) liability, especially in case of limited and insufficient capitalisation.[20]

Another—only difficult to solve—problem is that of the formal identification of the e-Person, which as a legal person must be recognisable to the outside world.[21] Some AI systems—but by no means all—are actually embodied, be it in a device, vehicle or robot. Nevertheless, it is questionable what exactly is to be identified and assigned as a legal subject. Because in the shell of a device, several autonomous subsystems can (inter-)act. Do they then form one or more e-Persons? If, on the other hand, a physical embodiment of the systems is not possible at all, it is unclear whether the algorithm in general or individual instances represent the e-Person.

The material allocation of an AI system to an e-Person is also difficult to solve: From what degree of autonomy should an AI system be endowed with legal capacity? This is especially difficult against the background of the already mentioned development starting from a weak AI today.[22]

3.3 The e-Person is Currently not a Solution

It therefore becomes apparent that, if one wanted to introduce the e-Person, an immense effort (such as the establishment of (electronic) registers of e-Persons

[18] Riehm (2020, para. 13).

[19] Leupold and Wiesner (2021, para. 103).

[20] Linke (2021, p. 202).

[21] Riehm (2020).

[22] On the aspect of formal and material identification Riehm 2020

and their respective liability masses) would have to be made and, in addition, a multitude of new problems and questions would be raised. It would require countless new regulations, which could hardly be seamlessly and consistently integrated into existing liability regulations.

It is also not currently desirable to transfer more responsibility to AI systems.[23] Because if and to the extent that employers remain directly liable for the consequences of using AI, they have a strong self-interest in controlling the AI and its results and minimising damage risks.

Ultimately, the e-Person cannot facilitate compensation or eliminate liability problems. Because the prerequisites for a liability claim still have to be presented and proven. And there would still have to be someone who represents the e-Person in legal transactions or takes out insurance for them and pays the insurance premium.[24] If a responsible person can be identified in this respect, one can and should also hold this (natural or legal) person directly accountable.

We have a legal system that—as of today—provides a liability regime by recourse to the employers behind the AI. It should also be acknowledged that fault-based liability sometimes also means that there is no absolute security and that there is a possibility that operators or employers may not be liable for damages that are not their fault.

The introduction of an e-Person is therefore superfluous today. Currently, the assignment of a separate legal personality for AI systems probably still goes beyond the technological reality and very likely would also have significant market failure problems as a consequence.[25]

4 Outlook on Employer Liability for AI in 2030

By 2030, AI will become more autonomous, adaptive and learning, and thus more unpredictable. It will therefore become more difficult to attribute the actions of an autonomous AI to a human being who is supposed to be responsible for them. This new degree of independence could require new legal concepts.[26] Because it

[23] Bitkom (2020, p. 40).

[24] Bitkom (2020, p. 40).

[25] Leupold and Wiesner (2021, para. 103).

[26] This is especially true if the current approaches of machine learning, such as deep learning using neural networks, are consistently further developed.

can also become impossible to trace damages back to a specific natural or legal person as the cause. This raises liability issues that go beyond the analogue understanding of the current legal system and push the common concepts of strict and fault-based liability to their limits. With increasing autonomy, a legal capacity of AI with its own liability of the e-Person could then be a suitable means despite all its complexity and the resulting questions and solve numerous future legal problems. An alternative could also be the extension of strict liability to AI systems.[27] The discussion about this should be initiated today in order to have enough time to find good and appropriate solutions and not to slow down the further development or use of AI in working life because of incalculable risks.

5 Summary and Practical Recommendations

Employers are liable for misconduct of AI and resulting damages to their employees. This corresponds to the current law and seems fair.

Perspectively, a separate liability of AI seems conceivable, the stronger it becomes. However, there is a long way to go until then. If it is introduced de lege ferenda one day, e-Persons may be liable and relieve the employer directly. Indirectly, however, the liability will still affect the employers. Because the e-Person does not fall from the sky with liability assets, but has to be equipped with such by the employer accordingly.

Employers are therefore well advised to carefully select AI systems, monitor them constantly in use, maintain them and anticipate and avert potential dangers for employees. They should also do everything possible to optimally secure themselves in terms of warranty and liability in the contracts with the manufacturers when purchasing AI.

References

Bitkom Bundesverband Informationswirtschaft, Telekommunikation und Neue Medien e. V. (2020). Stellungnahme der digitalisierten Wirtschaft: Haftung für Systeme Künstlicher Intelligenz. https://www.bitkom.org/sites/default/files/2020-10/bitkom-position-zu-haftung-fur-ki-mit-one-pager.pdf. Accessed: 13. Juli 2021.

[27] On the current discussion, see Neighbour (2020, para. 74).

Böglmüller, M., Günther, J. (2017). Künstliche Intelligenz und Roboter in der Arbeitswelt. Betriebsberater (pp. 53–58).

Keßler, O. (2017). Intelligente Roboter—neue Technologie im Einsatz | Voraussetzungen und Rechtsfolgen des Handelns informationstechnischer Systeme. MMR (pp. 589–594).

Linke, C. (2021). Die elektronische Person Erforderlichkeit einer Rechtspersönlichkeit für autonome Systeme? MMR (pp. 200–204).

Neighbour, K. (2020). § 8 Arbeitsrecht—Realität und Herausforderungen. In T. Sassenberg & T. Faber (Eds.), *Rechtshandbuch Industrie 4.0 und Internet of Things.* C.H. BECK in Gemeinschaft mit Vahlen, München.

Plattform Industrie 4.0. (2019). *Ergebnispapier Künstliche Intelligenz und Recht im Kontext von Industrie 4.0.* Bundesministerium für Wirtschaft und Energie.

Riehm, T. (2020). Nein zur ePerson! Gegen die Anerkennung einer digitalen Rechtspersönlichkeit. *RDi,* pp. 42–48x.

Schwab, B. (2016). Haftung im Arbeitsverhältnis—2. Teil: Die Haftung des Arbeitgebers. NZA-RR, pp. 230–234.

Stiemerling, O. (2020). Technische Hintergründe. In N. Kaulartz & T. Braegelmann (Eds.), *Rechtshandbuch Artificial Intelligence und Machine Learning* (pp. 15–31). Beck.

Leupold, A., & Wiesner, M. (2021). Zivilrechtliche Haftung bei Einsatz von Robotern und Künstlicher Intelligenz. In A. Leupold, A. Wiebe, & S. Glossner (Eds.), *Münchener Anwaltshandbuch IT-Recht* (pp. 1054–1115). Beck

The Co-Determination Right of the Works Council According to § 87 Para. 1 No. 6 BetrVG in the Use of AI Systems in the Company

An Overview of the Development of Case Law and Challenges in Practice

Gerlind Wisskirchen and Marcel Heinen

1 Introduction

The German Works Constitution Act (BetrVG) aims to achieve a reasonable balance between the co-determination rights of the works council on the one hand and the entrepreneurial freedom of decision on the other. A broad interpretation of the right of co-determination on the introduction of technical applications by the Federal Labour Court (BAG) poses great challenges for companies.

The two authors would like to thank the research assistants Ms Sandra Maas and Mr Moritz Herrmann for their assistance in preparing the article.

G. Wisskirchen (✉) · M. Heinen
CMS Germany, Köln, Germany
e-mail: gerlind.wisskirchen@cms-hs.com

M. Heinen
e-mail: marcel.heinen@cms-hs.com

I. Knappertsbusch and K. Gondlach (eds.), *Work and AI 2030*,
https://doi.org/10.1007/978-3-658-40232-7_11

2 Status Quo/Inventory and Case Studies

The German regulation contains some undefined legal terms (§ 87 para. 1 no. 6 BetrVG). The term "technical device" is understood to mean any device that is capable of operating independently, with or without human assistance.[1] The works council in Germany has a right of co-determination when introducing, using or adapting such systems if they are "intended for monitoring employees". However, there is no specific definition of the broad term artificial intelligence (AI systems).[2] AI is usually used to mean machine learning or big data analytics.[3] The European Commission has made an attempt to define it in Art. 3 no. 1 of its draft dated 21st April 2021 for a regulation laying down harmonised rules for AI: An AI system is a software that has been developed with one or more of those technologies and approaches listed in the annex to the European regulation and can produce results such as content, predictions, recommendations or decisions that affect the environment with which they interact. According to the case law of the BAG, any software that is operated with a computer already constitutes a technical device.[4] AI systems that process data, according to the definition discussed in the European legislative process, usually fall under the term technical device within the meaning of the law, leading to co-determination rights of the works council.

2.1 Interpretation in Case Law and Literature

The case law has extended the term "monitoring" so far that now a large number of—also partially automatic—processes in the use of AI systems are covered, as far as they allow predictions to be mades about the behaviour or performance of individual employees.[5] Even the mere storage of data falls under the term monitoring.[6]

---

[1] BAG 6.12.1983–1 ABR 43/81, NJW 1984, 1476; (Wisskirchen, et al., 2017, p. 2106).

[2] (Schiefer & Worzalla, 2021, p. 822): On the difficulty of a generally valid definition of AI.

[3] (Freyler, 2020, p. 284).

[4] BAG, Urt. v. 26.07.1994–1 ABR 6/94, NZA 1995, 185.

[5] BAG 13.12.2016–1 ABR 7/15, NZA 2017, 657; Richardi/*Richardi* Works Constitution Act with Election Regulations, 2018, § 87 margin no. 500.

[6] BAG 13.12.2016–1 ABR 7/15, NZA 2017, 657; BAG 06.12.1983–1 ABR 43/81, NJW 1984, 1476.

The BAG already ruled in the past that a technical device is "intended" for monitoring if it is objectively suitable for monitoring—it is irrelevant whether the employer actually pursues this goal or collects the data unintentionally.[7] The BAG argued that even the mere possibility interferes significantly with general privacy rights.[8] Thus, the use of AI systems is usually "intended" for monitoring. Even rudimentary software will be controlled by AI systems in the future, as standard programs are already significantly influenced by AI systems.[9] In the case of data processing by the recording components provided for by the EU regulation (e.g. cache, history functions), the objective suitability for monitoring the employees is therefore always present.

The interpretation of the intentional use of any data for monitoring overstretches the wording of the law (§ 87 para. 1 no. 6 BetrVG).[10] In fact, actual monitoring would be required to trigger the right of co-determination. A specific intention logically requires a purpose, while the abstract possibility is -independent of any purpose. Even the bill, which was introduced into the parliamentary procedure back in 1972, was based on this more restrictive understanding.[11]

In this respect, parts of the BAG already introduced a so-called "immediacy requirement" as a corrective element in 1975. According to this requirement, technical monitoring is only present if the system directly leads to the monitoring.[12] On the other hand, the co-determination right is excluded if the transmitted data do not allow any conclusions to be drawn about the behaviour of the employee and further measures are required for a monitoring effect.[13]

The "immediacy requirement" is a suitable corrective element, also to take into account the increasing implementation of artificial intelligence in companies. A link to collection must be denied in particular for data that are only stored manually.[14] The same applies if human behaviour is required as an additional measure for the specific execution of any monitoring.

---

[7] BAG 11.3.1986–1 ABR 12/84, NZA 1986, 526.

[8] BAG 11.3.1986–1 ABR 12/84, NZA 1986, 526; German Bundestag, 1971, p. 48 f. The explanatory memorandum of 1972 still spoke of the "personal sphere of the employees".

[9] (Niklas 2021).

[10] (Günther & Böglmüller, 2015, p. 1025; Wisskirchen et al. 2017, p. 2108).

[11] Cf. BT-Drucks. VI/1786 p. 48/49.

[12] Overall: BAG 9.9.1975–1 ABR 20/74, NJW 1976, 261.

[13] BAG, 9.9.1975–1 ABR 20/74, NJW 1976, 261.

[14] ArbG Heilbronn 8.6.2017–8 BV 6/16, NZA-RR 2017, 476, appeal pending under AZ. 4 TaBV 5/17; (Schiefer & Worzalla, 2019, p. 2017).

2.2 Examples of Implementation

According to the BAG, merely operating a Facebook page is not sufficient for the works council to have a co-determination right. However, manually inputting data within the Faebook comment function is sufficient for a co-determination right, because the data are then stored and can be accessed.[15]

The BAG's Facebook decision was heavily criticised by the German law community.[16] The telos of the co-determination right is not to unnecessarily thwart modern technologies.[17] This approach was also rightly pursued by the lower court (LAG (Regional Labour Court) Düsseldorf): Both the writing of the comment on Facebook and its evaluation required further decisions to be made by humans, meaning that data processing did not take place immediately.[18] A right of co-determination has to be rejected.

In 2018, the LAG Hamm even assumed that there was a right of co-determination in the introduction of standard programs such as Microsoft Office Excel.[19] The BAG rejected the non-admission complaint and added that a—actually very suitable—de minimis threshold did not matter.[20]

In a decision of the LAG Hamburg, a Twitter account set up by the employer was the subject of dispute. Regardless of the external storage location of the posts of other Twitter users, the employer created the possibility of writing posts about the company and thus also its employees through its own Twitter account.[21] Unlike the lower court, the LAG Hamburg unfortunately did not take the time to deal with the "immediacy requirement" in its decision.

---

[15] BAG 13.12.2016–1 ABR 7/15, NZA 2017, 657; BAG 15.12.1992–1 ABR 24/92, BeckRS 1992, 30.743.536.

[16] (Haußmann & Thierne, 2019, p. 1612; Ludwig & Ramcke, 2016, p. 2293; Wisskirchen et al., 2019, p. 1460; Schreiner 2019, p. 554; Giesen 2020, p. 73; Schiefer & Worzalla 2019, p. 2017).

[17] BAG 8.11.1994–1 ABR 20/94, NZA 1995, 313; BAG 10.12.2013–1 ABR 43/12, NZA 2014, 439.

[18] LAG Düsseldorf 12.1.2015–9 TaBV 51/14, NZA-RR 2015, 355; (Schiefer & Worzalla, 2019, p. 2017).

[19] LAG Hamm 10.4.2018–7 TaBV 113/16, BeckRS 2018, 27.857.

[20] BAG 23.10.2018–1 ABN 36/18, AP BetrVG 1972 § 87 Überwachung Nr. 50.

[21] LAG Hamburg 13.9.2018–2 TaBV 5/18, NZA-RR 2018, 655.

On the other hand, software updates only trigger a co-determination right according to the opinion of many legal expertsif there is a structural extension, an expansion to other individuals or an increase in monitoring.[22] The works council only has to be informed about the new update in this case.

3 Challenges and Solutions

When transferring the case law regarding the co-determination rights on implementing Facebook to services such as LinkedIn, XING, YouTube or Instagram, a public exchange of opinions without the prior consent of the works council is virtually impossible. The media platforms mentioned automatically allow criticism by third parties against employees or against content that allows conclusions to be drawn about individual employees.[23] BAG's argument that the employer was free to use Facebook as a digital medium is not comparable to the use of any websites.[24] A practical approach, on the other hand, is provided by the labour court in Heilbronn (ArbG Heilbronn): According to the court, a contact form available on the website is to be treated like a digital mailbox, for which there is no co-determination right of the works council.[25]

4 Preview of AI in the Context of Co-Determination Rights in 2030

Co-determination rights will have to be adapted in the coming years, especially in connection with the introduction of AI systems.

[22] (Müller-Glögge et al., 2021), BetrVG § 87 Rn. 59; cf. (Wiese et al., 2018, § 87 Rn. 595; Henssler. 2020, BetrVG § 87 Rn. 125; so in the result also: BAG 11.12.2018–1 ABR 13/17).

[23] (Gooren 2020, p. 114).

[24] Cf. BAG 13.12.2016–1 ABR 7/15, NZA 2017, 657.

[25] Cf. ArbG Heilbronn, 08.06.2017–8 BV 6/16, NZA-RR 2017, 476; appeal pending under AZ. 4 TaBV 5/17; LAG Düsseldorf on Facebook (overruled by BAG 13.12.2016 aaO) NZA-RR 2015, 355; (Fuhlrott 2017, p. 348).

A more literal interpretation ("determines") is required.[26] The introduction of a prohibition of evidence obtained in violation of co-determination rights, according to which any evidence obtained in violation of co-determination rights could not be introduced in any court proceedings between individual employees and employers, would already ensure adequate protection of the employees' interests.[27] However, the case law is cautious in this regard and affirms a prohibition of evidence only in individual cases.[28] Further anonymisation obligations and comprehensive information rights of the works council could adequately complement a corresponding prohibition of evidence.[29]

Moreover, the European and German data protection laws already provide sufficient protection requirements with regard to the processing of personal employee data.[30]

The obvious difficulties experienced in practice should also be an alarming indication for the legislator to amend German co-determination rights with respect to AI. In the context of the Works Council Modernisation Act, the legislator reacted to the increasingly demanding and complex issues that are associated with the introduction of AI. In particular, with regard to the issues concerning the technical possibilities and the functionalities of the systems to be introduced, works councils are to be entitled to rely on professional expertise while dealing with such issues. The commendable goal of the legislator to achieve faster decisions of the works council on the introduction of new AI systems[31], is accompanied by the significant costs at the expense of the employer. The federal government's assumption that the experts for the works council will be able to provide sufficient consultation and perform sufficient examination of the AI components within one day seems completely unrealistic.[32] However, it is still to be expected that employers will agree to the engagement of those experts as a precautionary measure in order to avoid further delays when implementing any kind of AI.

---

[26] (Günther & Böglmüller, 2015, p. 1027; Ludwig & Ramcke, 2016, p. 2293).

[27] (Günther & Böglmüller, 2015, p. 1025; Maschmann, 2002, p. 13); also LAG Tübingen, 06.05.1999–12 Sa 115/97, BB 1999, 1439; (Fischer 1999, p. 154).

[28] BAG 27.03.2003 -2 AZR 51/02, NZA 2003, 1193; LAG Hamm Urt. v. 25.1.2008 -10 Sa 169/07, BeckRS 2008, 53.207.

[29] (Schipp 2016, p. 177).

[30] (Giesen 2020, p. 73).

[31] BT-Drucks. 19/28.899 p. 23.

[32] BT-Drucks. 19/28.899 p. 17.

Legislative amendments would also be necessary for the procedures before the conciliation board (*Einigungsstelle*), which can take up to three years. As a result of such long delays the new technologies are often already out of date when they are introduced. Therefore, with respect to improved employer flexibility, the possibility of a temporary introduction of AI systems would be highly appreciated.[33] In parallel, the legislative process of the EU regulation on AI harmonisation will be interesting to see, especially with regard to whether the current version of the AI definition and the reference to different technologies and concepts will be implemented as planned. The focus of this initiative is to maintain a minimum level of trust in AI systems. This is to be achieved by a risk-based approach, according to which high-risk AI systems are prohibited and strict requirements are imposed on high-risk systems (e.g. documentation & information obligations, risk mitigation systems).

The use of AI systems in German companies is gaining ground, whether in semi-automated travel expense accounting via app or self-learning tutorial management systems, where seminars are suggested by comparable employees. Such systems are likely to be classified as low-risk AI according to the proposal of the EU Commission, but nevertheless those algorithms are able to be used for monitoring at least objectively anyway and thus regularly trigger the works council's right of co-determination according to the law (§ 87 para. 1 no. 6 BetrVG).

5 Summary and Recommendations for Employers

The interpretation of the co-determination right by the courts is no longer practical in view of the technical progress and prevents corporate digitalisation. The use of modern AI systems is now essential to meet a large number of legal requirements and to remain competitive at the same time. It should be kept in mind that the majority of AI programs primarily serve the interests of the employees as well.

It would be more useful if a co-determination right were only established in cases where the employer intends to monitor the performance or behaviour of the employees. Until legal amendments are implemented, it must be in the interest of all employers to consider a co-determination right of the responsible works council even at an early stage when implementing new AI infrastructure.

[33] (Bitkom.org 2021, p. 6).

References

Bieder, M. (2016). Die Entwicklung der betrieblichen Mitbestimmung in sozialen Angelegenheiten (§ 87 I BetrVG). *Neue Zeitschrift für Arbeitsrecht Rechtsprechungsreport,* 225–233.

Bitkom.org. (2021). Stellungnahme zum Referentenentwurf des Bundesministeriums für Arbeit und Soziales "Entwurf eines Gesetzes zur Förderung der Betriebsratswahlen und zur Stärkung der Betriebsräte (Betriebsrätestärkungsgesetz)". https://www.bitkom.org/sites/default/files/2021-01/bitkom-stellungnahme-referentenentwurf-betriebsrate-starkungsgesetz-bmas-19012021.pdf. Accessed: 14. Juli 2021.

Deutscher Bundestag. (1971). Entwurf eines Betriebsverfassungsgesetzes, Drucksache VI/1786. http://dipbt.bundestag.de/doc/btd/06/017/0601786.pdf. Accessed: 14. Juli 2021.

Fischer, U. (1999). Prozessuales Verwertungsverbot für mitbestimmungswidrig erlangte Beweismittel. *Betriebsberater,* 154–157.

Freyler,M. (2020). Rob-Recruiting, Künstliche Intelligenz und das Antidiskriminierungsrecht. *Neue Zeitschrift für Arbeitsrecht,* 284–290

Fuhlrott, M. (2017). Smartphone-App zur Kundenbefragung ist mitbestimmungsfrei. *Gesellschafts- und Wirtschaftsrecht,* 348.

Giesen, R. (2020). Materielles Betriebsverfassungsrecht und Digitalisierung. *Neue Zeitschrift für Arbeitsrecht,* 73–76.

Gooren, P. (2020). Kein Tweet ohne Betriebsrat?—Neues zur Mitbestimmung. *Neue Juristische Wochenschrift Spezial,* 114–115.

Günther, J., & Böglmüller, M. (2015). Arbeitsrecht 4.0—Arbeitsrechtliche Herausforderungen in der vierten industriellen Revolution. *Neue Zeitschrift für Arbeitsrecht,* 1025–1031

Haußmann, K., & Thieme, L.M. (2019). Reformbedarf und Handlungsoptionen in der IT-Mitbestimmung. *Neue Zeitschrift für Arbeitsrecht,* 1612–1620.

Henssler, M et al. (Eds.). (2020). *Arbeitsrecht Kommentar* (9. Edn.). Otto Schmidt.

Ludwig, D., & Ramcke, O. (2016). Verhaltens- und Leistungskontrolle nach § 87 Abs. 1 Nr. 6 BetrVG—Plädoyer für einen Neuanfang! *Betriebsberater,* 2293–2298.

Maschmanm, F. (2002). Zuverlässigkeitstests durch Verführung illoyaler Mitarbeiter? *Neue Zeitschrift für Arbeitsrecht,* 13–22.

Müller-Glöge, R., et al. (Eds.). (2021). *Erfurter Kommentar zum Arbeitsrecht* (21. Edn.). C.H. Beck.

Niklas, T. (2021). Kabinett beschließt Betriebsrätemodernisierungsgesetz. https://www.arbrb.de/blog/2021/04/13/kabinett-beschliesst-betriebsraetemodernisierungsgesetz/. Accessed: 16.Juli 2021.

Richardi, R. (Eds.) (2018). *Betriebsverfassungsgesetz mit Wahlordnung* (16. Edn.). C.H. Beck

Schiefer, B., & Worzalla, M. (2019). Moderne Arbeitswelt (Teil 2): Betriebsrat und Betriebsratsaufgaben—Mitbestimmungsrechte bei technischen Einrichtungen, Qualifizierungen, Arbeitszeit und Arbeitsweise sowie Betriebsratsarbeit und Betriebsratswahl –. *Der Betrieb,* 2017-2022.

Schiefer, B., & Worzalla, M. (2021). Das Betriebsrätemodernisierungsgesetz—Eine "Mogelpackung"? *Neue Zeitschrift für Arbeitsrecht,* 817-823.

Schipp, J. (2016). Industrie 4.0 und Mitbestimmung bei technischen Innovationen. *Der Arbeits-Rechtsberater,* 177–180.

Schreiner, P. (2019). Mitbestimmungsrecht des § 87 Abs. 1 Nr. 6 BetrVG unterliegt keiner Erheblichkeitsschwelle (§ 87 Abs. 1 Nr. 6 BetrVG). *Der Betrieb,* 54.

Wiese, G., et al. (Hrsg.). (2018). *Gemeinschaftskommentar zum Betriebsverfassungsgesetz* (11. Edn.). C.H. Beck.

Wisskirchen, G., Schiller, J. P., & Schwindling, J. (2017) Die Digitalisierung—eine technische Herausforderung für das Mitbestimmungsrecht aus § 87 Abs. 1 Nr. 6 BetrVG. *Betriebsberater,* 2105–2109.

Wisskirchen, G., Schiller, J. P., & Schwindling, J. (2019). Betriebliche Mitbestimmung bei IT-Implikationen—Ein deutscher Sonderweg und eine Innovationsbremse. *Betriebsberater,* 1460–1464.

Data Protection Assessment of Predictive Policing in the Employment Context

Legal Basis and Its Limits

Inka Knappertsbusch and Luise Kronenberger

1 Introduction

Police authorities have been using this form of data analysis for several years to find out where and with what probability which crime could be committed.[1] For this purpose, data from countless crimes are entered into a software, which then analyses them using artificial intelligence and then makes a prediction about where the next crime could be committed.[2] Artificial intelligence can reveal connections that are not or only recognisable by manual or tedious research.[3]

This principle could be extended to employment relationships in the future.[4] A comprehensive analysis of employee data using artificial intelligence could provide information on whether and, if so, with what probability an employee violates rules or commits a crime.[5]

[1] (Härtel, 2019, p. 54).

[2] https://www.goethe.de/prj/jad/de/the/ari/22082332.html

[3] (Hoch, 2020, p. 295).

[4] (Hinz, 2020, p. 543).

[5] (Riesenhuber, 2021, § 26 Rn. 139.1; Rudowski, 2019, p. 72).

I. Knappertsbusch (✉) · L. Kronenberger
CMS Germany, Köln, Germany
e-mail: inka.knappertsbusch@cms-hs.com

I. Knappertsbusch and K. Gondlach (eds.), *Work and AI 2030*,
https://doi.org/10.1007/978-3-658-40232-7_12

2 Status Quo

One application of artificial intelligence already in use in companies is fraud monitoring.[6] Such an application can determine fraud probabilities based on data analysis.[7] One use case is the evaluation of the user behaviour of the employee.[8] Statistical methods can filter out irregularities that may indicate cybercrime, such as data theft.[9]

In contrast, predictive policing starts one step earlier by detecting the risk of committing misconduct without a concrete suspicion against the affected employee.

3 Challenges and Solutions

Predictive policing is limited by data protection, as there must always be a legal basis for the processing of employee data. Possible legal bases for the use of predictive policing in the employment relationship are § 26 para. 1 s. 1 and the consent pursuant to § 26 para. 2 BDSG (German Data Protection Act).

1. General clause (§ 26 para. 1 s. 1 BDSG)
   It seems conceivable to use the general clause of employee data protection. Pursuant to § 26 para. 1 s. 1 BDSG, personal data of employees can be processed for the purpose of carrying out the employment relationship, if this is necessary for this.
   § 26 para. 1 s. 2 BDSG cannot be applied, however, because predictive policing pursues a preventive approach.[10] § 26 para. 1 s. 2 only aims at the detection of crimes already committed.[11] The prevention of crimes or other breaches of duty is, however, strongly distinguishable from the detection of crimes.[12]

[6] (Hentrich & Pyrcek, 2016, p. 1452).

[7] (Hentrich & Pyrcek, 2016, p. 1452).

[8] (Hentrich & Pyrcek, 2016, p. 1452).

[9] (Hentrich & Pyrcek, 2016, p. 1452).

[10] (Hinz, 2020, p. 543; Rudowski, 2019, p. 73).

[11] (Gola, 2019, § 26 Rn. 125; Riesenhuber, 2021, § 26 Rn. 130).

[12] (Riesenhuber, 2021, § 26 Rn. 130).

a) Carrying out the employment relationship
   Predictive policing is intended to prevent compliance violations by employees by analysing individual employees.[13] In businesses where several employees work, it is necessary for the purpose of carrying out the employment relationship that the employer takes measures to prevent breaches of duty by the employees.[14] This was already assumed by the legislator in the old § 32 para. 1 BDSG[15], which is continued in § 26 para. 1 BDSG.[16] The new version is intended to confirm the previous regulations on the permission facts.[17] Thus, predictive policing serves the purpose of carrying out the employment relationship within the meaning of § 26 para. 1 s. 1 BDSG.
b) Necessity
   However, it is questionable whether predictive policing is also necessary for the purpose of carrying out the employment relationship.
   Basically, it is first necessary to determine whether there is not also a less intrusive, but equally effective method to counteract violations of rules.[18] An open video surveillance, i.e. a surveillance that is easily recognisable for the employee,[19] could also prevent crimes and violations of rules by its deterrent effect.[20] However, such a surveillance is not a less intrusive measure and may not be carried out without a reason[21].
   Within the scope of necessity, a balancing of the conflicting interests must be carried out.[22] The interest of the employer in the data processing and the personal interest of the employee must be brought into balance.[23] In doing so, it must be taken into account that a balancing must generally only be carried out if the employee could commit the specific offence at

[13] (Rudowski, 2019, p. 73).
[14] (Riesenhuber, 2021, § 26 Rn. 138).
[15] BT-Drs. 16/13.657, 21.
[16] BT-Drs. 18/11.325, 96 f.
[17] (Riesenhuber, 2021, § 26 Rn. 57).
[18] (Gola, 2019, § 26 Rn. 16).
[19] (Bongers, 2019, Rn. 779).
[20] (Gola, 2018, Art. 6 Rn. 182).
[21] (Maschmann, 2020, § 26 Rn. 45a).
[22] (Riesenhuber, 2021, § 26 Rn. 113).
[23] BT-Drs. 18/11.325, 97.

all according to his or her activity.[24] This applies in particular to employees in positions susceptible to fraud with an increased risk of crimes such as money laundering.[25] A balancing is also excluded from the outset if the whole personality of the employee is to be illuminated in order to assess his or her compliance.[26] Rather, only analyses of individual character traits may be carried out.[27] A criterion for a balancing could be how likely a misconduct is and what damage the employer faces if it occurs.[28] The higher the likelihood of a criminal offence and the potential damage, the more likely this interest could outweigh the interest in processing the data. For employees who can only access the employer's assets to a limited extent, predictive policing is therefore not necessary for this reason already in principle. A decisive factor for a balancing in favour of the employee and against the necessity is the following: For surveillance measures such as video surveillance and the use of software keyloggers, it is required that a certain level of suspicion indicates a violation of rules.[29] An investigation "into the blue" cannot be necessary.[30] Something else only applies if the interference with the general privacy rights of the employee is only of low intensity.[31] This is the case, for example, with random checks of the history data of an internet browser.[32]

However, a data analysis predicting future misconduct interferes more strongly with the privacy rights than the mere checking of historical data. If one assumes that predictive policing has a similar control function as surveillance, it must also be based on the fact that at least the suspicion exists that a crime or a serious violation of rules has been committed by an employee. This is exactly what predictive policing does not aim at. It is intended to carry out an analysis without suspicion, even for susceptible employees, in order to take quick measures.

---

[24] (Rudowski, 2019, S. 74).

[25] (Hinz, 2020, S. 543).

[26] (Hinz, 2020, S. 543; Rudowski, 2019, S. 74).

[27] (Rudowski, 2019, S. 74).

[28] (Riesenhuber, 2021, § 26 Rn. 142).

[29] (Fuhlrott & Oltmanns, 2019, S. 1108; Maschmann, 2020, § 26 Rn. 45a u. 47).

[30] BAG, NZA 2017, 1327 o. BAG, Urt. v. 27.7.2017–2 AZR 681/16.

[31] BAG, NZA 2017, 1327 o. BAG, Urt. v. 27.7.2017–2 AZR 681/16.

[32] BAG, NZA 2017, 1327 (1330) o. BAG, Urt. v. 27.7.2017–2 AZR 681/16.

c) Intermediate result

§ 26 para. 1 s. 1 BDSG therefore excludes as a suitable legal basis for non-suspicion analyses due to lack of necessity.

2. Consent (§ 26 para. 2 BDSG)

The data analysis could, however, be justified by the consent of the employee. The problem in the employment relationship is especially whether a consent can be given voluntarily in the specific individual case due to the dependence.[33] There must always be a real choice.[34] The employee must therefore have the possibility to decide against the use of predictive policing at the workplace by refusing or revoking the consent.

The voluntariness is further made firm in the employment relationship in § 26 para. 2 s. 2 BDSG and can in particular exist if a legal or economic advantage is achieved for the employed person or if the employer and the employed person pursue similar interests.

a) Legal or economic advantage

A legal or economic advantage is achieved for the employee if the consent leads to an actual or potential improvement of the status quo of the person concerned, because his or her assets increase or his or her legal position improves.[35]

A low risk assessment could lead to increased trustworthiness and thus to more opportunities for the employee. This could enable the employee to obtain a position that requires, for example, a trustworthy handling of the employer's assets. Due to this positive risk assessment, other monitoring measures could also possibly be omitted. Consent to an analysis would at least be conceivable for employees who want to achieve such a responsible position.

The analysis can also result in the employees gaining further rights. Due to the trustworthiness, they could be allowed to take sensitive documents home. At the same time, they could be granted more extensive powers in payments and contract conclusions.

b) Pursuit of similar interests

In addition, pursuant to § 26 para. 2 s. 2 BDSG voluntariness can exist in particular if the employer and the employed person pursue similar interests.

---

33 (Nebel, 2018, S. 523).

34 (Gräber & Nolden, 2021, § 26 Rn. 28; Ströbel & Wybitul 2019, § 10 Rn. 61).

35 (Martini & Botta, 2018, S. 629).

The decisive factor in this respect is the cooperation between the employer and the employee in the sense of a corporate coexistence.[36]
Similar interests are pursued, for example, if a camera surveillance also serves to protect the employee.[37] Moreover, based on case law, the consent can be given voluntarily, if a video surveillance in the company leads to the fact that employees can exonerate themselves from an accusation.[38] This shows that consent is of particular importance especially in complianc matters.[39] In such a case, both the employee and the employer benefit from the camera surveillance.
This also applies to predictive policing. When employees are checked for their trustworthiness, a safe working environment is also created for the employee. He also has the possibility to exonerate himself from the beginning of accusations and thus possibly not even get into a situation where he has to defend himself.
For these reasons, similar interests can also be acknowledged regularly.[40]

c) Result
Thus, predictive policing can be based on the consent of the employee.

4 Outlook on Predictive Policing in 2030

Data analysis using artificial intelligence will be indispensable in 2030. Predictive policing will also be a good way to create a control option for employers in the modern working world. Mobile working and flexible working hours are here to stay. But this positive development on the part of the employee is directly connected with a loss of control for the employer. Employees can, for example, take sensitive documents anywhere and also the actual working time is difficult to verify. Predictive policing can give employees the benefit of the doubt and thus be used sensibly by employers. At the same time, unlike surveillance measures, no new data is collected, but only existing data is evaluated.

---

[36] (BT-Drs. 18/11.325, p. 97; Gräber & Nolden, 2021, § 26 margin no. 27).
[37] Riesenhuber, 2021, § 26 margin no. 47b; Zöll, 2019, § 26 margin no. 78.
[38] BAG, NZA 2017, 443 (447).
[39] (Zöll, 2019, § 26 margin no. 78).
[40] (contra Rudowski, 2019, p. 73).

One has to consider, however, that the employee could understand predictive policing as a basic mistrust of his or her person and thus fears could arise with regard to a negative working atmosphere. These concerns can, however, be at least partially dispelled, because in the future, employees will already be confronted with data analysis from the time of application, which will positively influence the acceptance of data analysis in general and predictive policing specifically in the long run.

Predictive policing is currently only possible on the basis of consent. Consequently, it always has to be checked in the specific individual case whether voluntariness can be acknowledged. This is not practicable, especially in large companies, which probably need such data analysis the most.

There may also be concerns that employees can revoke their consent at any time. However, in that case, the result of the predictive policing will usually already be known to the employer, so that the data processing is completed. Against this background, the employer's obligation to delete the processed data because of the revocation does not hinder a successful use of predictive policing.

Moreover, it has to be considered that a safe working environment can only be created if as many employees as possible, for example within a department, agree to such a data analysis. This is because if only a few employees agree, predictive policing provides only a low value in terms of a safe working environment.

5 Summary and Practical Recommendations

Predictive policing can also be used in the employment context in the future. With regard to data protection considerations, concerns can at least be partly dispelled. Based on the current legal situation, a consent by the respective employee is required. Data processing on the basis of § 26 para. 1 sentence 1 BDSG is not possible due to lack of necessity.

Especially in companies that work with sensitive data or in case employees come into contact with large sums of money, predictive policing can be a good way to make the work environment safer. It also represents an opportunity to quickly gain trust in an employee and thus qualify the employee for promotion. Especially in large companies, this could change the world of work.

With regard to the acceptance of the employees, the following points should be considered:

If there is a works council, it should be involved early and its co-determination rights must be respected. This includes, among others, the co-determination right from § 87 para. 1 no. 6 BetrVG due to the introduction and application of a technical device suitable to track the employees' behaviour.

The trust of the employees should also be strengthened. This could be done by a pilot project in which only interested employees participate.

In addition, investment should be made in educating the employees. For example, targeted training could reduce the fear of using artificial intelligence in the company.

Taking these measures into account, nothing should hinder a successful use of predictive policing in the employment relationship in 2030, at least in the framework of the current legal provisions.

References

Bongers, F. (2019). Überwachung durch IT und Datenschutz. In S. Kramer (Eds.), *IT-Arbeitsrecht. Digitalisierte Unternehmen: Herausforderungen und Lösungen* (pp. 713–871). C.H. Beck.

Fuhlrott, M., & Oltmanns, S. (2019). Arbeitnehmerüberwachung und interne Ermittlungen im Lichte der Datenschutz-Grundverordnung. *NZA*, 1105–1110.

Gola, P. (2018). *Datenschutz-Grundverordnung*. C.H. Beck.

Gola, P. (2019). § 26 BDSG. In P. Gola & D. Heckmann (Eds.), *Bundesdatenschutzgesetz*. C.H. Beck.

Gräber, T., & Nodeln, C. (2021). § 26 BDSG. In B.P. Paal & D. A. Pauly (Eds.), *Datenschutzgrundverordnung—Bundesdatenschutzgesetz*. C.H. Beck.

Härtel, I. (2019). Digitalisierung im Lichte des Verfassungsrechts—Algorithmen, Predictive Policing, autonomes Fahren. *LKV*, 49–60.

Hentrich, W., & Pyrcek, A. (2016). Compliance und Fraud Monitoring im Zeitalter von digitaler Transformation und Big Data. *BB*, 1451–1455.

Hinz, K. (2020). Auswahlentscheidung nach sog. Predictive Policing. In M. Kaulartz & T. Braegelmann (Eds.), *Rechtshandbuch Artificial Intelligence und Machine Learning*. (pp. 543–545) Beck.

Hoch, V. R. S. (2020). Big Data und Predictive Analytics im Gerichtsprozess. Chancen und Grenzen der Urteilsprognose. *MMR*, 295–300.

Lundschien, S. (2021). Predictive Policing. Wie Algorithmen bei der Verbrechensbekämpfung helfen. https://www.goethe.de/prj/jad/de/the/ari/22082332.html. Accessed: 30. Juni 2021.

Martini, M., & Botta, J. (2018). Iron Man am Arbeitsplatz?—Exoskelette zwischen Effizienzstreben, Daten- und Gesundheitsschutz. *NZA*, 625–637.

Maschmann (2020). § 26 BDSG. In J. Kühling & B. Buchner (Eds.), *Datenschutzgrundverordnung—BDSG*. C.H. Beck.

Nebel, M. (2018). Big Data und Datenschutz in der Arbeitswelt. *ZD*, 520–524.

Riesenhuber (2021). § 26 BDSG. In S. Brink & H. A. Wolff (Eds.), *BeckOK Datenschutzrecht*. C.H. Beck.

Rudkowski, L. (2019). “Predictive Policing” am Arbeitsplatz. *NZA*, 72–77.
Ströbel, L., Wybitul,T. (2019). Einwilligung in Datenverarbeitung im Arbeitsverhältnis. In L. Specht & R. Mantz (Eds.), *Handbuch Europäisches und deutsches Datenschutzrecht—bereichsspezifischer Datenschutz in Privatwirtschaft und öffentlichem Sektor.* C.H. Beck.
Zöll, O. (2019). § 26 BDSG. In J. Taeger & D. Gabel (Eds.), *Kommentar DSGVO—BDSG.* Recht Und Wirtschaft GmbH.

Legal Requirements for AI Decisions in Administration and Justice

Johannes Schmees and Stephan Dreyer

1 Introduction

An increasing number of administrative and judicial proceedings with a growing shortage of staff[1] as well as a steadily growing amount of data that can be used as a basis for legal decisions, drive impulses for comprehensive modernisation and digitalisation strategies of the administrations of the federal and state governments. The prerequisites for theoretical to specific considerations of a stronger use of AI systems thus also come into focus in areas of state administration and judicial proceedings. Only for administrative and judicial proceedings do special requirements apply that at least limit the use of AI. The aim of this article is to provide an overview of the existing constitutional requirements for the use of AI in the executive and judicial branches in Germany and to take a cautious look into the near future.

[1](These will certainly be exacerbated by the foreseeable shortage of staff in both administration and justice, see, among others, Eydlin, 2021).

J. Schmees (✉) · S. Dreyer
Leibniz-Institute for Media Research, Hans-Bredow-Institut, Hamburg, Germany
e-mail: j.schmees@leibniz-hbi.de

S. Dreyer
e-mail: s.dreyer@leibniz-hbi.de

I. Knappertsbusch and K. Gondlach (eds.), *Work and AI 2030*,
https://doi.org/10.1007/978-3-658-40232-7_13

2 Status Quo

For the implementation of digital systems that use artificial intelligence technologies[2], numerous potential areas of application are emerging in the state sector. For example, AI software systems are currently used in administration for data analysis, such as weather data or traffic data for traffic control systems[3], but also decision systems for decision preparation, such as for the automated issuance of tax returns (§ 155 para. 4 AO)[4]. Hypothetical areas of application in administration and justice extend beyond that to entire fundamental decision making processes up to the installation of a "robo-judge"[5]. Thus, many considerations oscillate between enthusiasm and suspicion, often assumptions instead of realistic assessments and sometimes exaggerations of what is currently technically feasible can be observed. In doing so, one can look critically at the hitherto rapid and optimistic (at least with regard to the technically realisable) developments in the field of artificial intelligence with good reason.[6]

First, however, the existing legal framework needs to be outlined: For the administration in Germany, the central provision is § 35a VwVfG (Administrative Procedures Act), which the federal German legislator created for the automated issuance of administrative acts. According to this provision, an administrative act can "be issued completely by automatic means, provided that this is permitted by legal provision and that neither discretion nor a margin of appreciation exists". In addition, there are other, general procedural requirements that apply in principle also to (partially) automated decision processes, including, for example, the requirement of justification according to § 39 VwVfG—which in turn raises issues of traceability, transparency and explainability of AI systems in state use.[7] Moreover, data protection law also provides a basic—with exceptions—prohibition of fully automated decisions.[8]

A provision corresponding to § 35a VwVfG for judicial proceedings does not exist in German law so far. Here, the legal framework is set by simple statutory requirements for court decisions, in particular the rules of procedure, and constitutional requirements such as the principle of judicial independence according to

[2] (For a definition of the term "artificial intelligence" see further Krafft et al., 2020; Yeung 2019, p. 6).

[3] (Djeffal, 2018, p. 497).

[4] (Martini 2017, p. 452; Mund 2020, p. 179).

[5] (Schwintowski, 2018, p. 1603; Kaspar et al., 2020, p. 50 f).

[6] (So Floridi, 2020, p. 2).

[7] (Instructive Wischmeyer, 2020, p. 73).

[8] (See Dreyer & Schulz, 2018; Nink, 2021, p. 245 ff.).

Art. 97 para. 1 GG (Basic Law; the German constitution), but also by a constitutional human-oriented concept of a judge.[9] So far, in Germany, unlike in the international comparison, research and pilot projects are carried out that primarily provide for decision-supporting implementation of AI in the judiciary, for example for the automated extraction of legally relevant information from electronic documents. In addition, AI systems are used rather low-threshold, such as in the form of speech recognition programmes as an alternative to classic dictation devices and subsequent transcription.[10] Nevertheless, it is to be expected that the use of AI will find widespread use in various facets both in administration and in justice (see sect. 4).

3 Challenges and Solutions

Law appears as a social practice.[11] A legal decision usually serves to achieve conformity with higher-ranking norms by substanciating and adapting to specific facts and persons and thereby—regularly—resolving conflicts.[12] In this respect, the decision is a method-driven act of legal application[13], a judicial or executive interpretation of a legal norm, which is brought into effect locally and situationally[14] and condenses into a fundamental rights-relevant intervention.[15] The legal decision appears, as it were, as a complexity-reducing fiction, which constructs a final act of a complex, law-applying decision process, which is creative, social and discursive.[16] Purely functional objections can already be derived from this background against the postulate of a human-equivalent AI in fully automated legal decision processes. Especially in instances of decision-making leeway, which have to be filled with factual, contextual, impact and implicit knowledge, it becomes apparent that such legal decisions presuppose a certain understanding of

[9] (Nink, 2021, p. 288 ff., 260 ff).

[10] (Scientific Service of the Bundestag, 2021, p. 5).

[11] (Morlok et al., 2000, p. 15; Boulanger, 2019, p. 173).

[12] (With further evidence cf. Hill, 2017, p. 437).

[13] (Larenz, 1991, p. 137).

[14] (Vesting, 2019, p. 116).

[15] (As a legitimate and coercive legal application, decisions are actually practical processes, legally sociological Baer, 2021, p. 246).

[16] (Darnstädt, 2019, p. 1580).

the world and seem hardly compatible with the mechanisms of an AI system—as of now.[17]

It is central to acknowledge that the law-applying executive or judicial decision is subject to increased constitutional requirements due to the direct binding of the state to fundamental rights guarantees. Thus, it is open whether and how, when implementing AI systems in state decision-making practice, the human dignity and autonomy[18], freedom and equality rights, legitimacy[19] and rule-of-law principles can be preserved, in particular, if state authorities de facto relinquish their decision-making competences by this means or temporally shift them to the area of AI programming or AI learning. For the legal admissibility of the use of AI in administration or in court, this results in high hurdles and fundamental questions. If AI is used to reduce the burden of action and decision-making, at least a suitable degree of (human) responsibility must be maintained and appropriate responsibility structures must be created.[20] In this context, the aforementioned upstream area of programming and modeling plays a central role, as this is where the preconditions are created that might have a significant influence on later AI decisions. Here, reservations regarding the technical as well as the legal feasibility have to be identified. Especially in the area of the judiciary, achieving an adequate data basis is difficult, which would be required for working AI systems for fully automated decision-making.[21] This problem is illustrated by an AI-based project for suicide prevention in prison: Due to the lack of real training data, actors had to be used, rendering the actual benefit of the AI-systems compared to humans questionable.[22] Another significant problem is the one of justifications for automated administrative acts, which are an indispensable element and precondition of the right to judicial protection under Art. 19 para. 4 GG and the rule of law principle under Art. 20 para. 3 GG and would accordingly have to be implemented in a suitable form for AI decisions.[23]

---

[17] (Dreyer & Schmees, 2019, p. 762).

[18] (Dreyer, 2018, p. 135).

[19] (Herold, 2020).

[20] (Rademacher, 2020, p. 67).

[21] (Dreyer & Schmees, 2019, p. 759; further Nink, 2021, p. 177).

[22] (Stukenberg, 2021).

[23] (Roth-Isigkeit, 2020, p. 1020; Wischmeyer, 2020, p. 73).

4 Outlook on AI Systems in Administration and Justice in 2030

The idea that by 2030 only fully automated machines will issue administrative acts and decide as "robo-judges" over Germany's citizens seems unlikely in view of the constitutional limitations of delegating legal decisions to AI systems. *With significant decisions*, the human actor will likely remain the central decision-maker. In fields of large quantity decision making areas with uniform procedures, however, corresponding systems will be used more and more extensively in administrations. Here, the practical benefit outweighs the legal risks, especially in the face of staff shortages in state organisations. Apart from that, there will be incremental changes in the executive branch, where the focus will be on the development and implementation of decision *supporting* systems.

Developments are also to be expected in the judiciary, which will aim to support judges in finding judgements. Given a rather poor record of digitisation of some court administrations and a traditionally independence-minded German judiciary, however, one can look with scepticism at too great technological leaps in the nearer future in this area of legal decision-making—even if quite sensible software systems, for example for automated file analysis or research, are conceivable.[24]

However, there will be a longer-lasting, extensive debate about the advantages and disadvantages of supposedly "more objective" decisions by AI systems. Here, further approximations to decision rationality, -justifiability and -traceability of algorithmic decision systems can be achieved.[25] But even if one follows the appeal of *Rademacher* to include the potential solution of some technical AI problems in considerations in perspective[26], the question of the scope and quality of data on which systems are trained remains. These will foreseeably and inevitably contain hidden biases. In addition, the structural problem remains that with the shift of the decision-maker towards AI, there is also a problematic temporal shift of decision-making power.[27]

Overall, with a progressive digitisation of justice and administration, there will be mainly new, hybrid decision configurations of humans and AI. In variants of

[24] (LegalTech in the judiciary currently focuses rather on issues of the electronic court file, cf. Starosta, 2020, p. 216).

[25] (Cf., among others, Koulu, 2021, p. 81; Rudin, 2019, p. 206).

[26] (Rademacher, 2020, p. 71).

[27] (f. Dreyer & Schmees, 2019, p. 763).

automated data preparation, decision support or decision recommendation, it will be crucial to create arrangements, structures and procedures that preserve the constitutional guidelines outlined above and which guarantee that not only "the" human, but also the "right" human actor exercises (decision-)power in a democratically legitimate, rule-of-law manner. Developing active design proposals and analysis tools for this is the fundamental task for jurisprudence, computer science and software engineering, requiring orientation towards innovation and interdisciplinarity.

5 Summary and Practical Recommendations

Do the legal professions in the judiciary and executive branch remain a special case when it comes to the use of AI? Probably yes, as far as fully automated decisions are concerned. This is due to the fact of the direct binding of fundamental rights of all state organs and thus also of their decision-makers. A complete automation of all administrative or judicial proceedings is not a legitimate option, especially for constitutional reasons. Even a partial automation that is deemed purposeful entails its legal pitfalls. Not too much decision-making power may be delegated away from human actors, despite all their real or presumed flaws, to AI systems. Court and administrative decisions will foreseeably in most cases continue to be made by humans; but state institutions have to get used to the fact that AI systems could support them in finding these decisions. This requires new forms of competence, expertise and experience.

References

Baer, S. (2021). *Rechtssoziologie—Eine Einführung in die interdisziplinäre Rechtsforschung*. Nomos.

Boulanger, C. (2019). Die Soziologie juristischer Wissensproduktion—Rechtsdogmatik als soziale Praxis. In C. Boulanger, J. Rosenstock, & T. Singelnstein (Eds.), *Interdisziplinäre Rechtsforschung: Eine Einführung in die geistes- und sozialwissenschaftliche Befassung mit dem Recht und seiner Praxis* (pp. 173–192). Springer.

Darnstädt, T. (2019). Die Suche nach dem richtigen Weg. *Neue juristische Wochenschrift,* 1580–1586.

Djeffal, C. (2018). Normative Leitlinien für künstliche Intelligenz in Regierung und öffentlicher Verwaltung. In: R. Mohabbat Kar, B. E. P. Thapa, & P. Parycek (Eds.), *(Un)berechenbar? Algorithmen und Automatisierung in Staat und Gesellschaft* (pp. 493–415).

Dreyer, S. (2018). Predictive Analytics aus der Perspektive von Menschenwürde und Autonomie. In W. Hoffmann-Riem (Eds.), *Big Data—Regulative Herausforderungen* (1. Aufl., S. 135–144). Baden-Baden: Nomos.

Dreyer, S., 7 Schmees, J. (2019). Künstliche Intelligenz als Richter?—Wo keine Trainingsdaten, da kein Richter—Hindernisse, Risiken und Chancen der Automatisierung gerichtlicher Entscheidungen. *Computer und Recht,* 758-764.

Dreyer, S., & Schulz, W. (2018). Was bringt die Datenschutz-Grundverordnung für automatisierte Entscheidungssysteme? Potenziale und Grenzen der Absicherung individueller, gruppenbezogener und gesellschaftlicher Interessen. Impuls Algorithmenethik #7. Bertelsmannstiftung (Eds.), Gütersloh. doi: https://doi.org/10.11586/2018011.

Eydlin, A. (2021). Deutscher Beamtenbund: Personallücke im öffentlichen Dienst um zehn Prozent gewachsen. Zeit Online, 29. März 2021. https://www.zeit.de/politik/deutschland/2021-03/deutscher-beamtenbund-oeffentlicher-dienst-personalluecke-kita-pflege-schule. Accessed: 14. Juni 2021.

Floridi, L. (2020). AI and its new winter: from myths to realities. *Philos. Technol.,* 1–3.

Herold, V. (2020). *Demokratische Legitimation automatisiert erlassener Verwaltungsakte.* Duncker & Humblot.

Hill, H. (2017), Die Kunst des Entscheidens. *Die öffentliche Verwaltung,* 433–443.

Kaspar, J., Höffler, K., & Harrendorf, S. (2020). Datenbanken, Online-Votings und künstliche Intelligenz—Perspektiven evidenzbasierter Strafzumessung im Zeitalter von „Legal Tech“. *Neue Kriminalpolitik,* 35–56.

Koulu, R. (2021). Crafting Digital transparency: Implementing legal values into algorithmic design. *Critical Analysis of Law,* 81–100.

Krafft, P. M., Young, M., Katell, M., Huang, K., & Bugingo, G. (2020). Defining AI in Policy versus Practice. Proceedings of the AAAI/ACM Conference on AI, Ethics, and Society, 72–78.

Larenz, K. (1991). *Methodenlehre der Rechtswissenschaft.* Springer.

Martini, M. (2017). Transformation der Verwaltung durch Digitalisierung. *Die öffentliche Verwaltung,* 443–455.

Morlok, M., Kölbel, R., & Launhardt, A. (2000). Recht als soziale Praxis. Eine soziologische Perspektive in der Methodenlehre. *Rechtstheorie,* 15.

Mund, D. (2020). Das Recht auf menschliche Entscheidung—Freiheit in Zeiten der Digitalisierung und einer automatisierten Rechtsanwendung. In R. Greve, B. Gwiasda, T. Kemper, J. Moir, S. Müller, A. Schönberger, S. Stöcker, J. Wagner, & L. Wolff (Eds.), *Der digitalisierte Staat—Chancen und Herausforderungen für den modernen Staat* (pp. 177–198). Nomos.

Nink, D. (2021). *Justiz und Algorithmen: Über die Schwächen menschlicher Entscheidungsfindung und die Möglichkeiten neuer Technologien in der Rechtsprechung.* Duncker & Humblot.

Rademacher, T. (2020). Künstliche Intelligenz und neue Verantwortungsarchitektur. In M. Eifert (Eds.), *Digitale Disruption und Recht: Workshop zu Ehren des 80. Geburtstags von Wolfgang Hoffmann-Riem* (pp. 45–72). Nomos.

Roth-, D. (2020). Die Begründung des vollständig automatisierten Verwaltungsakts. *Die öffentliche Verwaltung,* 1018–1026.

Rudin, C. (2019). Stop explaining black box machine learning models for high stakes decisions and use interpretable models instead. *Nat. Mach. Intell.,* 206–215.

Schwintowski, H.-P. (2018). Wird Recht durch Robotik und künstliche Intelligenz überflüssig? *Neue juristische Onlinezeitschrift,* 1601–1609.

Starosta, G. (2020). Die richterliche Unabhängigkeit im Zeitalter der Digitalisierung. *Die öffentliche Verwaltung,* 216–225.

Stukenberg, T. (2021). Gefängnis: Warum Künstliche Intelligenz Suizide nicht verhindert. netzpolitik.org 09. Juni 2021. https://netzpolitik.org/2021/gefaengnis-warum-kuenstliche-intelligenz-suizide-nicht-verhindert/. Accessed: 14. Juni 2021.

Vesting T (2019) Rechtstheorie: Ein Studienbuch. C.H. Beck, München

Wischmeyer, T. (2020). Künstliche Intelligenz und neue Begründungsarchitektur. In M. Eifert (Eds.), *Digitale Disruption und Recht: Workshop zu Ehren des 80 Geburtstags von Wolfgang Hoffmann-Riem* (pp. 73–92). Nomos.

Wissenschaftlicher Dienst des Bundestages (2021). Künstliche Intelligenz in der Justiz—Internationaler Überblick, WS 7—3000—017/21, 2021. https://www.bundestag.de/resource/blob/832204/6813d064fab52e9b6d54cbbf5319cea3/WD-7-017-21-pdf-data.pdf. Accessed: 14. Juni 2021.

Yeung, K. (2019). Responsibility and AI, A study of the implications of advanced digital technologies (including AI systems) for the concept of responsibility within a human rights framework. Council of Europe study DGI(2019)05. https://rm.coe.int/responsability-and-ai-en/168097d9c5. Accessed: 14. Juni 2021.

AI in the economic world of work

Intelligent IT Systems in Business Application

Control and Transparency as Means of Building Trust in AI

Alexander Rühr, Benedikt Berger and Thomas Hess

1 Introduction

The increasing availability of large amounts of data, the steadily growing computing capacities for processing this data and innovative algorithms have led to technological progress in the field of artificial intelligence (AI). Information technology systems (IT systems) that make use of this progress are referred to as intelligent or cognitive IT systems (Watson, 2017). These systems enable the automation of numerous tasks and can support humans in various work processes. An example of such a process is staff planning in a call centre, requiring a prediction of how many calls the call centre will receive on a given day. Using machine learning methods and based on existing data on calls from the past, an intelligent IT system can provide this prediction. A human can then proceed with the staff planning based on this prediction. The increasing amount and variety of automatable tasks means at the

A. Rühr (✉)
OMMAX - Building Digital Leaders, London, United Kingdom
e-mail: alexander.ruehr@ommax.de

B. Berger
Department of Information Systems, University of Münster, Münster, Germany
e-mail: benedikt.berger@ercis.uni-muenster.de

T. Hess
Institute for Digital Management and New Media, Ludwig-Maximilians-Universität München, Munich, Germany
e-mail: thess@bwl.lmu.de

I. Knappertsbusch and K. Gondlach (eds.), *Work and AI 2030*,
https://doi.org/10.1007/978-3-658-40232-7_14

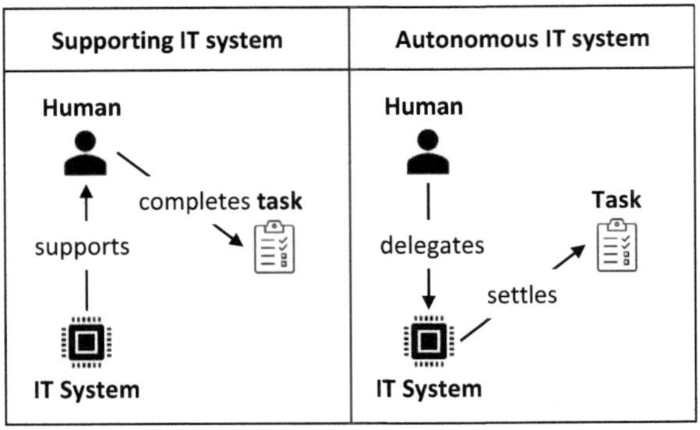

Fig. 1 Supporting and autonomous IT systems

same time that intelligent IT systems can take over larger parts of work processes, i.e. act more autonomously (Baird & Maruping, 2021). In the call centre example, this would mean that the intelligent IT system not only makes the prediction of the number of calls, but also determines the staff demand and assigns shifts. The increasing degree of autonomy changes the interaction between humans and IT systems. Instead of actively using an IT system as support for a task, users can instead delegate the entire decision or task to the system (see Fig. 1).

Whether companies benefit from the increasing automation potential depends on whether the responsible employees are willing to be supported by intelligent IT systems or to delegate tasks to them. Analogous to the delegation decision between two humans, trust in intelligent IT systems plays an important role for delegating tasks and decisions. Only if the users trust the IT system, can companies successfully exploit the automation potentials that arise from the developments in the field of AI. However, humans do not automatically trust intelligent IT systems. Possible consequences of a lack of trust can be that humans ignore the recommendations of an intelligent IT system, deliberately bypass the system or even feel threatened by the system. This chapter therefore deals with trust building in intelligent IT systems and the resulting implications for the world of work.

2 Status Quo

In order to analyse these implications, a common understanding of basic terms is necessary. The *automation* of a task or decision refers to the degree to which functions that could be performed by a human are taken over by an IT system

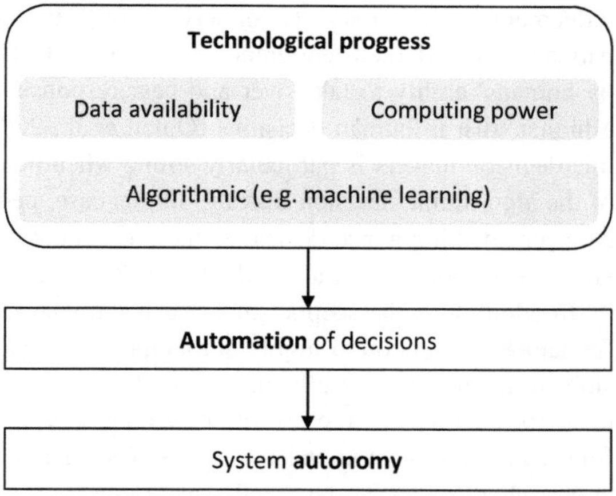

Fig. 2 Tendency towards automation and autonomy

(Parasuraman et al., 2000). Degrees of automation result from the extent of algorithmic conduct in information processing, decision making and the subsequent implementation within a task (Lee & See, 2004). At the workplace, an increasing automation is common in many areas. Automated production processes, for instance, not only increase the efficiency of the processes, but also provide important data that, in turn, enable further process improvements and predictive maintenance. Automation can also secure and accelerate processes in management, such as checking cash flows or selecting new employees. The degree to which an IT system can make certain decisions or perform tasks describes the *autonomy* of the system. The more diverse the decisions and tasks that an IT system can handle, the more autonomous this system is. While automation refers to a task, autonomy is a property of an IT system (see Fig. 2).

3 Challenges in Building Trust

The attitude of people towards automated tasks and decisions is subject to extensive behavioural research. Despite the potential for increasing efficiency, many people are sceptical about delegating decisions to algorithmic decision makers (Dietvorst et al., 2015). Instead, people tend to prefer human over algorithmic decisions. In studies on the application of automation in human resources, participants justified their aversion to algorithmic decision making by arguing that it was less professional and flexible (Diab et al., 2011) or did not have the intuitive judgement of a human (Highhouse, 2008). In medical delegation decisions, patients justified their preference for human decision makers over algorithms by humans' ability to take over and bear responsibility. This is also reflected in a higher trust in human decisions (Önkal et al., 2009). The aversion to algorithmic decision makers is particularly strong when people notice suboptimal results of the algorithmic decision making. In this case, people prefer even those human delegates making worse decisions than the suboptimal algorithmic decision makers. This phenomenon is also called *algorithm aversion* (Dietvorst et al., 2015).

In addition to the scepticism towards automated decisions, humans exhibit a fundamental aversion to highly autonomous delegates. The causes for this aversion are manifold. Basically, people tend to rely on themselves most. This is evident, for example, in the use of consulting services. People often only partially follow even the recommendations of professional and suitable consultants (Yaniv & Kleinberger, 2000). The tendency to preserve one's own influence on a final decision is based, among other things, on *overconfidence* (Tan et al., 2012). Since people systematically overestimate their own abilities, it seems less attractive to them to delegate decision-making competencies than would be optimally the case. In addition, delegation involves the loss of control over a decision process or an IT system that executes the decision. Such a loss of control is closely associated with the perception of risk. The more competencies a person delegates to an IT system, the greater the potential damage if the system fails or delivers undesirable results (Rijsdijk & Hultink, 2003). This perception of risk, in turn, impedes the development of trust in the IT system. Fig. 3 shows the tensions arising from the potentials of higher automation and system autonomy on the one hand and human behavioural responses on the other hand.

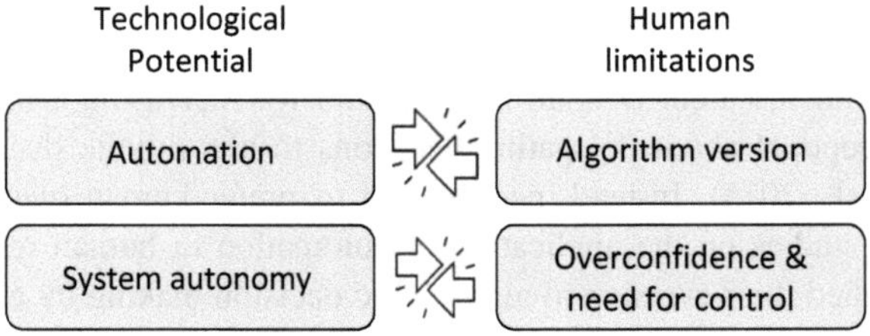

Fig. 3 Automation potentials and user behaviour

4 Possible Solutions: Control and Transparency

To reduce the scepticism towards intelligent IT systems due to algorithm aversion or perceived loss of control, an obvious strategy is to convey to people the *feeling of control* over the systems. The perception of having influence on an IT system reduces the risk associated with the system. This increases users' trust in the IT system and increases their willingness to delegate a decision to the system. This effect is consistent with the human tendency to overestimate one's contribution to the decision process and the resulting influence on a satisfactory outcome (Tan et al., 2012). In the context of highly autonomous IT systems, it can be assumed that the initial risk perception of users is strong and requires reduction to facilitate delegation (Złotowski et al., 2017).

One approach to convey the feeling of control over a highly autonomous IT system is to actually grant them the possibilities to actively intervene in the system's actions. In the introductory example of the call centre staff planning, a human could, for instance, override the intelligent IT system's prediction of the number of calls for a certain day. However, providing *control options* over autonomous IT systems comes at a price. While control options can appease users' concerns in the short term, the actual use of such options may impair the system's performance in the long term. Extensive research in computer science has shown that algorithmic decision making is superior to human decision making in many contexts (Kleinberg et al., 2018). Especially for tasks that require the analysis of large amounts of data and aim for an objective result, algorithms improve the outcome (Diab et al., 2011). Therefore, the active intervention of users in the task conduct of an intelligent IT system can worsen its performance. The IT system's performance, in turn, significantly influences the consideration of delegating tasks or decisions to the system. If an employee of the aforementioned call centre would repeatedly override the intelligent IT system's predictions, the staff planning based on these predictions could deteriorate and make the system's use less attractive. Consequently, it is helpful for the realisation of automation potential to assure users of their involvement in the decision process while at the same time limiting the actual influence.

A second approach to reduce the perceived risk of users without allowing for an active influence on the IT system is a transparent system design. The *transparency* of an IT system refers to its ability to convey an understanding of how the system works. For this purpose, IT system providers can, for example, provide explanations of the automated decision process or the underlying data basis (Wang & Benbasat, 2007). With the help of this information, users can better understand and comprehend the IT system. The higher comprehensibility facili-

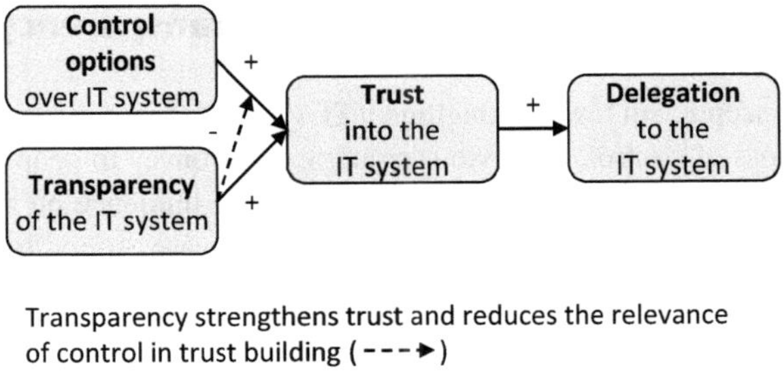

Fig. 4 Proposed effects of transparency and control on trust in IT systems

tates classifying the actions of the IT system and assessing its added value, which has a positive effect on trust (Wang & Wang, 2019). In contrast to active control options, transparency is a passive form of user participation in the decision process. The IT system can satisfy users' basic desire for control at least in part by granting them insights into its processes and allowing them to familiarise themselves with the system (Kahai et al., 1998). The possibility to observe or monitor the decision behaviour of the IT system thus partially replaces active control options and reduces their relevance in trust building as well as the potential disadvantages of human influence. Therefore, it would make sense for the intelligent IT system in the introductory example to provide clues for the prediction of the number of calls in the call centre on a certain day. Fig. 4 summarises the possibilities of influencing trust in IT systems and the willingness to delegate tasks to them.

5 Actionable Recommendations and Outlook

Despite the potential benefits of granting IT systems high autonomy in decision processes, users generally prefer to retain control over their decisions. IT systems that provide users with control options appear less risky in use. This reduced risk perception increases the willingness to trust an IT system and to delegate decision competencies to it. To realise automation potentials, the providers of IT systems should therefore create possibilities for active control, especially when the individual judgement of users can help to avoid inaccuracies. This applies, for example, to the interpretation of data or the development of solutions for complex problems. Moreover, companies should exploit automation potentials in routine tasks (Oeste-Reiß et al., 2021). In the development of video games, for example, the design of virtual landscapes represents such a routine task. Tasks in which a human intervention due to a strong desire for control could be explicitly detrimental, however, should be covered by particularly transparent IT systems. In this case, the communicative impact of transparent design elements, such as explanations, can mitigate the desire for control.

For the world of work in 2030, a broad spectrum of collaboration between humans and intelligent IT systems is likely to emerge. Intelligent IT systems may check human work and point out improvements, as is already possible today in text processing or software programming. Conversely, humans may check and potentially improve the work of intelligent IT systems. Both scenarios can result in mutual learning (Oeste-Reiß et al., 2021). The exact configuration of such collaborations, also referred to as *hybrid work*, depends not only on the task characteristics outlined above, but also on the context. Legal and ethical considerations, for instance, can influence the decision of how autonomously an IT system should be able to act. In addition to efficiency considerations, the configuration of human-AI collaboration should also be guided by the fulfilment and satisfaction of humans through their work as these outcomes also play an important role for the long-term success of a company.

References

Baird, A., & Maruping, L. M. (2021). The Next Generation of Research on IS Use: A Theoretical Framework of Delegation To and From Agentic IS Artifacts. *MIS Quarterly, 45*(1), 315–341.

Diab, D. L., Pui, S. Y., Yankelevich, M., & Highhouse, S. (2011). Lay Perceptions of Selection Decision Aids in US and Non-US Samples. *International Journal of Selection and Assessment, 19*(2), 209–216.

Dietvorst, B. J., Simmons, J. P., & Massey, C. (2015). Algorithm Aversion: People Erroneously Avoid Algorithms after Seeing Them Err. *Journal of Experimental Psychology, 144*(1), 114–126.

Highhouse, S. (2008). Stubborn Reliance on Intuition and Subjectivity in Employee Selection. *Industrial and Organizational Psychology, 1*(3), 333–342.

Kahai, S. S., Solieri, S. A., & Felo, A. J. (1998). Active Involvement, Familiarity, Framing, and the Illusion of Control during Decision Support System Use. *Decision Support Systems, 23*(2), 133–148.

Kleinberg, J., Lakkaraju, H., Leskovec, J., Ludwig, J., & Mullainathan, S. (2018). Human Decisions and Machine Predictions. *The Quarterly Journal of Economics, 133*(1), 237–293.

Lee, J. D., & See, K. A. (2004). Trust in Automation: Designing for Appropriate Reliance. *Human Factors, 46*(1), 50–80.

Oeste-Reiß, S., Bittner, E., Cvetkovic, I., Günther, A., Leimeister, J. M., Memmert, L., et al. (2021). Hybride Wissensarbeit. *Informatik Spektrum, 44*(3), 148–152.

Önkal, D., Goodwin, P., Thomson, M., Gönül, S., & Pollock, A. (2009). The Relative Influence of Advice from Human Experts and Statistical Methods on Forecast Adjustments. *Journal of Behavioral Decision Making, 22*(4), 390–409.

Parasuraman, R., Sheridan, T. B., & Wickens, C. D. (2000). A Model for Types and Levels of Human Interaction with Automation. *IEEE Transactions on Systems, Man, and Cybernetics, 30*(3), 286–297.

Rijsdijk, S. A., & Hultink, E. J. (2003). "Honey, Have You Seen Our Hamster?" Consumer Evaluations of Autonomous Domestic Products. *The Journal of Product Innovation Management, 20*(3), 204–216.

Tan, W. K., Tan, C. H., & Teo, H. H. (2012). Consumer-Based Decision Aid that Explains Which to Buy: Decision Confirmation or Overconfidence Bias? *Decision Support Systems, 53*(1), 127–141.

Wang, W., & Benbasat, I. (2007). Recommendation Agents for Electronic Commerce: Effects of Explanation Facilities on Trusting Beliefs. *Journal of Management Information Systems, 23*(4), 217–246.

Wang, W., & Wang, M. (2019). Effects of Sponsorship Disclosure on Perceived Integrity of Biased Recommendation Agents: Psychological Contract Violation and Knowledge-Based Trust Perspectives. *Information Systems Research, 30*(2), 507–522.

Watson, H. J. (2017). Preparing for the Cognitive Generation of Decision Support. *MIS Quarterly Executive, 16*(3), 153–169.

Yaniv, I., & Kleinberger, E. (2000). Advice Taking in Decision Making: Egocentric Discounting and Reputation Formation. *Organizational Behavior and Human Decision Processes, 83*(2), 260–281.

Złotowski, J., Yogeeswaran, K., & Bartneck, C. (2017). Can We Control It? Autonomous Robots Threaten Human Identity, Uniqueness, Safety, and Resources. *International Journal of Human Computer Studies, 100*, 48–54.

Successful Introduction of AI in the Company

Building Blocks for Change Management

Sascha Stowasser

> *"We can only master the transformation into a working world 2030 with AI technologies together. We have to think together about what this new working world should look like. For this, a change management towards a culture of change is necessary and to be lived in the companies every day."—Sascha Stowasser*

1 Introduction

AI not only offers a lot of opportunities for innovative business models of companies and institutions, but also brings about radical changes in the working world within companies. AI tools or learning (work) systems develop the work 4.0 to work 5.0. While work 4.0 is currently focused on the networked digitalisation and the flexibility of work location, time, organisation and freedom of action, work 5.0 will be enriched with intelligent assistance, learning robots and user-optimised information provision. For the employees, the use of AI means more flexibility, the performance of more demanding tasks, individually adapted information and relief from monotonous mental routine tasks.

S. Stowasser (✉)
ifaa—Institut für angewandte Arbeitswissenschaft, Düsseldorf, Germany
e-mail: s.stowasser@ifaa-mail.de
URL: https://www.arbeitswissenschaft.net

I. Knappertsbusch and K. Gondlach (eds.), *Work and AI 2030*,
https://doi.org/10.1007/978-3-658-40232-7_15

Since AI technology and learning systems have the potential to reduce or channel the flood of work-relevant information in the field of information and knowledge work, and learning robot systems and AI-based automation solutions can take over physically demanding activity components in the field of production work, there is the prospect that employees in a work world shaped by AI will tend to experience reduced stress. This has a positive effect on the risk assessments for the stress effects of the employees in the best case.

The use and implementation of new technologies are a familiar variable in the companies and in the work environment, which relies on a familiar change toolkit and the legal rules—such as co-determination and data protection. Nevertheless, new challenges for change processes arise from the specifics of AI, such as the learning aspect of machines, robots and software systems, the use of large amounts of data as a learning basis or the predictive analytics by AI systems. In addition, questions of discrimination by data and algorithms, personal rights or the relationship between human and machine—including the scope of action and the attribution of responsibility—become more prominent (cf. Terstegen et al., 2021).

2 Status Quo—Assessments and Fears Regarding the Introduction of AI

The working world is changing. Our own project analyses show that about 75% of the work systems will change. This affects all kinds of work systems, i.e. both production and knowledge work.

A survey within the project "Digital Mentor—Model and Testing of a Preventive Acting AI Helper (short en[AI]ble)"[1] reflects the assessment of AI technologies

---

[1] "Digital Mentor—Model and Testing of a Preventive Acting AI Helper (short en[AI]ble)", project duration: September 2020—September 2023, consortium leader: ifaa—Institute for Applied Work Science e. V. The project is funded by the Federal Ministry of Labor and Social Affairs (BMAS) within the framework of the Initiative New Quality of Work (INQA) and is accompanied by the Federal Institute for Occupational Safety and Health (BAuA) in terms of content. The project sponsor is the Society for Social Business Consulting mbH (gsub). Funding number: EXP.01.00008.20.

by different company stakeholders (Fig. 1). While people are willing to use AI in the private sector without much reservation, for example as a navigation aid or for music selection with their smartphone, many see it quite differently at the workplace. They fear the misuse of their personal data and are afraid of being spied upon by the new technologies.

We register three basic fears in surveys and in operational implementation projects, which are repeatedly addressed:

- What will happen to my job?
- What will happen to my personal data?
- And: Will I be able to keep up with the digitalisation? Am I skilled enough to work with an AI?

These fears must be taken seriously and therefore clarity must be created about the use of AI. Little helpful is an activist use of AI technologies, an unstructured imposition of new tools and the lack of consideration of the affected people. Rather, an employee-oriented, participatory introduction process emerges as a success factor. An intensive change participation of the employees and the works council includes

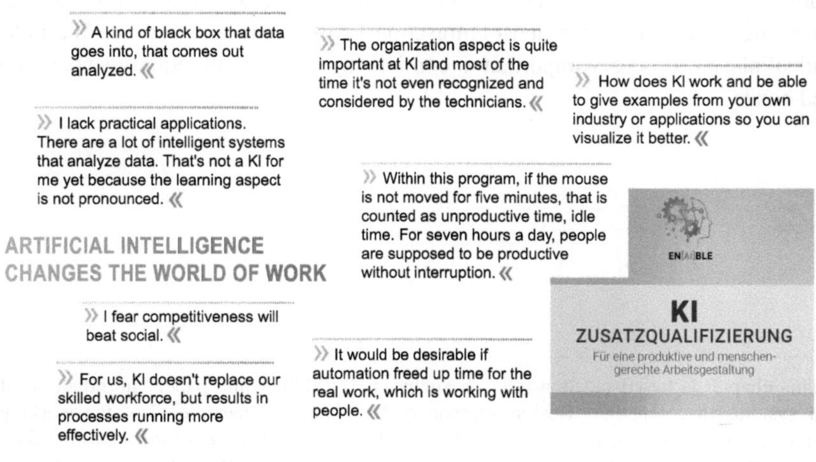

Fig. 1 Original quotes from entrepreneurs, association representatives and works councils on the assessment of AI in the working world (based on ifaa, 2021, p. 16–21)

- showing the benefits of AI and how it can be used,
- determining the need for qualification and implementing it,
- co-designing the new work systems.

Not least, the design of a changing relationship between human and technology through AI is in the foreground. It is important to emphasise the respective strengths of human and machine, in order to enable a productive cooperation and to support the human in his work activity.

3 Challenges and Solutions[2]

The introduction of AI technologies is just beginning in many companies, institutions and facilities in 2021. Today, therefore, is the right time to discuss possible opportunities, hurdles and limits of AI in the work environment, to explore design options and to use AI technologies for both the economic success of companies and a human-oriented work world in the sense of the employees by 2030 (cf. Terstegen et al., 2021).

When does the use of AI in the work world 2030 prove to be successful? In the following, the framework conditions and building blocks for successful change with AI will be outlined and possible solutions will be presented. Different phases of change processes are used as a basis (Fig. 2). Simplified, these phases are shown sequentially in the figure. In the operational—meanwhile agile—practice, there are always recursion and optimisation loops within and between the individual phases. It is very important to consistently consider all requirements in the red boxes.

[2]This chapter is based on Stowasser & Suchy et al. (2020) white paper "Introduction of AI systems in companies. Design approaches for change management" of the working group 2—Work/Qualification, Human-Machine Interaction of the Platform Learning Systems (https://www.plattform-lernende-systeme.de/ag-2.html#BT101). Many thanks for the kind permission and cooperation with the German Academy of Engineering Sciences e. V. (Acatech).

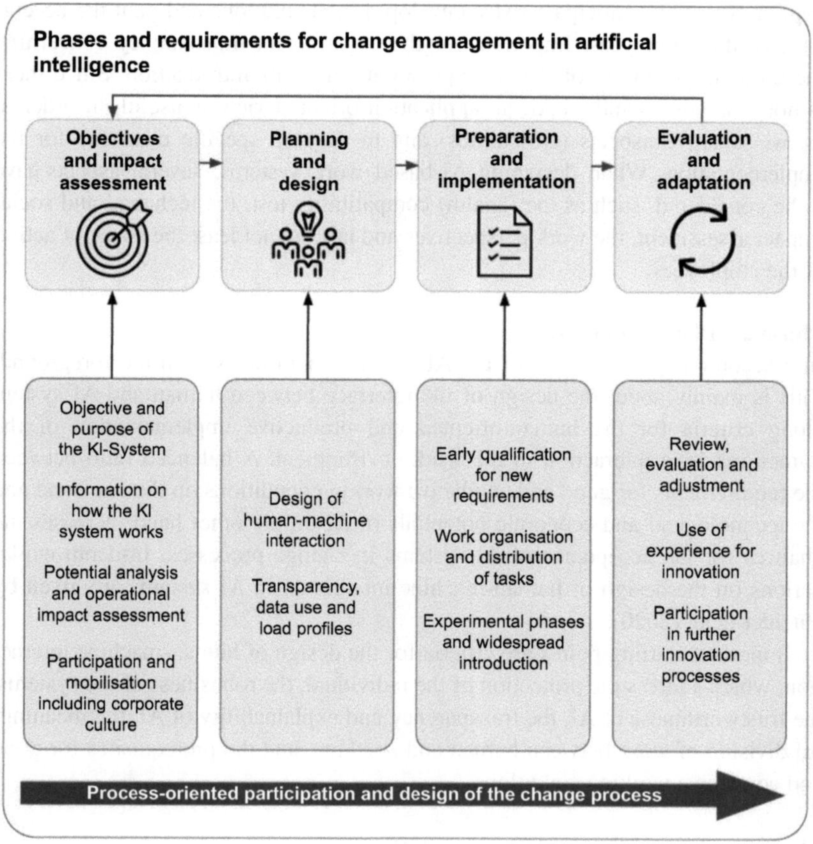

Fig. 2 Phases of and requirements for change management with AI (Stowasser & Suchy, et al., 2020)

Phase 1—Goal Setting and Impact Assessment

All actors who are responsible for the introduction of AI technology and the design of change processes—such as management and human resources, programmers and IT departments, as well as employees and works or staff councils—should cooperate before the introduction of AI systems in the company, in order to agree on the optimisation goals for the operational use and the requirements for the AI systems and to anticipate possible effects for the work design as early as possible. When introducing AI systems, a careful potential analysis and an operational impact assessment are necessary, in order to optimally exploit potentials, anticipate risks, develop design solutions and gain the acceptance of the employees. Important for this is sufficient transparency concerning the mode of operation of the AI application before its introduction. Since there is not "the" one standard AI, an application-oriented view is useful, in order to assess sensitive aspects (criticalities) and to develop specific measures for the implementation. When designing AI-based work systems, several aspects have to be considered, such as the (health) compatibility test, the technical and social impact assessment, the work perspectives and last but not least the scope of action of the employees.

Phase 2—Planning and Design
In a second step, the design of the AI systems themselves is in the foreground. This is mainly about the design of the interface between human and AI system along criteria for the human-oriented and productive implementation of the human-machine interaction in the work environment. A balanced ratio between the requirements for good and conducive working conditions on the one hand and the technological and economic potentials of AI on the other hand increases the chances for the acceptance of AI systems in change processes. In-depth explanations on the design of human-machine interaction in AI systems are given by Huchler et al. (2020).

Important starting points are criteria for the design of human-machine interaction, which address the protection of the individual, the robustness of the systems, the trustworthiness of AI, the transparency and explainability of AI, the meaningful division of work between human and machine, and the prerequisites for good and conducive working conditions.

Phase 3—Preparation and Implementation
The AI systems also have to be integrated in a suitable way into existing or new work processes and possibly into changed organisational structures. This means preparing the employees early for new tasks and initiating necessary qualification measures. It is also important to design new task and activity profiles for employees and to adapt the work organisation to a changed relationship between human and machine.

The change processes with AI should include more pilot and experimentation phases, in which experience values and best practice examples can be collected. They should also allow checking the effects and interfaces of the AI systems with regard to the objectives and in view of a humane work design, as well as to exclude undesirable effects as far as possible and to gain positive experience values with the systems in the work environment.

Phase 4—Evaluation and Adaptation

After the introduction of AI systems, a continuous review and evaluation of the AI use should take place, in order to ensure possible adjustments with regard to the design of the applications, the work organisation or the further qualification of the employees. In addition, the regular evaluation of the AI use can use the experience of the employees and initiate further innovation processes—both with regard to the further improvement of processes and in relation to new products and business models—together with the employees as designers of change.

4 Outlook on the Introduction of AI in Companies in 2030

When will AI succeed in the world of work 2030? For all the optimism, we can also exaggerate—if we design work in the future in such a way that the employees would only be an appendage of AI systems. Here I count on a moral debate that considers both the enormous benefits of AI and the human aspects of its use. This involves design opportunities for companies, employees, social partners and politics. If this debate is conducted in a human-oriented way, humans will continue to perform steering, executing and monitoring activities—a company without humans will then not occur.

As promising elements in change management, the following prove to be useful:

- Experimental spaces and pilot projects, which then radiate as lighthouses, prove to be expedient to a) test and evaluate new technologies in the field and b) achieve comprehensive readiness and acceptance of AI solutions.
- Transparent and legally secure works agreements involving the data protection officer are useful, which clarify the use of AI and especially the handling of employee data in a binding way.
- An awareness-raising offensive for a lived active data protection culture with the employees or the works council and the data protection officer proves to be positive.
- Early qualification campaigns to impart the necessary skills of employees and managers are to be carried out.

A comprehensible design of the use of AI and a process-oriented participation of the employees and the works representation in the change process are a central element to address reservations early or to find and negotiate constructive solutions for goal conflicts—for example with regard to data use. A suitable instrument can also be ethical guidelines of the companies within the framework of the existing recommendations—such as the EU High-Level Expert Group on Artificial Intelligence (2018)—and those along the existing (legal) framework conditions for the use, introduction and handling of AI systems (Stowasser & Suchy et al,. 2020).

5 Summary and Practical Recommendations

How the forms of employment and work activities in the future world of work will be shaped in detail cannot yet be clearly determined. Without doubt, the successful implementation of new work environments requires the appropriate corporate culture for a positive use of AI technologies.

AI introduction means: An entrepreneur should always ask himself why an AI method should be used in his company. So what benefit does the company expect from the use of an AI method or what added value does the use of such a method create for the customers and the employees? Once this question is clearly answered, it is a matter of introducing the methods in a structured way. As with any technological innovation, this involves questions of corporate strategy, process planning, data handling and technology procurement. When introducing the AI methods into the company, it is also important to keep an eye on the topic of corporate culture and leadership. The sketched four phases are action-bearing and should be fully completed by the company. Education and training are an important key here. The successful transition to the world of work in 2030 will only succeed with a trustworthy and human-friendly corporate culture that involves the employees.

References

EU-High-Level Expert Group on Artificial Intelligence (2018). Ethik-Leitlinien für eine vertrauenswürdige KI. https://ec.europa.eu/newsroom/dae/document.cfm?doc_id=60425. Accessed: 16. June 2021.

Huchler, N., Adolph, L., André, E., Bauer, W., Bender, N., Müller, N., Neuburger, R., Peissner, M., Steil, J., Stowasser, S., & Suchy, O. (2020). Kriterien für die Mensch-Maschine-Interaktion bei KI. Ansätze für die menschengerechte Gestaltung in der Arbeitswelt. Whitepaper aus der Plattform Lernende Systeme, München. https://www.plattform-lernende-systeme.de/files/Downloads/Publikationen/AG2_Whitepaper2_220620.pdf. Accessed: 16. June 2021.

ifaa (Hrsg.) (2021). KI-Zusatzqualifizierung—Für eine produktive und menschengerechte Arbeitsgestaltung. ifaa—Institut für angewandte Arbeitswissenschaft e. V., Düsseldorf. (Publikation aus dem Forschungsprojekt „Digital Mentor — Modell und Erprobung eines präventiv agierenden KI-Helfers (kurz en[AI]ble)"). www.arbeitswissenschaft.net/enaible_broschuere. Accessed: 15. June 2021.

Stowasser, S., & Suchy, O., et al. (Hrsg.) (2020). Einführung von KI-Systemen in Unternehmen. Gestaltungsansätze für das Change-Management. Whitepaper aus der Plattform Lernende Systeme, München. https://www.plattform-lernende-systeme.de/files/Downloads/Publikationen/AG2_Whitepaper_Change_Management.pdf. Accessed: 16. June 2021.

Terstegen, S., Suchy, O., & Stowasser, S., & Heindl, A. (2021). Bausteine für das Change-Management bei der Einführung von KI-Systemen in Unternehmen. In GfA (Hrsg.), *Arbeit HumAIne Gestalten. Bericht zum 67. Kongress der Gesellschaft für Arbeitswissenschaft vom 03.—05. März 2021. ISBN 978–3–936804–29–4,* GfA-Press.

Responsible and Robust AI in Companies

How to Manage AI-Related Risks Against Bias and Discrimination

Claudia Pohlink and Sebastian Fischer

1 Introduction

AI is spreading rapidly and permeating almost all areas of our lives. To ensure long-term business success, hardly any company can afford to ignore the application of AI today. AI also finds more and more fields of application at Deutsche Telekom AG. On the one hand, AI is used for the management of the telecommunications network, both in planning, such as the expansion of the fibre optic network, and in operation, by predictive maintenance to prevent failures in advance. Also in the area of customer communication, AI applications such as chatbots are increasingly used. In addition, other internal group functions benefit from AI: For example, the areas of finance and human resources use AI models to improve planning processes.

All these application scenarios have in common that machines take over tasks that were previously reserved for humans. Thus, more and more business processes are partially or fully automated by the use of AI. With the increased use of AI and the associated risks, the demand for *robustness of AI,* which should prevent both unwanted distortions of the results and deliberate manipulations,

C. Pohlink (✉)
Berlin, Germany
URL: https://www.linkedin.com/in/Claudia-Pohlink/

S. Fischer
Berlin School of Economics and Law, Berlin, Germany
e-mail: sebastian.fischer@hwr-berlin.de
URL: https://www.linkedin.com/in/dr-Sebastian-fischer/

I. Knappertsbusch and K. Gondlach (eds.), *Work and AI 2030*,
https://doi.org/10.1007/978-3-658-40232-7_16

increases in order to secure the trust in the results of AI models. In the following sections, we explain why trust in AI is so important and what path we take at Deutsche Telekom to achieve robust AI for whatever case we choose to use it.

2 Status Quo and Case Studies

As described, AI takes over an increasing part of the decision-making in companies. With this development, there is also an increase in the associated risks. This is not only about deliberate attacks on AI systems. Numerous publicised examples show how *AI bias*—the unconscious distortion of decisions—can have serious consequences.

The company Amazon uses AI to support its human resources department to an increasing extent. The software is used by the company not only for administrative activities, but also for monitoring its own employees and suppliers. This resulted in a recently reported example that a package delivery driver of the company received a termination letter written by the algorithm, because tours were not completed fast enough. According to the former employee, the algorithm penalises things that he had no influence on, such as locked building complexes, poor road conditions or waiting times when loading (Soper, 2021). In this context, it is reported that Amazon uses AI systems to monitor, evaluate and even fire people—without human control (Gershgorn, 2018).

Another example for the existence of AI risks, especially of AI bias, are facial recognition systems of large technology companies. In a study, Timnit Gebru, former AI ethics researcher at Google, together with Joy Boulamwini analysed the facial recognition systems of Microsoft, IBM and Megvii. The study shows that facial recognition worked less accurately in identifying people of darker skin colour than people of lighter skin colour. There were also differences between the genders: facial recognition systems recognised men more reliably than women. One of the systems examined showed a reduced accuracy of around 35% for dark-skinned women compared to light-skinned men (Buolamwini & Gebru, 2018).

One thing becomes clear: supposedly intelligent systems make mistakes, just like humans. These mistakes can have serious consequences for a company. The more decisions that are taken over by AI, the more serious the possible consequences. This can be losses in revenue, legal consequences or even damage to brand and reputation. The fact is that the development and application of AI systems pose numerous challenges for companies. What these are and what possible solutions look like are explained in the following section.

3 Challenges and Solutions

The mentioned case studies show that supposedly artificially intelligent systems create the basis for decisions (partial automation) or make them independently (complete automation). They thus have a direct impact on key entrepreneurial competencies, such as the management of customer relationships, and support investment decisions based on evidence. Often it is clear why an AI came to a decision, or what the most important influencing factors for an AI decision were. Such clear AI systems are referred to as "black boxes". Often, methods of machine learning (ML) are used in these systems to provide the actual intelligence. Here, one distinguishes between ML methods that are interpretable by themselves, and those that still need to be "made explainable". Interpretable approaches are, for example, linear regression and classification algorithms, from which one can directly read which influencing variable had what influence on an AI decision. In contrast, there are neural networks. These approaches are the ML technique with the best performance for many use cases, but they offer only limited interpretability. It requires additional ways of doing things that take into account the desire for best performance and highest transparency of AI systems. It is therefore not surprising that *explainable AI (XAI)*, i.e. the comprehensible and transparent AI, which makes the steps to decisions explainable, is gaining more and more importance as a research area.

Explainability is only one aspect on the way to responsible use of AI. Overall, we are talking about requirements for the *robustness of AI,* which should prevent both unwanted distortions of the results and deliberate manipulations, in order to secure the trust in the results of AI models.

On the way to robust AI, companies must be aware that AI does not make the "right" decisions by itself. It must be "right" or better programmed and trained. And possible sources of error must be identified and eliminated from the outset, in order to minimise costly corrections in operation. By constantly learning with new data, AI systems change their behaviour over time and thus errors such as bias can manifest, which is why a one-time check is not enough.

In summary, we note that companies must manage to develop, implement and use AI solutions that are both morally responsible and legally flawless and economically goal-oriented. This can only succeed if companies look at the topic of "responsible and robust AI" holistically on different levels and do not lose sight of this neither in the initial design of AI applications nor in the operation of the systems.

4 Outlook on Mastering AI-Related Risks in 2030

The question now arises as to how companies can master AI-related risks in the future. What aspects will be decisive for using AI responsibly and sustainably? We take a look into the future and show which four dimensions will play an essential role from our point of view.

4.1 Strategic Anchoring

Companies need to address the topic of responsible AI at a strategic level. While a large part of the executives are aware of the importance and potential benefits of AI, many lack the strategic focus and the attention to crucial success factors such as trust, transparency and ethics. In a study published by Cognizant, only about half of the 1000 surveyed companies indicated that they had established guidelines that address the ethical aspects of AI application and implementation (Cognizant, 2018). Deutsche Telekom has taken up this topic early on and created guidelines that describe how the company wants to deal with AI and develop AI-based products and services in the future. These include the following nine principles: Responsible, careful, supportive, transparent, secure, reliable, trustworthy, cooperative and explanatory (Telekom, 2018).

But not only for their own development of AI systems do guidelines set the necessary framework. Guidelines become particularly important when AI systems are purchased from third-party providers. Often there is a lack of transparency and algorithms are like a "black box". Therefore, we recommend companies to clearly define explainable and robust AI by means of specific requirements and to oblige third-party providers to comply with corresponding guidelines. Only in this way can it be ensured that the systems used meet the standards of trustworthy AI expected by the company. Deutsche Telekom has published these and other thoughts and solutions in its white paper on *Robust AI Assessment* (Telekom, 2021).

4.2 Data and Training

In addition to strategic anchoring, companies should pay high attention to data quality. Data plays a much more important role for dynamically learning systems in AI than for traditional static software. In order to recognise patterns and trends and to make decisions based on them, an AI system depends on data sets from which it can learn the "right behaviour". If an AI system is trained with data that are not representative of the environment of the intended use case or if they even reflect inherent biases, this can have a direct influence on the decisions of the AI system. True to the motto *"garbage in—garbage out"*, poor quality of the training material leads to poor output (behaviour).

This means, conversely, that a lack of quality in the data leads to algorithms making decisions that can be discriminatory against certain groups or economically suboptimal in their results. But how can this risk be minimised?

The challenge is to train AI without discrimination. For this, it is important to carefully check the data sets used in training for possible bias and, if necessary, enrich or weight them differently. Only in this way is it possible to avoid underrepresentation in the data set having serious consequences. The example mentioned in the chap. "Practical guide to AI = Collaborative and Interdisciplinary" of the lower face recognition rate of dark-skinned women compared to light-skinned men was largely due to the fact that there was no balance between these two groups in the training data set. The system had thus received more light-skinned men as examples to learn from than dark-skinned women. Such imbalances can be easily corrected in advance—if developers know that such distortions can occur.

Companies must also be aware that diversity is an important success factor in the context of AI. With a clear commitment to more diverse development teams, companies can proactively counteract application problems or unintentional discrimination by AI systems (Fountaine et al., 2019).

4.3 Processes and Governance

A third building block of success will be to formulate rules within an AI governance model that determine a clear framework for dealing with AI systems. This includes essentially the definition of audit and validation processes. Similar to, for example, security tests or user tests in other software development projects, a dedicated project component in the AI context should be the *robustness test*.

It is important that tests are carried out regularly not only during development, but also during updates and retraining of the systems. The goal must be to continuously ensure that the results do not contain any bias and that the systems are adequately prepared for new threat scenarios. In addition to testing mechanisms, it is also necessary to define responsibilities and set up standardised change logs that record any modification of data. Only in this way can AI models be made secure and transparent.

In a research project "Robust AI" at the Telekom Innovation Laboratories, essential foundations were developed from 2018 to 2021 with the compliance department and the Ben-Gurion University (Telekom, 2021).

4.4 Technical Tests and Automation

The testing of AI models is complex and not really feasible manually with the large number of AI applications deployed. Therefore, it is important to automate the testing mechanisms as much as possible and test AI models regularly. Depending on how critical AI applications are, a continuous monitoring of the models may also be appropriate, in order to quickly identify possible deviations based on defined metrics.

In this context, a monitoring system could be set up, which measures the behaviour of AI systems during the life cycle based on predefined KPIs (Mehrabi et al., 2019). Taking as an example the targeted customer approach by playing out personalised advertising and offers. In this case, an AI would support the decision to address customers who have a high willingness to switch to the competition—but who would refrain from this decision due to targeted special offers. There is a risk here that customers are unintentionally discriminated against based on protected characteristics such as gender or age (they are not proactively offered ongoing offers). To measure such a gender or age bias, a monitoring system would check whether the AI used behaves differently for women than for men. The degree of difference in behaviour could be determined, for example, by a static measure such as accuracy or false positive rate. This is done by comparing the model result with a previously made selection of representative existing customers (test sample). If the statistical measures detect differences in behaviour that exceed a predefined threshold, the monitoring system could signal to the person responsible that a bias has occurred and that it should now be adjusted. Experiments by the Telekom Innovation Laboratories in cooperation with Ben-Gurion University have shown in an internal research and development project "Testing ML / Robust AI" that bias can be identified and adjusted in this way, without having to accept limitations in the accuracy (performance) of the model.

5 Summary and Practical Recommendations

New technologies bring opportunities and risks—AI is no exception. The important thing here is to weigh the risks and benefits, in order to achieve economic goals and at the same time maintain the corporate responsibility towards the different stakeholders affected. The first step here is to deal intensively with the associated risks, in addition to exploiting the advantages of AI technology to improve business performance. Companies should also consider the ethical questions that the use of technology raises and develop skills to use AI both effectively and ethically in a responsible manner.

For this, it will be crucial in the long term to anchor the topic of responsible AI in the top management and agree on binding AI guidelines as a basis for further development. In the foreseeable future, the use of AI will probably be more influenced by regulations and requirements, so that dealing with them will become a mandatory task for every company.

The action fields shown here do not represent a revolution in the development of software, but bring in another dimension of non-functional and quality-assuring requirements for software, in addition to fields such as security or user centricity. These must be taken seriously at all levels of strategy, development, operation and assessing results.

References

Buolamwini, J., & Gebru, T. (2018). Gender shades: Intersectional accuracy disparities in commercial gender classification. PMLR. http://proceedings.mlr.press/v81/buolamwini18a/buolamwini18a.pdf. Accessed: 12. July 2021.

Cognizant. (2018). *Making AI responsible – – – And effective*. https://www.cognizant.com/whitepapers/making-ai-responsible-and-effective-codex3974.pdf. Accessed: 12. July 2021.

Fountaine, T., McCarthy, B., & Saleh, T. (2019). *Building the AI-powered organization*. Harvard Business Review. https://www.kungfu.ai/wp-content/uploads/2019/07/HBR-Building-the-AI-Powered-Org.pdf. Accessed: 12. July 2021.

Gershgorn, D. (2018). *Companies are on the hook if their hiring algorithms are biased*. Quartz. https://qz.com/1427621/companies-are-on-the-hook-if-their-hiring-algorithms-are-biased/. Accessed: 12. July 2021.

Mehrabi, N., Morstatter, F., Saxena, N., Lerman, K., & Galstyan, A. (2019). *A survey on bias and fairness in machine learning*. arXiv preprint. https://arxiv.org/pdf/1908.09635.pdf. Accessed: 12. July 2021.

Soper, S. (2021). *Fired by bot at Amazon: 'It's you against the machine'*. Bloomberg. https://www.bloomberg.com/news/features/2021-06-28/fired-by-bot-amazon-turns-to-machine-managers-and-workers-are-losing-out. Accessed: 12. July 2021.

Telekom. (2018). *Leitlinien für Künstliche Intelligenz.* https://www.telekom.com/de/konzern/digitale-verantwortung/details/ki-leitlinien-der-telekom-523904. Accessed: 12. July 2021.

Telekom. (2021). *Whitepaper zum Robust AI Assessment.* https://www.telekom.com/de/konzern/management-zur-sache/details/implementierung-von-digitaler-ethik-sichert-nachhaltigen-unternehmenserfolg-630942. Accessed: 12. July 2021.

AI as a Driver of Hybrid Forms of Employment

The Future of Intelligent Job Matching: Potentials and Challenges of AI Tools in Recruiting and Talent Selection

Daniel Barke

1 Introduction

Non-standard employment is gaining in importance. At the same time, science, society and economy are in a phase of transition. The change in values and digitalisation open up new possibilities and are drivers for more flexibility and *new work*.

To reflect the needs of employees and progressively positioned employers in new constructs, atypical models of working time flexibility are needed. Hybrid forms of employment (a mix of freelancer and permanent employment), job and employee sharing or crowdworking could replace the traditional concept of permanent full-time employment in the future. Because by scaling the use of flexible workers, companies increase their efficiency and competitiveness. This model is now made possible for the first time by the use of AI as a tool for intelligent matching.

However, working time flexibility means much more than just a dynamic allocation of working time. The vision looks like this: Each contractor works for several employers at the same time. While this is going on, companies share talents and benefit from their diverse skills and abilities. In human resource management,

D. Barke (✉)
WorkGenius GmbH, Hamburg, Germany
e-mail: db@workgenius.com
URL: https://www.linkedin.com/in/dbarke

I. Knappertsbusch and K. Gondlach (eds.), *Work and AI 2030*,
https://doi.org/10.1007/978-3-658-40232-7_17

the focus is on skill-based matching. Instead of the categorical staffing according to job profiles, the skills of a talent will be in the foreground in the future. For this, a fundamental rethink for the future of learning and education takes place. The rigid boundaries between freelancers and permanent employees are thus dissolved.

Working with freelancers is still a challenge for many companies. Because selection, management and processing of flexible workers lead to a significant additional effort—on the part of the companies and the talent. By using AI, this additional effort can be eliminated in the future. In which areas does the implementation make sense? And how can it promote equal opportunities?

2 Status Quo/Case Studies

The Digital Future

Freelancing is a trend that has been growing for years. In 2019, 1.4 million people worked as freelancers in Germany (cf. Volini et al., 2020). This means an increase of 36% in the number of self-employed in the liberal professions in the last ten years. In addition, we are in the midst of digital transformation. Not only the Covid-19 pandemic has accelerated the digital transformation and the freelance revolution. Also, consumer behaviour is moving in many areas of social life towards the on-demand economy. From the constantly available offer of music, art and movies to the paradigm shift in the world of work: The present is characterised by the desire for flexibility, availability and networking.

The change in family structures, the pursuit of self-fulfillment and the change in values of Generation Y and Generation Z show that the need for individuality and meaningfulness has been growing for years. Hybrid, AI-driven earning constellations enable to meet these wishes for purpose-driven, independent work. A look at the USA also confirms: Freelance work has been a persistent, growing megatrend for decades. According to expert forecasts, there will be more freelancers than employees there by 2027 (cf. Pruchniewski, 2020). Especially for Generation Z and Millennials, freelancing is becoming increasingly important here as well: 46% of 18- to 39-year-old Americans can imagine working freelance.

For the company side, the digital revolution also makes it easier than ever to build a global network of experts. Working with a freelancer instead of filling a position permanently saves resources and brings competitive advantages. Flexible workforces offer opportunities for more entrepreneurial freedom and the reduction of financial risk—regardless of the development of the economic situation.

Overhead, Employee Sharing and AI Use

Recruiting, processing and managing external workers creates additional time. If a flexible worker is needed for a project for 40 hours of work, an overhead of an average of 11 hours is created. This is working time that does not flow into the product or result, but into the choice of the freelancer, the management and the processing.

This is where intelligent, digital tools such as online marketplaces for acquisition make efficient collaboration with freelancers possible. One approach to bring people together in more agile working models and new work topics are job sharing platforms. Employees are networked for job sharing tandems using an intelligent matching algorithm, based on professional and human competencies (cf. Tandemploy, 2021). That such a concept also has great potential at the top management level is proven by successful co-leadership models, in which two women have been sharing a leadership position for ten years (cf. Sturm, 2020). AI also provides a blueprint in job matching. On online marketplaces, freelancers and companies are networked. The use of AI can help here to match the client with the optimally fitting flexible talent—based solely on skills, reputation system and professional quality (cf. WorkGenius, 2021).

Equality and (Anti-)Discrimination

Discrimination is still commonplace in the application process (cf. Federal Anti-Discrimination Agency, 2021). That is why the use of AI-driven technology makes sense especially when it comes to talent selection. In the subjective selection, it can happen that recruiters are unconsciously guided by the first impression or personal preferences. A look at demographic values, age, gender or the application photo can be an exclusion criterion to the disadvantage of the applicant. This not only creates unequal opportunities with equal qualifications and personal disadvantage of qualified talents; it also leads to economic damage. Anonymising application procedures to one hundred percent is made possible by the use of technology. With AI as an incorruptible, objective help, the fair staffing of diverse teams is possible.

3 Challenges and Solutions

Permanent Employment Equals Risk Minimisation?

While the project-based use of freelancers in the creative industry has long been common practice, traditional sectors still have reservations. This shows how much traditional project management shapes the economy. Breaking away from the associated traditional mindset is likely to be the biggest challenge, because companies sometimes perceive working with freelancers as a risk. The flexible workforce could jump off at short notice or be poached by another client. Hiring a permanent employee, on the other hand, suggests security to the employer. Employees are available at any time and bound to the company. It is clearly organised how much weekly working time the talent spends in the company—regardless of the actual workload and results.

Labour Law and New Terminology
Health insurance, social security and tax benefits: State security systems make permanent employment much more attractive than self-employment. But there is still a lot to do: Currently, it is not legally possible to work for more than two employers at the same time. Internal confidentiality agreements put additional obstacles in the way of the model "employee sharing". To pave the way for hybrid employment relationships, new labour law regulations and security systems for freelancers are also needed. Abolishing the systematic distinction between self-employment and permanent employment would be a step towards the work of the future.

Technical Challenges
Talent will be much more fluid in the future. With the right technical infrastructure, workers could log in to an interface in the future and assemble their orders for the month by drag-and-drop from a matrix. The data—skills, order history and work results—would be visible to every employer on the other side.

Can the use of AI and technology increase transparency and equal opportunities? Yes—social selling and subjective self-marketing would thus become irrelevant; the focus would rather be on the actual competencies. But if technology replaces personal recruiting, there are various requirements: The intelligent tool must have the ability to recognise the real potential of a job candidate. And beyond their order history, based on soft skills, rational and psychological aspects. It must, for example, recognise whether a career starter is capable of growing with his tasks/with the company.

4 Outlook on AI as a Driver of Hybrid Forms of Employment in 2030

Future-Proof Through AI Use
The use of AI-supported algorithms and technologies makes employment models of the future possible—by minimising friction. The threshold for traditional companies to use flexible workers disappears.

In the future, AI tools will solve the planning and transparency in the collaboration—on the client and talent side. Digital tools provide relief and simplify processes. Especially when it comes to handling projects with temporary workers or expanding the team with a specific skill set. With the help of technology, the use of freelancers can be scaled. Regardless of the number of employed freelancers, the effort, process and mechanisms remain the same. By automating certain processes, the company is relieved of administrative work and saves time and costs—while maintaining the same quality standard. The technology replaces manual processes and guarantees to find an available, qualified freelancer at any time and immediately. The tool manages the use of permanent employees as well as flexible workers. Even when working internationally.

Meaningfulness and Experience Values

Freelancers have their finger on the pulse. If a worker is active in several companies, fields of activity and professions, he or she gains new ideas, experience values, networks and cross-industry knowledge. The talent (worker) grows on diverse tasks and learns new tools and agile ways of working. The client also benefits from this. The future is characterised by more openness and mutual learning. The focus shifts from competitive thinking to cooperation—also between companies.

Because with the motive of fulfilment and individuality, the interests of people become more diverse. Employees learn different professions or study several times throughout their lives. Having multiple ways to earn a living is the norm. Job sharing thus not only increases employer attractiveness, but also satisfaction and productivity of the contractor as an opportunity for development.

Also, the desire of the young generation for networking and collaboration plays a significant role. Hybrid forms of employment are the answer to the difficulties of retaining the young generation. The value change of Generation Y and Generation Z brings new ideals: Where once career and monetary values counted, now freedom, autonomy, work-life balance and fun at work matter. With the help of technology, it is possible to earn a living two days a week in a permanent position in a company and invest the rest of the time flexibly in a meaningful heart project and individual fulfilment.

Employee "Sharing": Companies Without Permanent Employees?

In the future, it will be more likely that employees will work for more than one company. Driven by the demand of the talents, the possibility also arises on the employer side to share talents with other companies—both permanent employees and freelancers. The majority of the workforce is thus flexibly active. The number of employees will be reduced in this way and concentrated on a core team.

In an ideal, coming world of work, this means a system that is structured like a cooperative—like a hub, in which several companies can access a pool of talents. The model also makes it possible to align the demand for labour according to seasonality and economic cycles. Thus, high-demand industries can situationally "borrow" workers from low-demand industries. This requires new designations and headings: On the way from traditional concepts such as the 40-hour week and attendance obligation to trust culture and cooperation.

The idea that a permanent employee only works conscientiously on site is not up to date. The future will be determined by trust cultures. Due to the influence of the Covid-19 pandemic, traditionally led industries such as the industry were forced for the first time to rely on telework and remotely led teams (cf. Streim & Eylers, 2021). This digitisation boost in factories currently proves that the presence of a fixed team is not decisive for the work success. AI also comes into play here: Automated workflows, intelligent robots, predictive maintenance and networked production facilities enable working from home.

Skills Shortage or Matching Shortage

Fifty percent of German companies see a possible skills shortage as the biggest threat to their business development. And the demographic change fuels this fear even more (cf. Anonymous, 2021). Finding and retaining good employees with matching skills is essential. But German companies have problems filling positions. The use of artificial intelligence can counteract the shortage of experts—or show that there is no shortage of experts, but a lack of intelligent matching possibilities. In the future, it will be easier for the client to identify which skill set is needed to fill a position. AI-driven tools do the rest and find the suitable worker. Subjective matching is the effective answer to the challenges in global talent acquisition.

AI Against Language Barriers

Globalisation and the international talent market are on everyone's lips. But language is still a big barrier in the globalisation process and makes it difficult to recruit suitable talents across national borders. Obstacles in communication delay the nuanced selection process, even with the help of a lingua franca. As long as there are global differences in education systems and university education, the uniform evaluation of flexible workers remains complicated.

The use of neural translation machines almost completely eliminates the hurdle of the language barrier. AI-driven translation tools are making great leaps in development and are becoming more reliable. Thus, the AI identifies the worker who best matches an assignment based on his or her skills—regardless of location and language. The tool not only favours global work and barrier-free communication. It marks a big step on the way to an international talent market. Thus, it offers a possibility to eliminate the skills shortage globally.

5 Summary and Practical Recommendations

New forms of work are on the rise and there is no way around the change to diversity and flexibility. AI is one of the major drivers for the future of work. It will not only pave the way for hybrid employment models, but also radically transform social and economic life. The working world of tomorrow will be shaped by the new technical possibilities: On the one hand, with dynamic models, on the other hand, by the need for purpose-driven work. For the contractor, meaningfulness and individuality come to the fore.

The question of strengths and personal potential gains more weight than job titles. What skillset is really needed to fill the position of a marketing manager? What kind of person is needed to complete the assignment? Which soft and hard skills are essential? Here, AI becomes a significant engine. As a new tool, it counteracts discrimination in recruiting and creates equal opportunities through objective matching.

The digital transformation and the possibilities of using AI enable for the first time the realisation of hybrid, non-precarious employment models. It is now time to adapt obsolete ways of thinking and have the courage to implement new structures. A rethink will also take place with regard to the classification of permanent and freelance workers—because a job rotation between industries increases competitiveness.

The goal is flexibility with constant job security. If job sharing does not start at the companies, workers will take the first step and organise themselves. Whether at the management level or in freelancer management, job and employee sharing will be the focus in the future. With the focus on joint growth, trust culture, happiness and meaningfulness.

References

Anonymous. (2021). Fachkräfte für Deutschland. Bundesministerium für Wirtschaft und Industrie. https://www.bmwi.de/Redaktion/DE/Dossier/fachkraeftesicherung.html. Accessed: 21. June 2021.

Antidiskriminierungsstelle des Bundes. (2021). Unsere Projekte zu (Anti-)Diskriminierungsthemen. https://www.antidiskriminierungsstelle.de/DE/was-wir-machen/projekte/projekte-node. Accessed: 21. June 2021.
Pruchniewski, E. (2020). Der große Freelancer Vergleich [sic!]: USA vs. Deutschland. https://www.freelancermap.de/blog/freelancer-usa-deutschland/. Accessed: 21. June 2021.
Streim, A., & Eylers, K. (2021). Corona führt zu Digitalisierungsschub in der deutschen Industrie. https://www.bitkom.org/Presse/Presseinformation/Corona-fuehrt-zu-Digitalisierungsschub-in-der-deutschen-Industrie. Accessed: 21. June 2021.
Sturm, A. (2020). Zwei Frauen machen bei Unilever vor, wie Job-Sharing funktioniert. https://www.horizont.net/marketing/nachrichten/chan-zwei-frauen-machen-bei-unilever-vor-wie-job-sharing-funktioniert-185122. Accessed: 21. June 2021.
Tandemploy. (2021). https://www.tandemploy.com/de/. Accessed: 21. June 2021.
Volini, E., Schwartz, J. et al. (2020). Die soziale Organisation bei der Arbeit: Paradox in die Zukunft. 2020 Deloitte Human Capital Trends—Kurzfassung. https://www2.deloitte.com/content/dam/Deloitte/at/Documents/human-capital/at-hc-trends-2020-deutsch-kurzfassung.pdf. Accessed: 21. June 2021.
WorkGenius. (2021). https://www.workgenius.com/de/. Accessed: 21. June 2021.

Digital Finance—The Future of Financial Planning in Companies

Heinrich Kögel, Martin Spindler and Helmut Wasserbacher

1 Introduction

Digital finance refers to the digitisation of the finance function in companies. This creates new opportunities, but also technological, organisational and strategic challenges. Although digital finance is high on the agenda of many CFOs, there are still only few successful use cases, which are mostly limited to highly transaction-oriented areas such as invoice processing. In this article, we describe how digitisation and especially artificial intelligence (AI) will change financial planning in companies by 2030.

Creating forecasts and developing financial plans are two central tasks of the finance function in companies. However, the approach encountered in practice often has many shortcomings. It is usually time-consuming, resource-intensive, and prone to subjectively negotiable predictions and assumptions.

This text only reflects the personal opinion of the author, not that of Novartis or the management of Novartis.

H. Kögel (✉)
Economic AI GmbH, Regensburg, Germany
e-mail: koegel@economicai.com

M. Spindler
Universität Hamburg und Economic AI, Hamburg, Germany
e-mail: martin.spindler@uni-hamburg.de

H. Wasserbacher
Novartis International AG, Basel, Switzerland
e-mail: helmut.wasserbacher@novartis.com

I. Knappertsbusch and K. Gondlach (eds.), *Work and AI 2030*,
https://doi.org/10.1007/978-3-658-40232-7_18

Data-driven methods, based on AI and machine learning, have the potential to meet these challenges and fundamentally transform the finance function towards digital finance.

In the following, we describe the current challenges for the finance function in the area of planning and forecasting, which arise especially in the context of digitisation, as well as on the way to AI-driven financial planning. We also outline the changes and challenges that could arise from AI in this area by the 2030 horizon (in fact, some of them have already materialised). Our solutions are intended to help companies master the way into the future.

2 Status Quo

Forecasting, planning and analysis in the finance area are of central importance for companies. Therefore, most larger companies have special staff for these tasks within the finance department, sometimes grouped under "Financial Planning & Analysis" (FP&A). The overarching goal of FP&A is to support management decision-making. The development of financial plans also plays an important role in translating the strategic priorities of a company into specific, operational measures.

The analysis of the business environment and the business dynamics is an integral part of FP&A. The insights gained flow into the development of the forecasts and financial plans. Based on data from the different business areas, plans and forecasts are created in what are often "political negotiations" that are time- and resource-intensive. This process can be influenced by the strategic and personal interests of the parties involved, for example when a business area is interested in low forecasts in order not to receive too ambitious goals for the upcoming year. In addition, the process can also be influenced by unconscious biases of human perception.

Another challenge arises from the general advancement of digitisation of markets and business processes, which are changing faster and more dynamically as a result. To remain competitive, companies need to be able to respond flexibly and quickly to new conditions. For this, it is necessary to have current forecasts available in a timely manner and to be able to adjust plans flexibly. This is difficult to achieve with the traditional approach.

Digitisation also increases the amount and diversity of available data. In addition to more traditional data, companies today have access to entirely new data sources, such as audio files of customer conversations, movement patterns on social networks or digitally collected information from marketing and sales. The amount and speed of this "big data" make it increasingly impossible for humans to include insights from these data in the traditional approach.

3 Challenges and Solution

On the way to AI-driven financial planning and forecasting within digital finance, there are some challenges for companies that need to be addressed with appropriate measures. The main challenges can be divided into three areas: data & infrastructure, algorithms, and organisation & culture.

Data & infrastructure. Digitisation provides access to ever-increasing amounts of data. Big data will form the backbone of AI-supported financial planning in the future. For efficient use of the data, they must be merged, processed and standardised. A worthwhile investment is the establishment of company-wide data standards. A special challenge that has gained a lot of attention due to Corona is the "outlier detection", i.e. the identification and correction of outliers due to particular events affecting the company's environment. In addition to internal data, external data sources must also be added and checked for their reliability.

Appropriate infrastructure must be set up to store and process these large amounts of data. Due to the increasing importance of company data and the fact that, in addition to personal data, financial information is also highly sensitive, a solution is required that meets high security standards. Depending on the type and size of the company, either the establishment or expansion of company-owned infrastructure or a cloud solution from a commercial provider with appropriate precautions may be suitable.

Algorithms. As part of the company transformation, data preparation represents a large investment, especially at the beginning. Data have no value in themselves. The decisive factor is what conclusions can be drawn from them in order to arrive at better decisions. Accordingly, the development and implementation of algorithms also play a central role. In general, two types of tasks can be distinguished that can be solved with the help of data-driven methods and that are of particular importance for financial planning and forecasting (Wasserbacher & Spindler, 2021).

The first area consists of so-called prediction tasks, which also include the creation of financial forecasts. For pure prediction problems, it is sufficient to learn patterns from historical data and then apply them to the future. This is the strength of machine learning methods as a subfield of AI. Most of these algorithms were developed specifically for prediction based on pattern recognition (Taddy, 2019). A challenge with forecasts is often that the time series to be predicted are not very long and the number of possible prediction factors exceeds the number of data points of the time series. This is called a "high-dimensional problem", where classical methods from statistics are not applicable. Machine learning methods are very well suited for such situations and can often provide accurate predictions.

The second area involves tasks of causal inference to identify causal relationships. This is crucial, for example, within financial planning for efficient resource allocation. Determining causality reliably is usually much more difficult than making pure predictions. Instead of passive pattern recognition, causal inference is about understanding the consequences of an active intervention in a system, e.g. a market, in order to make investment recommendations and decisions. Although it is often known that "correlation is not causation", traditional machine learning methods are often mistakenly used to determine causal effects in current practice. Instead, methods from, for example, econometrics or causal machine learning should be used for this purpose. Powerful software for causal machine learning is now also available (Bach et al., 2021).

Organisation & culture. In addition, organisational and cultural adjustments are also required within the company.

In a first step, the organisation must commit to internal transparency of the data. A "data culture" must be established, in which easily accessible central data storage combined with intuitively usable dashboards and apps play a central role. In addition, skills must be built up in the company to develop, train and maintain algorithms. A basic understanding of data-driven methods must be developed among FP&A employees. This understanding is important so that they can effectively operate in an AI-driven environment.

Another prerequisite for success is the acceptance by the users of the largely automatically generated predictions and recommendations. Reliability of AI -based predictions and efficiency of the AI-driven resource plans will contribute to this. But two other aspects are almost more important. First, the close involvement of the end users as subject matter experts during the creation of the AI models. Second, a certain degree of interpretability and explainability of the AI results. Ideally, AI models do not run as a complete "black box", but instead provide a minimum level of transparency.

4 Outlook on AI-based Financial Planning by 2030

The way FP&A works will change fundamentally in the future. Within financial planning and forecasting, large parts of the processes will be data-driven and automated using AI algorithms. Some companies already use AI methods in this area today. However, even large companies are still at the beginning of the journey to AI-supported financial planning and forecasting. Nevertheless, the degree of AI-usage will have increased significantly by 2030. AI will have penetrated the financial area in most companies and thus significantly expanded the possibilities and fields of application of FP&A as part of corporate management.

The foundation for the spread of AI will be a centralised data base, in which data from various internal and external sources flow together and are updated in real time, automatically combining different data types and linking information from different origins. This enables a holistic overview of the company. The information is made available to decision-makers in real time via apps and dashboards. The tools themselves allow the flexible execution of detailed analyses at a more detailed levels by planners and analysts.

Reactive control thanks to AI-supported monitoriing of the company will represent only the first step. To move to proactive control, forecasts and scenarios will be generated in real time by AI. The algorithms will draw on the full spectrum of available data and thus enable optimal use for the most accurate predictions, with the necessary degree of detail. If negative developments are predicted, an alarm is automatically triggered and it is shown how, for example, a predicted price increase for a production factor—e.g. due to an extreme weather event—affects various operational metrics.

Going beyond pure forecasting and prediction, AI algorithms will also play a central role in the allocation of corporate resources within financial planning. Algorithms will thus no longer be used only for the purpose of passive-reactive prediction of the "existing world". They will also be used for active interventions to achieve the best possible outcomes. For example, a fixed marketing budget is automatically distributed across different countries and products of a company, so that certain market share or profit targets are maximised. The optimal allocation takes place not only once a year during the annual budgeting cycle, but also dynamically during the fiscal year based on the latest insights, which are automatically obtained from data in real time. By changing the marketing expenditures, potentially both the behaviour of the buyers and the competitors changes. Thus, an active intervention in the market takes place, which creates a "new world" (Taddy, 2019).

However, to be able to successfully use AI for active market interventions, an understanding of causal relationships is necessary (Pearl, 2019; Pearl & Mackenzie, 2018). Causal relationships go far beyond mere correlations.[1] Planners and analysts will therefore increasingly deal with applying suitable AI methods to causal questions relevant for the company, working together with technical specialists.

Due to the automation of a multitude of activities and the spread of AI to causal questions, the role of employees in FP&A will change. Time-consuming forecasting activities that require the collection and consolidation of figures from different business areas will be eliminated. Financial analysts and planners will use the time gained to work more in depth on the company strategy and to include additional business areas in the AI analysis network. FP&A will become more cross-functional and cross-departmental. For example, if a deviation from the established plan is predicted due to changed market conditions, the planner can alert her colleagues in marketing to this development and support them in finding appropriate countermeasures; at the same time, production gets also involved, as e.g. a new packaging size is part of the AI-predicted solution. In addition to the increased communication with other business areas, FP&A will also interact more with data scientists, data engineers and machine learning ops experts as well as other roles from the data and AI environment. In the world of digital finance, FP&A takes on a decisive bridging function between technically specialised colleagues and the various business functions and areas for which AI-based methods can be applied. Through their cross-functional competencies, FP&A employees will enable the development and successful use of algorithms for business optimisation.

[1] The following example illustrates the difference between correlation and causality. An ice cream seller only runs ads in the local newspaper during summer. At the same time, he notices that his ice cream sales are higher during summer than during winter. In the data, there is thus a positive relationship, or a positive *correlation*, between the advertising measures and the ice cream sales. From this, one could naively conclude that the advertising measures lead to an increase in ice cream sales. On closer inspection, however, it could also be that the ice cream sales increase due to the warmer weather during summer and not due to the advertising measures. In this case, the advertising measures would have no *causal* effect on the ice cream sales and it would be advisable not to run any ads. To distinguish correlation from causality, special methods for estimating causal effects must be used. This applies especially to methods of machine learning.

5 Summary and Practical Recommendations

Creating forecasts and financial plans plays a central role in modern businesses. It represents an important task of FP&A within the finance function. The traditional approach to planning and forecasting is often a lengthy and costly process. It is difficult to create timely forecasts and maintain flexibility for adjusting plans due to changing market conditions. Another challenge arises from the increase in data. With traditional approaches, the majority of data does not flow into the decision process.

To successfully shape the transformation to digital finance and AI-supported financial planning and forecasting, companies need to consider three core areas. In the area of data & infrastructure, it is essential to create a central, harmonised data base. This is fundamental, as data is the backbone of all AI-supported applications. For this, appropriate infrastructure must be set up. In the second area, algorithms, it is important to use the right methods. For prediction tasks, classical machine learning methods are suitable. For tasks in the area of causal inference, such as resource allocation, however, methods specifically developed for this purpose must be used, for example from the fields of econometrics and causal machine learning. Randomised experiments also offer an innovative way to estimate causal relationships. In the third area, organisation & culture, companies need to make tranformative organisational and cultural adjustments.

Financial planning and forecasting will change fundamentally in the wake of digital finance by 2030. With the help of AI, large parts of the processes will be automated and data-driven. The basis of these changes will be a central data base that provides both internal company data and external data in a broad, linked and timely manner. For predictive purposes, AI generates fully automated forecasts in real time to guide the company. Algorithms will also play a central role in the allocation of resources. AI suggests investments for R&D or distributes marketing budgets. Algorithms also allow for rapid adjustments when conditions change. And by automating many tasks, employees in the FP&A area will increasingly focus on working on strategic issues.

The transformation to an AI-driven finance function as digital finance requires significant changes in companies. However, it also offers enormous potential to relieve people, save costs and act flexibly in an accelerated world.

References

Bach, P., Chernozhukov, V., Kurz, M., & Spindler, M. (2021). DoubleML—An Object-Oriented Implementation of Double Machine Learning in R. Papers 2103.09603, arXiv.org, revised Jun 2021.

Pearl, J. (2019). The seven tools of causal inference, with reflections on machine learning. Commun. *ACM, 62,* 54–60.

Pearl, J., & Mackenzie, D. (2018). *The book of why: The new science of cause and effect.* Penguin Verlag.

Taddy, M. (2019). *Business data science: Combining machine learning and economics to optimize, automate, and accelerate business decisions*. McGraw-Hill Education.

Wasserbacher, H., & Spindler, M. (2021). *Machine learning for financial forecasting, planning and analysis: Recent developments and pitfalls*. Working paper.

AI in Banks

The Bank of the Future

Daniel A. Schmidt

1 Introduction

It is undisputed that technology is changing banking. In traditional banks, basic transactions are shifting from physical to digital channels, leading to major changes, such as banks reducing their branch network. In many developed markets, financial technology companies (fintechs) are taking market share from traditional banks, for example in payments or consumer lending. We want to take a look at the working world of the banks of tomorrow in this chapter and try to outline the potential changes that will come to the banks by 2030.

2 Status Quo

Digital and intelligent algorithms already help today to reliably detect fraud attempts, to raise the alarm in case of money laundering, to process financial transactions faster, to advise customers more comprehensively and to develop new business models. In particular, the employees play an important role in this constellation, because AI can support them in their daily work. Thus, complex business processes are accelerated by AI and employees are relieved of time- and resource-intensive tasks (Accenture, 2018; Deloitte, 2021; McKinsey Global Institute, 2018; PricewaterhouseCoopers, 2020). At the same time, many banks also have difficulties moving from experimenting with selected use cases

D. A. Schmidt (✉)
Frankfurt am Main, Germany

I. Knappertsbusch and K. Gondlach (eds.), *Work and AI 2030*,
https://doi.org/10.1007/978-3-658-40232-7_19

to scaling AI technologies across the enterprise. The reasons for this are often diverse, such as a unclear AI strategy, an outdated technology core or fragmented data sets. However, the banks agree that AI algorithms will play an important role as the primary customer interaction method in the future (according to the survey from (Accenture, 2018) about 76% of the study participants believe this). This situation poses ever greater challenges for the traditional banks, which we want to analyse below.

3 Challenges and Solutions

The sketched challenges of the banks due to a historically grown business model are reinforced by external influences. Thus, more and more technology companies are encroaching on the business fields of the banks. For example, the payment service provider PayPal from the USA is worth more than any other US bank except JPMorgan Chase (about 390 billion US$) with a market value of about 275 billion US$ at the end of 2020. In Asia, the customer migration in payments is even more dramatic, the e-commerce provider Ant Financial (Alibaba Group) has risen to become the market leader in mobile payments in a very short time. Banks already play only a minor role in this segment today. Other technology giants such as Google, Amazon, Facebook and Apple are also pushing into this market. In contrast to traditional banks, however, it is easy for a company like PayPal to implement and use AI technologies on a broad front, as all customer data and transactions can be used together to develop intelligent algorithms.

It is important for banks to acknowledge these new challenges and at the same time draw the right conclusions from them. Because the tech companies are essentially so successful because they operate very close to the customer. All operations and services that the companies offer are provided by them via the smartphone of the respective customer. For banks, this means that they have to quickly and comprehensively develop in the area of digital distribution channels and use artificial intelligence wherever possible. A new breed of banks, the so-called neo-banks (smartphone banks), are already taking this path today. The branch network of the future will consist only of a few flagship stores (representative bank branches). The substantial retail banking business will be essentially handled online and shaped by AI algorithms, which support bank employees in digital customer advice.

4 Outlook on AI in Banks in 2030

We now want to look at the front office, middle office and back office areas of a bank and discuss the potential for the use of AI technologies.

The term front office refers to all areas that work directly with products or services for the customer. These include, for example, the areas of capital markets, investment banking, asset management and wealth management, but also the lending business as well as all sales activities of banks.

The development of artificial intelligence in the front office of a bank depends on the area. For example, by 2030 there will probably be hardly any pure commission business (buyside/sellside trading) in the capital markets area, which is specialised only in executing stock market transactions. Here, the AI technologies and automation will come into full use. The banks that were designed as complete online banks already have no traditional commission business today, all orders are executed automatically without the intervention of traders. As more and more banks rely on online distribution or self-service distribution (e.g. through robo-advisers), AI will also play a big role in the sales area for private customers. Banks that offer research in addition to pure commission business, such as stock analysts, and thus conduct an expanded broker business, will probably still exist in 2030 in a similar way as today. The new breed of brokers, which specialise entirely on customer contact via smartphone, the so-called neo-brokers, also execute the orders completely automated.

For the other areas in the capital markets business, there will probably be less change due to AI technologies, as hedging derivative portfolios to hedge bank positions, as well as issuing capital market products, still require (product-dependent) a lot of human know-how. In addition, the regulation of banks in this area will probably not allow such sensitive and risky areas of a bank to be controlled by an AI (Financial Stability Board, 2017).

Investment banking will probably change only very marginally by the use of AI by 2030. Deals in investment banking are standardised products to some extent, but always with a very individual customer-oriented design. Therefore, in investment banking, one will use one or the other new tool, but the (data) analysts, who are already part of the investment banking teams today, will integrate them into their work without having to fear an impact on their working world.

When we look at private wealth management, there are essentially two distinctions: on the one hand, the very wealthy clients and on the other hand, the normal private investors. The wealthy private clients will rely on the personal contact with their wealth manager just like today in 2030. Managing large wealth is a matter of trust and will therefore probably be as little affected by the topic of AI as investment banking. Some new tools will find their way into private wealth management, but they will probably not have a disruptive impact on this business segment. The situation will probably look different in the private customer business. On the one hand, robo-advisors, i.e. automated wealth managers (algorithms), some of which also rely on AI, and on the other hand, smartphone banks, which rely entirely on the smartphone as a "customer advisor" or on self-service, are pushing into the market for private investors. At the current time, robo-advisors do not have a significant market share in the private customer business. Depending on the source (see, for example, BearingPoint & niiio finance group, 2018), the assets managed by all German robo-advisors together are between three and seven billion euros, which is a very small managed asset compared to other German asset managers such as DWS (793 billion euros), Union Investment (368.2 billion euros) or Deka-Gruppe (339.2 billion euros). However, the market share of robo-advisors will probably grow and thus the topic of AI will also become more important in this area. The topic of smartphone bank is also an important growth market in the private customer segment for the areas of asset management and credit business, with many possible applications of AI. Here, the focus is primarily on the data that smartphone banks have about their customers (trades, savings plans, loans, payments, etc.) and the possibility of training an AI system with them to make tailor-made offers to the customers.

To conclude the front office topics, we want to take a look at the fund business. Certainly, actively managed funds will also experiment with AI strategies and possibly be able to achieve excess returns in individual time periods. However, the market efficiency hypothesis (see Fama, 1970) can be applied to "AI fund strategies" on the stock market regardless of the strategy. This means that the topic of AI will also not have a disruptive impact on the equity fund business, as according to the market efficiency hypothesis, no strategy can outperform all other strategies in the long run. The arsenal of strategies will probably be supplemented by AI strategies, but this will probably not displace the other approaches.

The middle office of a bank typically includes units such as risk controlling, the compliance function, the finance area and various other units that support the front office in their daily work, as well as the IT area. The working world of the middle office will also develop differently depending on the area with regard to artificial intelligence.

The compliance area deals, among other tasks, with the prevention of money laundering, the fight against terrorist financing, the enforcement of financial sanctions and the prevention of other criminal acts in which a bank could be involved. Today, the prevention of the described criminal acts is often supported by deterministic software solutions. These software solutions check predefined cases that indicate potential criminal acts. The potentially problematic business transactions are then examined more closely by employees of the compliance unit. The clear disadvantage of such systems is that all cases that are not predefined cannot be found. This is exactly where AI algorithms will play a decisive role and change the work in the compliance area significantly. For the working world in 2030, this will mean that there will be fewer and fewer employees in the compliance area who have no connection to data analysis or AI. This is because the clerical activity in the compliance unit will lose importance and will probably be increasingly replaced by employees with data analysis skills. The compliance area is one of the areas that will probably change the most by 2030. The challenge here is to recognize the need for change and initiate it in time.

In risk controlling, probably less major changes are to be expected, as the models used there are already largely very sophisticated today. In addition, the facts that are dealt with there are mathematically often "non-linear" problems, with which AI algorithms often have problems (depending on the complexity of the non-linearity). These problems can only be solved by very large amounts of data, which are usually not available in the required quantity in risk controlling. Furthermore, the banking supervision will also not allow here, as in parts of the front office, that the risks of a bank are monitored by an AI (current considerations of the banking supervision in Financial Stability Board, 2017)). The explainability of the actions in risk controlling is very important and often not given when using an AI. There will certainly be applications for AI algorithms in risk controlling by 2030 (e.g. for market data control or market conformity), but the working world for risk controlling will probably not change disruptively.

In the areas of IT and finance, there are also many possible applications of AI algorithms. For example, in IT, there is the possibility of using AI-supported process mining to make IT processes more efficient. In principle, the possibilities of using AI in the IT or finance area of a bank do not differ significantly from those in other companies. Therefore, we will not go into more detail on these areas here.

To conclude the triad, we will now look at the back office of a bank. The back office of a bank includes all units that deal with the settlement of transactions, human resources and marketing/press, as well as the internal audit area. We want to focus mainly on the settlement and the internal audit area, as the other areas mentioned also occur in other companies.

Essentially, the employees of the settlement units check which obligations exist between two parties in securities transactions or transactions of balances and ensure their trouble-free fulfillment. The biggest challenges for the settlement units are that the information comes from many different interfaces of business partners and has to be booked in the bank's internal system and reconciled with the business partners. For this purpose, a lot of human know-how is still used today, as especially the handling of non-standardized interfaces is a big challenge for deterministic automation. This is exactly where intelligent, AI-supported process automation (e.g. robotic process automation) will come in. As soon as AI-supported process automation can be implemented cost-effectively for such processes, these techniques will cause a disruptive change in the back office. There will also continue to be specialists in this area who are familiar with and deal with the specifics of individual transactions, but the workforce in this area will shrink significantly by 2030.

The internal audit is an independent, objective audit and consulting unit. The internal audit is part of the corporate oversight function and the risk-oriented control system in banks. It conducts audits of the bank's internal processes, which usually last between three and nine months. Audits can be compared to projects that run for three to nine months and end with an audit report with potential weaknesses of the audit object. The audit routine in the internal audit is already undergoing change today. The classic case-by-case audit is increasingly giving way to an audit of large amounts of data (many business transactions) and their assessment by the auditors. The use of data analytics and artificial intelligence is finding more and more use cases and is used to uncover weaknesses in the internal control system. The working world will change in this respect, that more and more employees with data analysis skills will be needed and less classic auditors. The image of the auditor will change significantly by the year 2030, but the audit function will not be replaced by an AI. However, the audit units that do not rely on data analytics and artificial intelligence will provide the company with a much lower added value than those that have invested in this field. Just like in the area of compliance, a targeted investment in human know-how and technology is required.

5 Summary and Practical Recommendations

In summary, it can be stated that banks, just like all other areas of society and economy, are affected by the changes triggered by artificial intelligence. The most important insight for banks is that they invest in human know-how and technology in the banking areas that are likely to be strongly affected by the change through artificial intelligence. Only with the selection of the right workforce in combination with a viable AI strategy and the necessary technical prerequisites can the challenges of the coming decade be mastered. The time of the widespread branch banks is coming to an end and new AI-supported ways of customer acquisition and retention have to be established to cope with the challenges until 2030.

References

Accenture. (2017). Accenture technology vision banking. https://www.accenture.com/_acn-media/pdf-47/accenture-banking-technology-vision-2017.pdf. Accessed: 24. June 2021.

Accenture. (2018). Future Workforce Studie—Mit Künstlicher Intelligenz zu neuem Wachstum im Bankensektor. https://docplayer.org/114833820-Future-workforce-studie-mit-kuenstlicher-intelligenz-zu-neuem-wachstum-im-bankensektor-neue-schlag-kraft-fuer-banken-team-mensch-maschine.html. Accessed: 31. July 2021.

BearingPoint & niiio finance group. (2018). Robo-Advisory. Wertpapierberatung digital gestalten. https://www.bearingpoint.com/files/Robo_Advisory.pdf?download=0&itemId=549063. Accessed: 24. June 2021.

Deloitte. (2021). CFO 2030+ in der Finanzbranche. https://www2.deloitte.com/content/dam/Deloitte/de/Documents/financial-services/CFO-Brosch%C3%BCre-Digital-Final_1.pdf. Accessed: 24. June 2021.

Fama, E. (1970). Efficient capital markets, a review of theory and empirical work. *Journal of Finance., 25*, 1970.

Financial Stability Board. (2017). Artificial intelligence and machine learning in financial services. https://www.fsb.org/2017/11/artificial-intelligence-and-machine-learning-in-financial-service/. Accessed: 24. June 2021.

McKinsey Global Institute. (2018). Notes from the AI frontier: Modeling the impact of AI on the world economy. https://www.mckinsey.com/~/media/McKinsey/Featured%20Insights/Artificial%20Intelligence/Notes%20from%20the%20frontier%20Modeling%20the%20impact%20of%20AI%20on%20the%20world%20economy/MGI-Notes-from-the-AI-frontier-Modeling-the-impact-of-AI-on-the-world-economy-September-2018.ashx. Accessed: 24. June 2021.

PricewaterhouseCoopers. (2020). How mature is AI adoption in financial services. https://www.pwc.de/de/future-of-finance/how-mature-is-ai-adoption-in-financial-services.pdf. Accessed: 24. June 2021.

AI in the industrial world of work

Potentials of AI for Production

Obstacles, Potential Applications and a Scenario-Based Outlook for 2030

Marco Huber, Christian Jauch and Klaus Burmeister

1 Introduction

Making production more productive, employee-friendly and also more sustainable—these goals are obvious, but hard to achieve. AI technologies are supposed to enable this. The production environment has experienced technology trends in the past decades that were supposed to work towards one or more of these goals—from the human-free factory and "lean" concepts to networking and digitalisation around Industry 4.0. Many indications suggest that AI is partly based on or has to build on what has been achieved so far (no AI-based data analysis without data), but is substantially different in terms of effectiveness.

A human-free factory is not to be expected, on the contrary: The human being will accompany and orchestrate the implementation of AI-based applications and

M. Huber (✉)
Fraunhofer Institute for Manufacturing Engineering and Automation IPA, Stuttgart, Germany
e-mail: marco.huber@ipa.fraunhofer.de
URL: https://www.ipa.fraunhofer.de/ki

C. Jauch
Fraunhofer Institute for Manufacturing Engineering and Automation IPA, Stuttgart, Germany
e-mail: christian.jauch@ipa.fraunhofer.de

K. Burmeister
Founder and managing director foresight lab, Berlin, Germany
e-mail: burmeister@foresightlab.de

I. Knappertsbusch and K. Gondlach (eds.), *Work and AI 2030*,
https://doi.org/10.1007/978-3-658-40232-7_20

often AI will support the human being where pure automation is economically or technically unsuitable. This is at least the vision. However, we are still a long way from this desirable symbiosis today.

2 Status Quo

Overall, a diffuse picture emerges regarding AI in production. There is not *the* production and the proliferation both in terms of the possibilities of use and implemented AI applications in this environment is large. The potential uses of AI in production are already diverse today.

However, an exchange between human and AI is usually not or only rudimentarily present. Instead, AI and human coexist in production: AI takes care of a task, shows the result and the human has to accept the decision. As long as AI is working on a task, the human is out of the picture, and the same applies vice versa. AI looks at the results that the human has produced and uses this information as input data for the next decision making.

Moreover, the human being is currently indispensable for production, but is usually forgotten; optimisations are aligned to the process. Time-consuming training sessions are conducted for employees so that they can find their way around the submenus of the machine. The operation of the private cell phone or the own smart home, on the other hand, is intuitive and does not require training. Here, there is a gap between the private sphere and the job that needs to be closed.

However, there is potential in involving the human being in the optimisation. No one knows a process as well as the worker who performs it daily. The integration of the employee in the optimisation can leverage this potential. One then speaks not only of a *process*-oriented, but of a *human*-centred optimisation. This involves using the employee's prior knowledge and considering the employee with all his or her individuality, e.g. also ergonomic issues, preferences in handling e.g. between left-handed and right-handed people and motor skills.

3 Challenges and Solutions

With the diverse potentials, one might think that AI has already arrived comprehensively in the industrial world of work. But if at all, then mostly large companies have implemented first AI applications. What are the problems and what are the solutions?

Obstacles such as the lack of time, money or knowledge or skilled workers are also known from other innovations. Other barriers exist specifically against AI. Because this is or should be used to a large extent in factories that already exist for a long time. Such an installation in existing structures is associated with limitations and compromises. A previously AI-free production must be gradually made "AI-capable".

This includes that the digitalisation of production forms, the basis for the currently most widespread AI subfield of machine learning (ML). However, this is often not advanced enough. The lack of data, which is also the basis for various ML methods, also falls into this context. Currently, almost every manufacturer wants to keep their data for themself, but this limits or even makes a continuous digitalisation of production and AI solutions based on it impossible. Accordingly, open, uniform data structures in the sense of a "data governance" are important: data would then be provided or managed continuously and securely in high quality.

Also, the black-box character of typical ML algorithms is often disadvantageous. *How* they arrive at a result is not always clear even to experts. Many standards and regulations oppose this non-transparency and prevent the use of AI. Not without reason, the research field of "Explainable AI" (xAI) [1] is becoming increasingly important and is supposed to ensure that AI is used more widely in terms of legal aspects as well as trust and acceptance.

Currently, research is concerned with how to better use the potentials of the human being and also introduce this more widely, i.e. the integration costs should be kept low by means of as generalised systems as possible, in order to offer a low-threshold entry also for small and medium-sized enterprises (SMEs). An enabler for intelligent assistance and production control systems is the so-called automated activity recognition, which passively observes the worker and thus digitally captures even manual activities. This way, it can automate non-value-adding activities such as booking processes, documentation and the training of new employees. Further potentials are offered by the automated detection of errors. It is expected that in the next few years algorithms will be available that are specifically tailored to the activity recognition of humans in production and that have the necessary performance for industrial use. To reduce the amount of required training data for the ML algorithms, various solution strategies are particularly suitable in the industrial context. These include hybrid algorithms that combine analytical and data-based models. Or the training in the simulation with subsequent fine-tuning of the ML algorithms for the real application using a few additional training data.

With the help of this recognition, manual activities can be better integrated into the process and manual and automated processes can be closely linked.

4 Outlook on AI in Production in 2030

The use of AI in the industrial world of work is still at the beginning of a new data- and algorithm-based production paradigm [2]. The sketched potentials are large and diverse. However, in order to estimate the further development until 2030, a networked future analysis is needed, which also includes contexts and framework conditions in addition to production.

In this regard, the following sections draw on the results of a foresight and scenario study published in 2019 [3]. A set of 24 key factors from technology, the world of work, economy and companies, society and values, and politics and regulation were identified as the basis for the scenarios (Fig. 1). The result shows a map of six consistent and plausible scenarios of "AI-based working environments 2030".

All scenarios are based on common assumptions until 2030: A digital infrastructure for basic product and process innovations and digital value creation processes is created. Industry 4.0 technologies penetrate and shape all sectors. This entails a structural change of occupations and activities, which implies a high demand for qualification. AI supports the advancing automation processes in the industry and aims at the core to automate knowledge-based processes and activities.

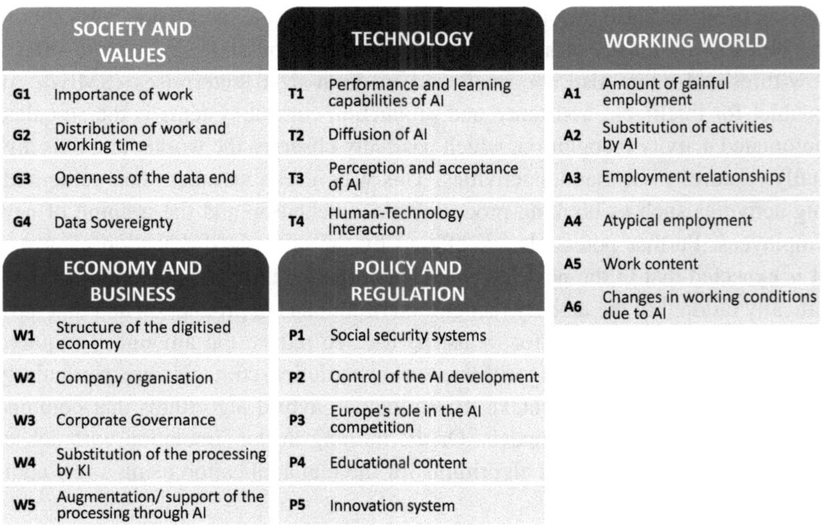

Fig. 1 Catalogue of key factors. (Source: Foresightlab)

4.1 The Scenario Map of AI-Based Work Environments 2030

The portfolio of scenarios (Fig. 2) outlines the target corridor of possible developments until 2030. It becomes clear that the industrial work environment faces important decisions and also disruptive changes. In a nutshell, the question is whether the work environment remains trapped in the shell of the *industrial* -shaped work environment (scenario 4) or whether it is possible to enable a holistic productivity leap (scenario 5). The latter is a pointed assessment and opportunity-oriented perspective, which is supported by the results of an expert survey [4]. The following brief presentation of the scenarios focuses on these two scenario dimensions.

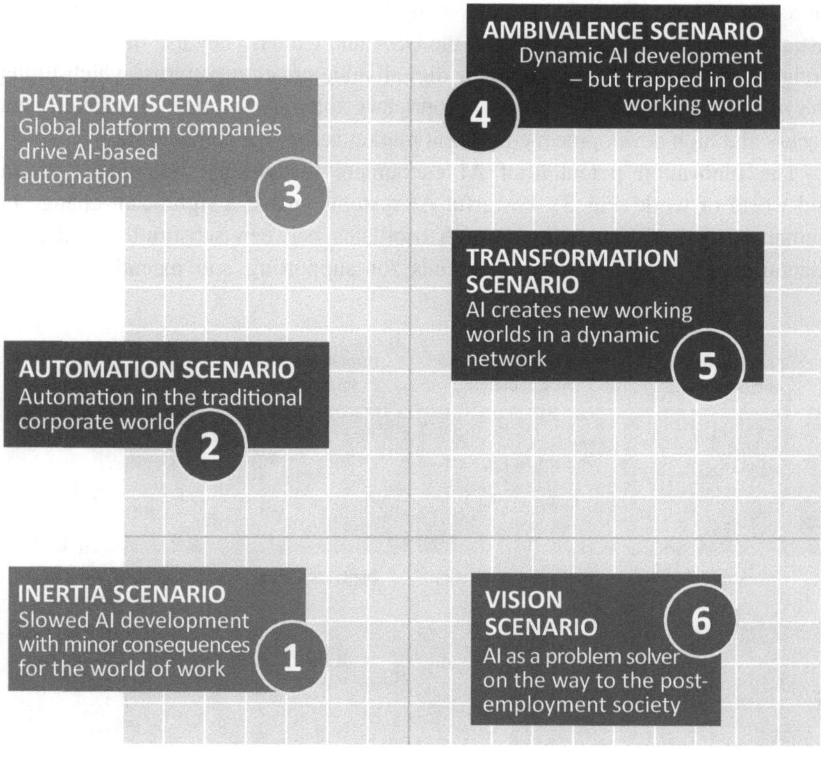

Fig. 2 Map of AI-based work scenarios 2030. (Source: Foresightlab)

Scenario 1: Inertia scenario – *Slowed down AI development with low impact on the work environment.*

The scenario describes a future of missed opportunities. The work environment is little affected by the developments in AI, because the necessary framework conditions change only slowly. Many potentials of AI remain unused.

Scenario 2: Automation scenario – *Automation in a classic corporate world.*
It outlines a development in which the industrial landscape continues to be shaped by the prevailing automation paradigm. There is a "business as usual" with old centralist leadership styles. The use of learning-capable AI systems is slowed down and their potentials are only realised in individual domains.

Scenario 3: Platform scenario – *Global platform companies drive AI-based automation.*
Here, the change dynamics are driven by the further establishment of monopolistic, global platform companies in the USA and China. They use their data sovereignty to further differentiate their digital value creation patterns, which limits the broad diffusion of AI systems. Proprietary software solutions, restricted data access and high development costs limit market access for competitors.

The innovation potential of AI encounters institutional boundaries of an old world of work (Fig. 3). Powerful AI systems enable a significant change in human-technology interaction, which oscillates between substitution and augmentation [5]. However, the potentials for supporting and extending human

AI development and data world	**Data sovereignty and AI acceptance**	**Human-machine interaction**	**Processing**
Dynamic, cross-domain AI development, but with limited access.	Low data sovereignty. Polarised public perception of AI	Augmentation of numerous activities by AI. Automation of routine tasks only.	Strong substitution of simple and medium clerical work. Augmentation mainly for efficiency improvement
Working world	**Company organisation**	**Digital economy**	**Control of the KI development**
Improvements in increasing a typical employment, but negative image remains	Rapid change in the context of closed concepts and centralised, cl- under- user leadership	Network effects drive platform competition	Intensive cooperation between states and global corporations

Fig. 3 Scenario 4: Core premises. (Source: Foresightlab)

activities in the sense of augmentation by AI are still hampered by the rules of the game and the mindset of a technology-centred work and business world. As a result, the substitution of manual activities and cognitive routine activities dominates and an increase in efficiency prevails over the improvement of human-technology interaction. The entrapment in the old world of work paradigm is reflected in the business concepts. These enable rapid change, but are still strongly influenced by traditional decision structures, which are effectively supported by AI.

Scenario 4: Ambivalence scenario *– Dynamic AI development, but trapped in an old world of work.*

The innovation potential of AI comes up against the institutional limits of an old working world (Fig. 3). Powerful AI systems enable a significant change in human-technology interaction that oscillates between substitution and augmentation [5]. However, the potential for supporting and expanding human activities in the sense of augmentation through AI continues to be held back by the rules of the game and the mindset of a technology-centred working and corporate world. As a result, the substitution of manual activities and cognitive routine activities dominates and an increase in efficiency prevails over the improvement of human-technology interaction. Being trapped in the old paradigm of the world of work is mirrored in corporate concepts. Although these allow for rapid change, they are still strongly characterised by traditional decision-making structures that are effectively supported by AI.

AI development and data world	**Data sovereignty and AI acceptance**	**Human-machine interaction**	**Processing**
Dynamic AI development with cloud-based applications and easy access with compartmentalised data use	Low data sovereignty. Nevertheless, trust in digital world and problem-solving potential of AI	Augmentation of numerous activities through AI as well as overall Strong automation	Substitution of even complex processing. Augmentation to improve efficiency AND results
Working world	**Company organization**	**Digital economy**	**Control of the AI development**
Gig economy with positive image, but poorer protection	Dynamic network world with AI-supported decentralised moderating leadership	Network effects drive platform competition	Intensive cooperation and promotion of broad innovation alliances

Fig. 4 Scenario 5: Core premises. (Source: Foresightlab)

***Scenario 5: Transformation scenario** – AI creates new work environments in a dynamic network.*
The scenario describes a future in which the potentials of AI are consistently used for a digital and ecological transformation in the sense of an energy-efficient and resource-efficient production (Fig. 4). Although digital sovereignty is also low here, there is high confidence in the digital transformation. Cloud-based AI applications ensure easy and cheap, cross-domain access for many medium-sized enterprises and start-ups. The emerging upheavals in the world of work are characterised by the substitution of even complex, knowledge-based activities. The development follows a human-centred augmentation in human-technology interactions, which promotes the emergence of new fields of activity. This reconfiguration of the worlds of work provides the basis for a productivity boost, which induces increased requirements for the qualification of the employees. Such a dynamic also enables an agile corporate world, in which the management—supported by AI—fosters self-organisation processes of employees and departments.

***Scenario 6:** Vision scenario – AI as a problem solver on the way to the post-work society.*
It points beyond 2030 and has the character of a specific utopia. The scenario sketches the contours of a society that is in transition to a "post-work society".

4.2 A Design Dilemma

Crucial for the future of AI in the industrial world of work will be whether and how a transition from the technology-centred ambivalence to a human-centred transformation scenario [6] can be designed. The differences between both scenarios (Fig. 3 and 4) point to the importance of open data worlds embedded in open innovation ecosystems and a change-ready business environment. In a new configuration of human-technology interaction in the transformation scenario, there are potentials for unleashing significant productivity gains that lie dormant in the ambivalence scenario. To unlock them requires a systemic understanding that addresses and rearranges the connection between work content and organisation, business organisation and leadership. In the transformation scenario, the upheaval is initiated by 2030. It takes place in the different production worlds each *industry-specific*. For this, innovation alliances will be necessary, supported by the employees, the social partners and the companies. Since there is no blueprint, experimental design arenas will have to become the norm.

5 Summary and Practical Recommendations

The industrial worlds of work are undergoing a structural change until 2030 and beyond. The use of AI in the world of work is still in its infancy. Not a massive destruction of jobs, but a profound transformation of occupations and activities characterises the development. In particular, a considerable potential for change is expected for analytical and interactive cognitive "non-routine tasks". Socially compatible answers are needed to the question of what happens to employees whose activities disappear. Furthermore, the question of which skills will be required in the future and who will train them must be answered.

The outlined upheavals in the worlds of work are an expression of the transitions to a new production paradigm. Visions, such as those that exist with Industry 4.0, provide an orientation that needs to be translated and tested for the operational day to day. The challenge is to realise the new paradigm in an asymmetric, platform-driven competitive environment. Approaches are needed for an open data economy, as pursued by the GAIA-X project [7]. The opportunity lies in redefining the overall context of production and distribution, based on the industrial core competence and embedded in an open European innovation ecosystem. A data-based and algorithm-based production system will inevitably require an open, agile and learning-capable corporate environment. The desired productivity progress will be successful in the target corridor of the great transformation when technical and social innovations are not opposites.

Finally, a small guide on how to approach the use of AI in the learning environment of production today [8]:

- Start small, think big: Work on individual process steps first. Small progress will provide initial knowledge of AI, confidence in the technology and thus also arguments in the company for bigger challenges.
- Start early: Identify useful use cases early on. Use short development cycles to make quick progress and be able to agilely counteract difficulties. This procedure enables a fast creation of prototypes.
- Pay attention to the benefit: This is ultimately what an AI project depends on. It must provide added value for the company. Let the business department drive the topic and not necessarily the IT.
- Take everyone along: Do not forget the employees in the project. If they are convinced of the benefits of an AI project, this prevents unnecessary hold-ups.

References

Burkart, N., & Huber, M. F. (2021). A survey on the explainability of supervised machine learning. *Journal of Artificial Intelligence Research (JAIR), 70,* 245–317. https://doi.org/10.1613/jair.1.12228.

Vergleiche hierzu u. a. BMWi. (2021). Plattform Industrie 4.0, Fortschrittsbericht 2021. https://www.plattform-i40.de/PI40/Redaktion/DE/Downloads/Publikation/2021-fortschrittsbericht.pdf?__blob=publicationFile&v=14. Zugegriffen: 20. Juli 2021; Lernende Systeme – Die Plattform für Künstliche Intelligenz. (2021). KI im Mittelstand – Potenziale erkennen, Voraussetzungen schaffen, Transformation meistern. https://www.plattform-lernende-systeme.de/files/Downloads/Publikationen/PLS_Booklet_KMU.pdf. Accessed: 20. July 2021.

Burmeister, K., Fink, A., Mayer, C., Schiel, A., & Schulz-Montag, B. (2019). *Szenario-Report: KI-basierte Arbeitswelten 2030*. Fraunhofer Verlag.

a. a. O. Seite 80 ff. und 132 ff.

Augmentierung meint hier die Assistenz von wissensbasierter Arbeit durch KI-Systeme und daraus resultierende neue Formen der Mensch-Maschine-Interaktion.

Die Unterschiede der beiden Szenariodimensionen zeigen Abb. 4.3 und 4.4.

https://www.data-infrastructure.eu/GAIAX/Navigation/EN/Home/home.html. Accessed: 20. July 2021.

Diese Handreichungen wurden bereits mehrfach an anderer Stelle veröffentlicht, zuerst in einem Fachartikel: Huber, M. (2019). Machine Learning: Daten sind Schlüssel für maschinelles Lernen. automationspraxis. https://automationspraxis.industrie.de/industrie-4-0/machine-learning-daten-sind-schluessel-fuer-maschinelles-lernen. Accessed: 20. July 2021.

The Grassroots Movement of AI

Data Governance and Servitisation as Drivers of the Digitalisation of Physical Infrastructures in the Energy Industry

Lars Michael Bollweg

> *"Data doesn't matter. It's what you derive from it. It's what you do with it that matters."*
>
> *Dr. Lars Michael Bollweg—Data Officer*

1 Introduction

The terms "digitalisation" and "artificial intelligence" have long since arrived in the vocabulary of top management of large and small energy companies in Germany in the 2020s. Executives in the energy industry no longer need to be convinced of the potential of digital transformation. But as in many other traditional industries, the energy industry still finds it difficult to keep up with the high speed of digital development or even actively shape it. The great and largely unresolved challenge for the energy industry is the operationalisation of digital transformation in an environment characterised by manual process chains for planning, construction and operation of large physical infrastructures for energy supply. In other words, the widespread use of artificial intelligence and other digital developments in the energy industry today fails not because of strategic planning, but rather because of operational implementation (Bollweg, 2021).

L. Bollweg (✉)
Westnetz GmbH, Dortmund, Germany
e-mail: lars.bollweg@westnetz.de
URL: https://www.linkedin.com/in/dr-lars-michael-bollweg-7823b886

I. Knappertsbusch and K. Gondlach (eds.), *Work and AI 2030*,
https://doi.org/10.1007/978-3-658-40232-7_21

The sentence fragments “process chains characterised by manual work” and “physical infrastructures” summarise the existing challenges for the digitalisation of processes and business models in the energy industry very well. Unlike pure service companies, the companies in the energy industry are not only faced with the challenge of automating self-contained information flows and processes. Like other traditional industries (e.g. manufacturing), the energy industry must also develop the “cyber-physical” (software and mechanics/technology) and the “human-computer” interaction along the existing value streams in high process quality in order to benefit sustainably from digitalisation (Baines, 2013). This so-called "handshake" between the digital and physical world poses particularly high demands on the supporting information flows (data quality) and the technologies used (e.g. sensors, robotics, etc.). Due to these high digital and technical requirements, the implementation complexity for value-added digital solutions and the use of AI is significantly increased and the sustainable digital development of companies from the energy industry is slowed down (Hadaya & Gagnon, 2017).

This article shows how the energy industry can operationalise digital transformation through a data-centric development of business processes (*Data Governance*) and the associated structured search for and targeted development of automation and service potentials (*Servitisation*) and establish it as an integral part of continuous digital business development.

2 Status Quo

The energy industry is currently facing a multitude of turning points. Hardly any industry is exposed to as many fundamental transformations as the energy producers, transmission and distribution network operators. An incomplete overview: The energy generation is changing, e.g. as a result of renewable energies. The consumer and the consumption patterns are changing, e.g. due to e-mobility. The physical infrastructures and their technologies are changing, e.g. with regard to smart grids. And finally, the legal and regulatory framework is changing, e.g. the EEC surcharge is noteworthy here.

Each of these changes alone already questions the established business model and the existing processes in the energy industry. But on the whole, the requirements for the digital capabilities of the companies increase exponentially with each of these changes. Under the pressure of change, a variety of digital tools are developed in isolation, e.g. to comply with information obligations, to control investment costs in a targeted manner, or to adapt to the communication with increasingly detailed customer groups and their needs. This leads to a fragmentation of the IT landscape in the energy industry, which becomes a problem whenever cross-cutting information chains are needed, for example, as a basis for prediction algorithms and the use of AI.

With the pressure of change from so many directions, the energy industry has no choice but to embrace and actively drive the digital transformation. The energy industry as a whole has recognised digitalisation and the use of artificial intelligence as an opportunity. But like so many industries and sectors, it struggles with the operationalisation of digital transformation and fails to use artificial intelligence in day-to-day business.

3 Challenges and Solutions

From theory to practice: The challenge for the energy industry with a view to the near future in 2030 lies in the operationalisation of digital transformation. This means that the companies in the energy industry still find it difficult to further develop their own business by using digital tools and applications within the traditional processes and procedures.

By "operationalisation of digital transformation" we do not mean the implementation and execution of large digital flagship projects, which are often implemented in so-called "speedboats" alongside the existing organisation, but rather the operationalisation of digital transformation along the grassroots of business development in the strategic (e.g. business development, process management, etc.) and operational units (e.g. network planning, network operation, etc.), which operate and shape the day-to-day business of the energy industry, both in municipal utilities and in large corporations. The focus on the "grassroots development" is important, because the strategic and operational units, which run the core business of the energy industry, hold the key to the successful digital transformation in their hands. They know their own business with all its strengths and weaknesses. They know the degree of detail and diversity of the complex processes along the value creation of the energy industry. They know the potential for improvement and above all they know the customers. The strategic and operational units are the ideal transformation units, because they can assess whether a digital solution after development also provides the promised and needed benefit for the respective business processes and business objectives (Erlach, 2020). A broad and deep digital transformation cannot be sustainably implemented from the outside in speedboats, it must be driven from the inside, from the strategic and operational units, to be successful.

As much as the strategic and operational units of the energy industry are also the decisive factor in solving the operationalisation of digital transformation, they are also part of the current problem. The operationalisation of digital transformation is a multi-layered and complex task, for which they must be prepared with the development of digital competencies. If the development of digital competencies is not recognised as a targeted and sustainable task, the widespread use of artificial intelligence will continue to fail at the internal competence barrier.

Two core competencies have emerged as crucial for digital business development: 1) The implementation of a continuous data-driven process development within the framework of a *Data Governance* and 2) the strict alignment of the data-driven process work on the identification and development of digitally enabled automation and service potentials (*Servitisation*). In the following, both the basics of these core competencies as a grassroots movement of digital transformation are briefly explained, in order to then provide some hints on their possible applications to promote artificial intelligence and digital solutions (Bollweg, 2021).

a. Data Governance

The term and the practical work of a *Data Governance* is very abstract for many employees in traditional companies at first glance. Often, the everyday experience each person has with the topic of data management and data-driven process development is lacking, to be able to grasp the added value for one's own work directly. Therefore, it always makes sense to start with a simple definition of *Data Governance* and to use it to show the potentials of a professional handling of data.

> "Data Governance is the structured embedding of the practices (procedures and methods) of data management in the organisational structure and processes of an enterprise." (Bollweg, 2021).

A *Data Governance* is therefore a tool to apply the practices of data management for the professional handling of data in the company ("How do we work with data?"). In addition, *Data Governance* is also a tool to anchor the application of the practices of data management in the breadth, i.e. in all departments, and in the depth, among all employees, of the organisational structure, by linking the data work with the process-oriented units (strategic and operational development) (Bollweg, 2021; DAMA International, 2017).

By creating *Data Governance* roles in the company, which are equipped with clearly defined responsibilities and tasks, *Data Governance* as a whole has a positive impact on the development of digital competencies in the organisational structure. The organisational structure, in turn, uses the *Data Governance* roles as multipliers for the development of the competencies of the process organisation (Gluchowski, 2020).

Data Governance thus not only supports the handling of data, but above all enables the employees in the strategic and operational units to recognise the challenges of digital transformation, i.e. to identify the existing process potentials and to transform them into digital solutions.

b. Servitisation

In contrast to *Data Governance*, *Servitisation* is much less a standalone craft than a paradigm and perspective shift. When a company looks at its own service portfolio through the "*Servitisation* glasses", the path leads away from the sole focus on value creation through the use of physical infrastructures and products (goods), towards the development of combination goods, i.e. a mix of goods and (digital) services (Vandermerwe & Rada, 1988). The term *Servitisation* accordingly describes a strategic extension of the company's processes and products by (digital) service offerings. The *Servitisation* logic works end to end. Companies that understand *Servitisation* only as an extension of the interface to the external customer and consumer miss the opportunity to also optimise the internal customer and supplier relationships along the business processes by developing the digital service mindset.

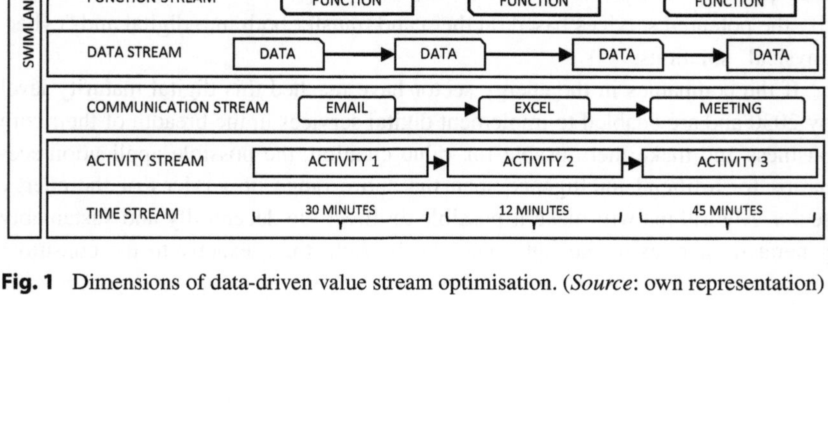

Fig. 1 Dimensions of data-driven value stream optimisation. (*Source*: own representation)

Together, *Data Governance* and *Servitisation* form the grassroots movement of digital transformation. Through the data-driven process work, with a clear focus on automation and service potentials, the digitalisation potentials are identified and their implementation prepared along the business processes (Urbach & Ahlemann, 2016). Thus, by using a service-oriented *Data Governance*, e.g. it becomes very easy to see, between which physical process steps, e.g. communication between humans and systems or systems and systems, a digital interface would be able to automate this communication as a digital service. The same consideration is also possible at the level of data flows, system functions, user interfaces or even the overarching IT systems. A service-oriented *Data Governance* works along all digital development levels (Bollweg, 2021) (see Fig. 1).

It can be summarised: The connection of *Data Governance* and the *Servitisation* logic leads the data-driven process work in each optimisation iteration to new digital automation and service potentials and thus into a continuous digital improvement process (Kersten, 2019). This permanent development process, from the digital grassroots upwards to the overarching corporate development, forms the basis for the professional operationalisation of digital transformation.

4 Outlook on AI in the Energy Sector in 2030

The high pressure for change on the energy sector (changed energy generation, energy consumption patterns, technologies as well as legal and regulatory framework conditions) accelerates the digital development of the industry and increases the expectation pressure regarding the use of digital technologies and modern algorithms (AI). Whether the industry will really be able to meet these expectations and exploit the benefits of digital transformation and modern algorithms by 2030, however, depends to a considerable extent on whether the energy sector will be able to optimise the information flows and information structures (e.g. data quality, events stored as time series, etc.) along the business processes, identify the potential services based on them and transfer both into digital and "cyber-physical" solutions.

If the companies in the energy sector have reached this digital maturity level by 2030 and are enabled to implement digital services in the breadth of their core business and make them usable for value creation, the possible application scenarios for artificial intelligence span the entire range of services of the energy sector. Algorithms will make it possible to direct the decentrally and sustainably generated energies to the right place at the right time, exactly to the consumer who needs them (whether households, industry, commerce or e-mobility), and to bill these energies flexibly and directly with the consumers via intelligent metering points. For the state and its regulatory mandate, new control possibilities will arise from energy generation to energy consumption, e.g. to steer the development of desired (e.g. LED lamps and energy-saving TVs, etc.) and undesired consumers (e.g. outdated electric heaters, refrigerators and freezers, etc.) by means of discounted or increased energy prices. As a consequence, the planning, operation and management of the physical infrastructures will also change fundamentally. From the fully automated planning process to intelligent maintenance (predictive maintenance), modern algorithms will automate and economically optimise the processes and workflows within the traditional value streams of the energy sector. In addition to all these big lines, however, it will be the automation of the small-scale process steps by the use of algorithms that will enable large productivity gains for the energy sector in total.

5 Summary and Practical Recommendations

Even in 2030, the use of artificial intelligence in the energy sector is more than a decision that a company can make at short notice by pressing a button. The use of modern algorithms and value-adding predictions requires an early structured and organised preparation of the data, processes and systems in the companies (Kotusev, 2018). Besides the necessary empowerment of the employees, the development of the appropriate data bases will be crucial for success. In addition, the potential services along the business processes must be identified in a structured way and implemented with a focus on a real contribution to value creation. Only the companies that understand today that a professional handling of data, systems and processes is the basis for the data-driven future and the successful implementation of the digital transformation will benefit from digitisation and the use of artificial intelligence in 2030.

Companies that want to actively steer the digital transformation today are well advised to establish an effective *Data Governance,* with a clear focus on the operationalisation of the digital transformation from the middle of the company and a focus on the identification and implementation of automation and service potentials along the business processes (*servitisation*). *Data Governance* and *Servitisation* offer companies the necessary tools with which they can develop the digital foundation—the grassroots movement—of the digital performance in 2030 already today.

References

Baines, T., & Lightfoot, H. (2013). *Made to serve: How manufacturers can compete through servitization and product service systems*. Wiley.

Bollweg, L. (2021). *Data Governance für Manager—Datengetriebene Prozess- und Systemoptimierung als Taktgeber der digitalen Transformation*. Springer.

DAMA International. (2017). *DAMA-DMBOK: Data management body of knowledge* (2. Ed.). Technics Publications.

Erlach, K. (2020). *Wertstromdesign: Der Weg zur schlanken Fabrik (VDI-Buch)*. Springer.

Gluchowski, P. (2020). *Data Governance: Grundlagen, Konzepte und Anwendungen*. Dpunkt.

Hadaya, P., & Gagnon, B. (2017). *Business Architecture: The Missing Link in Strategy Formulation*. ASATE Publishing, Montreal.

Kersten, M. (2019). *Project to product: How to survive and thrive in the age of digital disruption with the flow framework*. IT Revolution PR.

Kotusev, S. (2018). *The practice of enterprise architecture: A modern approach to business and IT alignment*. SK Publishing.

Urbach, N., & Ahlemann, F. (2016). *IT-Management im Zeitalter der Digitalisierung: Auf dem Weg zur IT-Organisation der Zukunft*. Springer.

Vandermerwe, S., & Rada, J. (1988). Servitization of business: Adding value by adding services. *European Management Journal, 6*(4), 314–324.

Employment Effects and Changes in Work Organisation Arising from AI

Werner Widuckel and Lutz Bellmann

1 Introduction

Digititalisations has led to major changes in many areas in recent years. This applies even to professional activities that until recently were considered irreplaceable by humans and that can now be digitised and automated. At the same time, the number of novel robots has increased (Deng et al., 2021). However, these collaborative robots no longer need to be equipped with protective devices, so that direct cooperation with humans is possible (Dengler & Matthes, 2018a). However, the potential for replacing human activity by machines cannot be directly derived from the labour market effects, because the specific design of the new technologies is shaped by ethical, legal, social, cultural and economic factors. In addition, there are cost reductions, which lead to a higher demand for goods and services as well as an increased demand for labour (Dauth et al., 2017, 2021). Furthermore, effects on the structure of employment and work organisation are to be expected, which also become visible through the emergence of new jobs or activities.

W. Widuckel (✉)
Friedrich-Alexander-Universität Erlangen-Nürnberg, Erlangen, Germany
e-mail: wiso-personalmanagement@fau.de

L. Bellmann
Friedrich-Alexander-Universität Erlangen-Nürnberg, Erlangen, Germany
e-mail: Lutz.Bellmann@iab.de

I. Knappertsbusch and K. Gondlach (eds.), *Work and AI 2030*,
https://doi.org/10.1007/978-3-658-40232-7_22

2 Status Quo: Use and Effects of AI

Based on the IAB Establishment Panel 2019, Deng et al. (2021) examined the use of robots in German establishments. While 8.2% of the establishments in the manufacturing sector use robots, only 0.9% of the establishments in the other sectors dare to do so. Although the share of all establishments that use robots was surprisingly low at 1.6%, the number of employees in these establishments was 3.2 million and the share of all employees was 8%.

What impact does the (partial) takeover of routine tasks, which are easy to programme and automate, by computers and computer-controlled machines have? According to calculations by Frey and Osborne (), the probability of digital automation of their activity is greater than 70% for 47% of the employees in the USA. In a similar study for Germany, Brzeski and Burk (2015) even determine a share of 59% of the jobs in Germany that could be replaced by machines in the next decades. Bonin et al. (2015) give a significantly lower value of 42%. In a further analysis step, the authors consider the level of activities instead of occupations and calculate a share of only 12% for the jobs that could be eliminated.

Dengler and Matthes (2015, 2018a, 2018b) have determined the substitution potential of occupations based on the expert database BERUFENET[1] of the Federal Employment Agency. It shows that the share of employees subject to social security contributions who work in an occupation with a high substitution potential increased from 15% in 2013 to 25% in 2016. This corresponds to a number of almost 7.9 million employees.

Particularly interesting is the consideration of the size and development of the substitution potentials for different requirement levels: For helpers and skilled workers, it was significantly higher in 2016 (58 and 54%, respectively) than for specialists and experts (40 and 24%, respectively).[2] The increase compared to the years 2013 and 2016 was 12% for the helpers, 8% for the skilled workers, 7% for the specialists and 6% for the experts. The substitution potential is therefore

[1] Contains, among other things, information on the tasks to be performed, the work equipment used, the design of working conditions and the necessary training requirements for the individual occupations.

[2] Experts are university graduates with at least four years of study, specialists are graduates of a master's and technician school, vocational school, professional or technical academy or bachelor's degree programs. Skilled workers have a minimum of two years of vocational training or a vocational qualification from a vocational or technical school. Helpers have no vocational training or a one-year vocational training.

greater the lower the level of requirements is, and also increases less with the level of requirements.

To determine the long-term employment effects of digitisation and especially the use of robots, Dauth et al., (2017, 2021) use data at the sector level of different countries for the period 1994–2014. In this period, the number of robots in Germany has increased significantly. According to their calculations, however, the 275,000 jobs that were lost in the manufacturing sector in Germany were compensated by the newly created jobs. This result differs significantly from that for the USA, which the authors attribute to the fact that in Germany not only robots are used, but also produced. This result is all the more significant as significantly more robots per employee have been installed in Germany than in the USA or in other European countries in the establishments of the manufacturing sector and their suppliers.

3 Challenges and Solutions

In the Linked Personnel Panel employee survey 2014/15, the employees were asked about the consequences of technological change (Arnold et al., 2016). The need to further develop their skills and competencies was mentioned by 78% of the respondents as the first priority. A higher workload was mentioned by 56% and the need for multitasking by 65% more often than physical relief (29%), greater decision-making freedom (32%) and reduced requirements (15%). Multiple answers were possible.

Becker et al. (2017) have shown for the occupation of mechatronics technician that the challenges of digitilisation result in great challenges at the level of those involved in the order of vocational training occupations and contents, the training and the trainees.

4 Outlook on Employment Effects and Changes in Work Organisation Arising from AI by 2030

The future perspectives of AI and digitalisation in the world of work are part of a comprehensive social transformation process, in which the fundamental upheavals of companies or organisations are embedded. It is therefore not only a *technological* change and also not gradual "change processes". Rather, companies fundamentally change their business models, organisational goals and purposes, processes and structures, work organisation and work culture in this context

(Widuckel, 2015, 2020). In addition, they are under pressure to make effective and measurable contributions to climate protection and sustainability. This is also a central element of the transformation, which forces companies to shape fundamental changes without having a ready-made "blueprint" for this. Moreover, this transformation has no discernible end point.

The behaviour of managers can contribute to creating or strengthening trust in the sense of the transformation or to disappointing it. However, this trust cannot simply be demanded, but must be based on a connection to the expectations, preferences and reflections of employees, who in turn deal with the transformation (Widuckel, 2018). Here, for example, expectations of fairness or expectations of the justifiable factual rationality of the transformation play a central role (Kratzer et al., 2015), which must not simply be glossed over with "change rhetoric". Rather, the participation of employees in the process of transformation is an essential prerequisite for the development of trust as a stabiliser and for the inclusion of those affected as persons with expertise in their situation. Participation also includes the involvement of works councils, which can have a significant influence on the emergence of trust in transformation processes. The co-determination at company level forms one of the central institutional frameworks of the transformation, which can have an innovative, enabling and opening effect on transformation processes (Schwarz-Kocher et al., 2011; Widuckel, 2020).

From our point of view, it can be expected that participatory and cooperative paths of digital transformation in companies have the greater chances of coping with this challenging, open process. This is supported by the factors of perceived self-responsibility, lifelong learning and the growing importance of team-based work organisation for complex work tasks, which depend on a positive reference to the motivation of employees and cannot simply be determined by others. For this reason, the structural and psychological empowerment of employees comes more into focus in the scientific and practical debate on digital work (e.g. Gül et al., 2020). This conceptual approach aims to strengthen the action and influence potentials of employees by designing digital work and thus to define a human-centred approach that has innovation, creativity and health-promoting potentials through dimensions such as meaningfulness, influence, competence and self-efficacy. However, this also entails greater dangers of possible overload, overstrain or the blurring of work boundaries, as e.g. studies on working from home show (e.g. Lott, 2019). It should be noted here that this conceptual approach has been mainly investigated for work in software development.

In addition to these different cultural conditions, further qualitative empirical studies in companies also show that the design of digital transformation in companies follows different paths, which depend on the company size, industry affiliation and the resulting differentiated functional assignments of digital technology (Hirsch-Kreinsen, 2018, Ittermann & Niehaus, 2018, Ittermann & Falkenberg, 2019). Conceptually, these differences have been condensed to structural patterns (scenarios) ranging from "devaluation", to "upgrading", to "polarisation" (low and high skilled activities grow) to "structure-conservative stabilisation" of industrial work or "simple work". As a result, it can be stated that no uniform development pattern of digital work can be discerned in the industrial companies examined. Rather, digital technologies are used for automation as well as for assistance and for the purpose of networking and organisational design of processes (Ittermann & Falkenberg, 2019). This shows that the digitalisation of work allows for design variations and by no means necessarily has to follow a "one best way" for technological functional reasons. A key requirement for a meaningful design appears to be not to disenfranchise and thus incapacitate employees through digitalisation, making them unable to react to unplannable events and deviations (Pfeiffer & Suphan, 2018).

5 Summary and Practical Recommendations

The potential of new technologies for replacing human activities, as determined in various empirical studies, cannot be directly derived from labour market effects or even the emergence of mass unemployment. Rather, the specific design of the changes caused by the new technologies is determined by ethical, legal, social and cultural factors. Digitisation and the associated transformation processes are embedded in fundamental social changes that should take place on the basis of stable social processes. No uniform pattern of digital work can be identified in qualitative empirical studies.

From these considerations we derive the following theses:

1. The digital transformation is an open upheaval process for companies without a recognisable end point. It requires trust as a social stabiliser. Transformation management is also trust management, which must be anchored in the corporate culture, the leadership behaviour and the direct participation of the employees as well as the co-determination of the works council.
2. The design of digital transformation in companies must include the expectations, preferences and expertise of the employees and be compatible with them. This must not be glossed over with change rhetoric.

3. The structural and psychological empowerment of employees is a decisive basis for the design of digital work, because this integrates motivational and health-relevant factors for employees and companies into the design of work.
4. The digitisation of industrial work moves in a corridor of alternative design possibilities. This corridor not only includes potentials for automation and the reduction of jobs, but also offers possibilities for supportive digitisation, to reduce stress or to make processes more transparent. In addition, human intervention possibilities in digitised production processes should be maintained, in order to avoid a disenfranchisement of the employees and to ensure the controllability of these processes.

References

Arnold, D., Butschek, S., Steffes, S., & Müller, D. (2016). *Digitalisierung am Arbeitsplatz (Forschungsbericht 468)*. Bundesministerium für Arbeit und Soziales.

Becker, M., Spöttl, G., & Windelband, L. (2017). Berufsprofile für Industrie 4.0 weiterentwickeln: Erkenntnisse aus Deckungsanalysen am Beispiel des Ausbildungsprofils Mechatroniker/-in. *Berufsbildung in Wissenschaft und Praxis, 46*(2), 14–18.

Bonin, H., Gregory, T., & Zierahn, U. (2015). Übertragung der Studie von Frey/Osborne (2013) auf Deutschland: Endbericht. (Forschungsbericht/Bundesministerium für Arbeit und Soziales, FB455; Zentrum für Europäische Wirtschaftsforschung GmbH). https://www.zew.de/publikationen/uebertragung-der-studie-von-frey-osborne-2013-auf-deutschland. Accessed: 10. Oct. 2021.

Brzeski, C., & Burk, I. (2015). Die Roboter kommen. Folgen der Automatisierung für den deutschen Arbeitsmarkt. *INGDiBa Economic Research*. https://www.ing.de/binaries/content/assets/pdf/ueber-uns/presse/publikationen/ing-diba-economic-analysis-die-roboter-kommen.pdf. Accessed: 31. July 2021.

Deng, L., Plümpe, V. & Stegmaier, J. (2021). Robot adoption at German plants VoxEU. https://voxeu.org/article/robot-adoption-german-plants. Accessed: 10. July 2021.

Dengler, K. & Matthes, B. (2015). Folgen der Digitalisierung für die Arbeitswelt – In kaum einem Beruf ist der Mensch vollständig ersetzbar. *IAB-Kurzbericht, 24*. https://www.iab.de/194/section.aspx/Publikation/k151209304. Accessed: 31. July 2021.

Dengler, K., & Matthes, B. (2018a). Wenige Berufsbilder halten mit der Digitalisierung Schritt. *IAB-Kurzbericht*, 4/2018. https://www.iab.de/194/section.aspx/Publikation/k180213301. Accessed: 31. July 2021.

Dengler, K., & Matthes, B. (2018b). The impacts of digital transformation on the labour market – substitution potential of occupations in Germany. *Technological Forecasting and Social Change, 137*, 304–316.

Dauth, W., Findeisen, S., Südekum, J., & Woessner, N. (2017). *German robots – The impact of industrial robots on workers*. CEPR Discussion Paper 12306.

Dauth, W., Findeisen, S., Südekum, J., & Woessner, N. (2021). The adjustment of labor markets to robots. *Journal of the European Economic Association, online first,* 1–52.

Frey, C. B., & Osborne, M. A. (2013). The future of employment: How susceptible are jobs to computerisation. https://www.oxfordmartin.ox.ac.uk/downloads/academic/future-of-employment.pdf. Accessed: 10. July 2021.

Frey, C. B., & Osborne, M. A. (2017). The future of employment: How susceptible are jobs to computerisation? *Technological Forecasting and Social Change, 114,* 254–280.

Gül, K., Boes, A., Kämpf, T., Lühr, T., & Ziegler, A. (2020). Empowerment – ein Schlüsselkonzept für die agile Arbeitswelt. In A. Boes, K. Gül, T. Kämpf, & T. Lühr (Hrsg.), *Empowerment in der agilen Arbeitswelt* (pp. 17–30). Haufe-Lexware.

Hirsch-Kreinsen, H. (2018). Einleitung Digitalisierung industrieller Arbeit. In H. Hirsch-Kreinsen, P. Ittermann, & J. Niehaus (Hrsg.), *Digitalisierung industrieller Arbeit* (2. Ed., pp. 13–32). Nomos.

Ittermann, P., & Niehaus, P. (2018). Industrie 4.0 und Wandel der Arbeit – revisited, Forschungsstand und Trendbestimmungen. In H. Hirsch-Kreinsen, P. Ittermann, & J. Niehaus (Ed.), *Digitalisierung industrieller Arbeit* (2. Ed., pp. 33–60). Nomos.

Ittermann, P., & Falkenberg, J. (2019). Funktionsweisen digitaler Technologien und Szenarien digitaler Einfacharbeit. In H. Hirsch-Kreinsen, P. Ittermann, & J. Falkenberg (Ed.), *Szenarien digitaler Einfacharbeit* (pp. 37–67). Nomos.

Kratzer, N., Menz, W., Tulius, K., & Wolf, H. (2015). *Legitimationsprobleme der Erwerbsarbeit.* edition sigma.

Lott, Y. (2019). *Weniger Arbeit, mehr Freizeit – Wenn Mütter und Väter flexible Arbeitsarrangements nutzen.* WSI-Report Nr. 47.

Pfeiffer, S., & Suphan, A. (2018). Industrie 4.0 und Erfahrung – das unterschätzte Innovations- und Gestaltungspotenzial der Beschäftigten im Maschinen- und Automobilbau. In H. Hirsch-Kreinsen, P. Ittermann, & J. Niehaus (Ed.), *Digitalisierung industrieller Arbeit* (2. Ed., pp. 275–301). Nomos.

Schwarz-Kocher, M., Kirner, E., Dispen, J., Jäger, A., Richter, U., Seibold, B., & Weifloch, U. (2011). *Interessenvertretungen im Innovationsprozess.* edition sigma.

Widuckel, W. (2015). Arbeitskultur 2020 – Herausforderungen für die Zukunft der Arbeit. In W. Widuckel, K. de Molina, M. J. Ringlstetter, & D. Frey (Ed.), *Arbeitskultur 2020* (pp. 27–44). SpringerGabler.

Widuckel, W. (2018). Kompetent führen in der Transformation. In K. de Molina, S. Kaiser, & W. Widuckel (Ed.), *Kompetenzen der Zukunft – Arbeit 2030, Freiburg* (pp. 205–234). Stuttgart, Haufe.

Widuckel, W. (2020). Arbeit 4.0 und Transformation der Mitbestimmung. In V. Bader & S. Kaiser (Hrsg.), *Arbeit in der Data Society – Zukunftsvisionen für Mitbestimmung und Personalmanagement* (pp. 17–34). Wiesbaden.

Opportunities of AI for Work Design in the Manufacturing Industry

Challenges and Potentials Using the Example of the Metal and Electrical Industry

Tim Jeske and Sebastian Terstegen

1 Introduction

Artificial intelligence is associated with as diverse associations as opportunities. The opportunities of AI arise in all areas of life due to advancing or advanced digitalisation and an associated availability of data and information. These can be evaluated and used with very different software tools. The basic mechanisms of these software tools are often borrowed from or designed to imitate natural processes (e.g. artificial neural networks). From the multitude of opportunities, arise for work design. This initially concerns support in dealing with large amounts of data and in identifying structures and relationships in these data. Similarly, AI-based assistance systems can be developed and used for the entire spectrum of human work activities—from mental or creative activities in development to assembly activities in production to physical activities in transport and provision of components and workpieces. The range of potential applications of AI in the world of work allows an impression of the potentials associated with their

T. Jeske (✉)
ifaa - Institut für angewandte Arbeitswissenschaft, Düsseldorf, Germany
e-mail: t.jeske@ifaa-mail.de
URL: https://www.arbeitswissenschaft.net/jeske

S. Terstegen
ifaa - Institut für angewandte Arbeitswissenschaft, Düsseldorf, Germany
e-mail: s.terstegen@ifaa-mail.de
URL: https://www.arbeitswissenschaft.net/terstegen

I. Knappertsbusch and K. Gondlach (eds.), *Work and AI 2030*,
https://doi.org/10.1007/978-3-658-40232-7_23

industrial use. Thus, significant productivity gains and corresponding increases in efficiency and competitiveness of companies are expected. Likewise, AI is associated with significant contributions to dealing with skills shortage and climate change as well as securing jobs and prosperity. AI is of correspondingly great importance for Germany's largest industrial sector, the metal and electrical industry with its around four million employees and an annual turnover of over 1000 billion euros (Körtel et al., 2020).

2 Status Quo/Inventory and Case Studies

The use of artificial intelligence can currently be interpreted as the highest level of development of digitalisation (see Fig. 1). Accordingly, information about the states and positions of machines, objects or workpieces must be made available as well as information about employees in the respective area of consideration or work system. The further stages of development are based on this foundation. Thus, available information can be used rule-based for interactions; for example, work tables can be automatically adjusted in height depending on the people working on them and the respective work task. Finally, artificial intelligence requires the two stages of development information and interaction to contribute to decision-making and to control objects and machines.

The stages of development of digitalisation show how information and information flows can be made available and used step by step. In some cases, this is also associated with automation, which means that information flows are faster. If inefficient processes are automated, they are executed faster, so that their inefficiency increases. Therefore, the information flows and the underlying processes must be designed according to the principles of lean management, lean and as waste-free as possible.

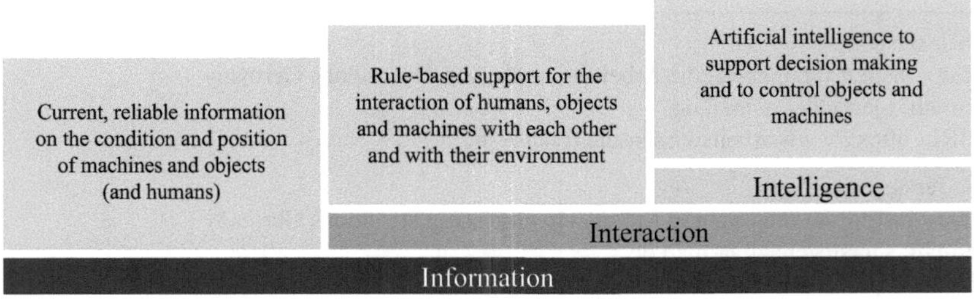

Fig. 1 Typical stages of development of digitalisation. Own representation based on Fraunhofer IAO (2015)

Survey results from the German metal and electrical industry from the years 2015 (ifaa 2015) and 2019 (Jeske et al., 2020) show that the methods of lean management are seen by more than 70% of the respondents as a requirement for the introduction/use of digitalisation measures—with an increasing tendency (see Fig. 2). Accordingly, these methods are already used by a large and increasing majority of respondents (around 85% in 2015 and around 95% in 2019) and are intended for future use with even greater and also increasing consent (around 95% in 2015 and around 100% in 2019).

Results from the surveys mentioned from the years 2015 and 2019 (ifaa 2015, Jeske et al., 2020) also show how the importance of digitalisation has developed across all business areas (see Fig. 3). Thus, in 2015, almost 70% of the respondents rated the importance as high or very high; in 2019, this share was larger, more than 80% of the respondents came to this estimation.

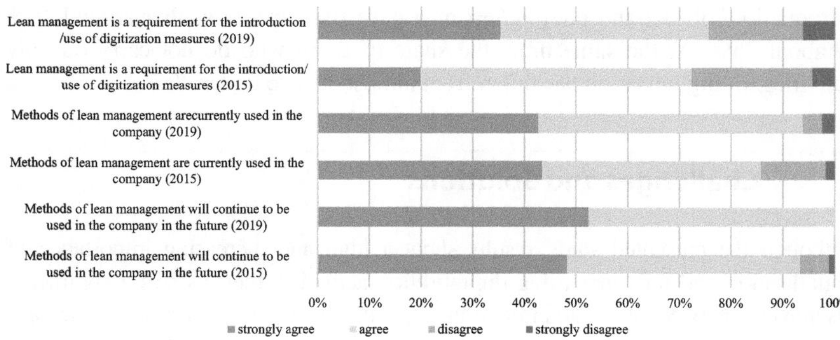

Fig. 2 Effects of digitalisation on methods of lean management (2015: n= 282–288; 2019: n = 82; ifaa 2015, Jeske et al., 2020)

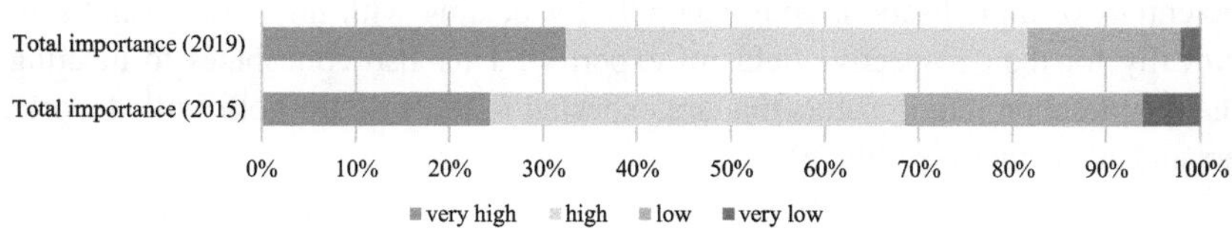

Fig. 3 Importance of digitalisation across all business areas (2015: n= 349–370; 2019: n= 79–89; Jeske et al., 2021)

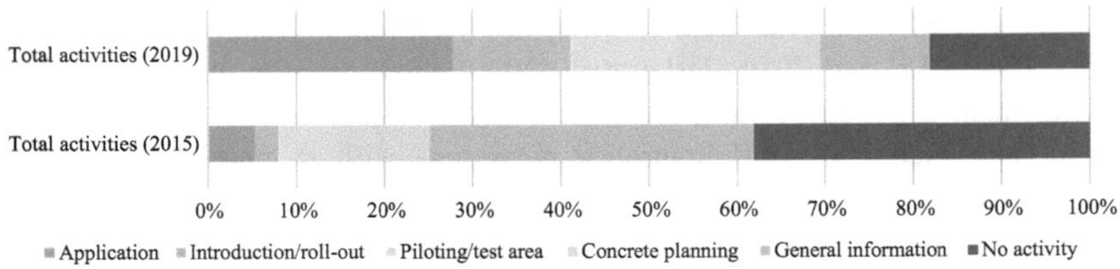

Fig. 4 Activities to implement digitalisation measures across all business areas (2015: n=286–309; 2019: n= 62–73; Jeske et al., 2021)

The higher estimated importance of digitalisation is evident from the activities for the implementation of digitalisation measures. These have increased significantly between the years 2015 and 2019 (see Fig. 4). Thus, the share of digitalisation measures already in use has risen from under 10% to almost 30%, introduction/roll-out and piloting/test area have risen from together around 10% to about 25%. At the same time, the share of those who do not carry out any activities for digitalisation has fallen from almost 40% to under 20%.

3 Challenges and Solutions

Although the presented study results show a high and increasing importance of digitalisation and thus increasing digitalisation activities, there is still a significant discrepancy between the stated importance and the level of implementation. Accordingly, there is great potential to improve the technical prerequisites for the use of AI by further advancing the implementation of digitalisation measures—starting with the design of lean processes according to the principles of lean management.

In this context, qualification measures can help to show the employees the potentials of digitalisation, both generally for dealing with information and specifically for their respective fields of expertise. This also contributes to meeting the qualification requirements that are expected to increase for both skilled workers and academics (see Fig. 5).

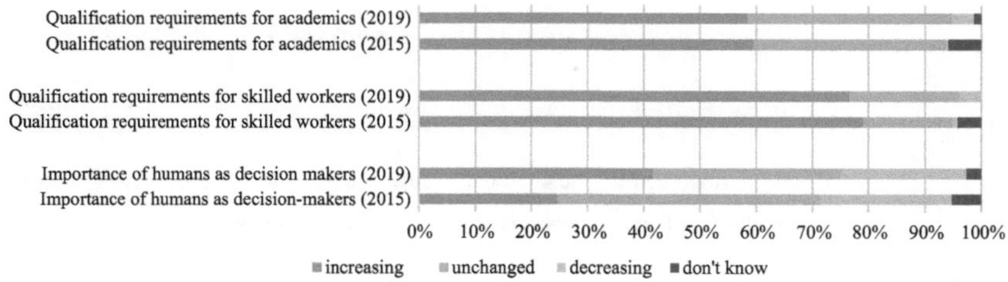

Fig. 5 Effects of Industry 4.0 or digitalisation on employees (2015: n= 305–310; 2019: n= 77; ifaa, 2015; Jeske et al., 2020)

Figure 5 also shows that due to digitalisation, an increasing importance of humans as decision-makers is expected. In this context, changes for managers also have to be considered. For example, 80% of the respondents of an ifaa study stated that the requirements for managers have changed due to the introduction of digital technologies (Jeske et al., 2020). In line with this, 69 % of the respondents in the same study stated that they wanted to change the leadership culture in their company. For this purpose, measures can be used that change the leadership understanding in such a way that a trustful and open corporate culture is created that enables changes (Frost et al., 2019). Thus, qualification measures for managers can additionally support the digitalisation and thus the use of AI.

4 Outlook on AI in the Manufacturing Industry in 2030

Since the use of AI requires the previously described steps of digitalisation, the expectations associated with digitalisation can be used as an orientation for the further development of digitalisation using AI.

For example, survey results from 2019 largely expect that shift work will decrease due to digitalisation—both in terms of night shifts and weekend shifts (Jeske et al., 2020, see Fig. 6). This is advantageous for both employees and companies, because strains (ergonomic and monetary) decrease and consequently increasing performance and competitiveness can be expected. These expectations can be further supported by the use of AI and the associated opportunities for process monitoring and control.

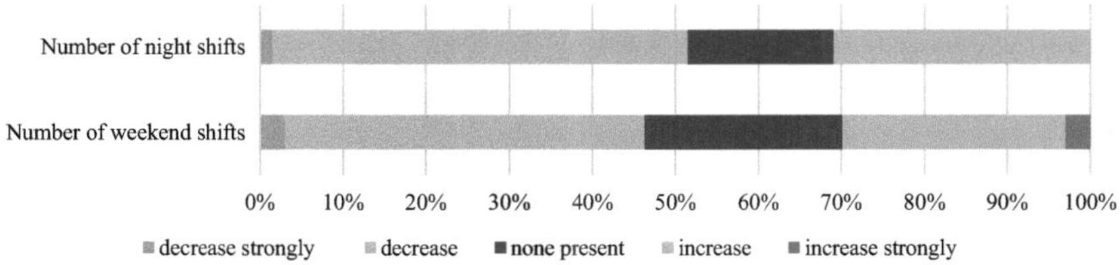

Fig. 6 Effects of digitalisation on the number of weekend/night shifts (n= 67–68, Jeske et al., 2020)

Survey results on the effects of digitalisation on spatial, temporal and content-related/professional flexibility could be compared between the years 2015 and 2019 (ifaa, 2015; Jeske et al., 2020). In all three dimensions of flexibility, the expectations of the respondents regarding the opportunities for employees have increased, while the expectations regarding the requirements for their flexibility have decreased (Würfels & Jeske, 2021, see Fig. 7). Here, too, it can be assumed that an increasing use of AI will lead to a continuation of this trend.

Further potentials of digitalisation and AI for shaping the future of work are systematically described e.g. in Terstegen & Jeske, 2021.

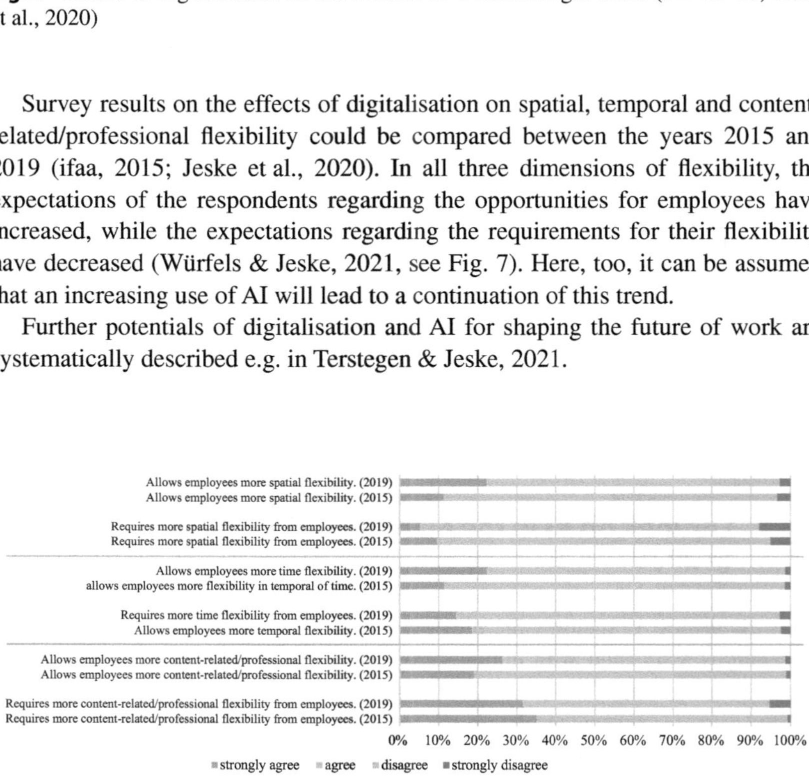

Fig. 7 Effects of digitalisation on the flexibility of employees (2015: n= 296–305; 2019: n= 76; ifaa 2015; ifaa 2015, Jeske et al., 2020)

5 Summary and Practical Recommendations

The opportunities that arise from digitalisation and the use of artificial intelligence for shaping the world of work are manifold. The use of AI requires both, processes designed according to lean management principles and a sufficient degree of digitalisation.

Study results show that the importance of digitalisation from the perspective of companies in the German metal and electrical industry has increased significantly in recent years; there is also progress in implementation. However, the implementation does not yet correspond to the expressed importance, so there is still a need for action. In this context, qualification needs also have to be addressed and the leadership and corporate culture have to be further developed. In this way, the opportunities of digitalisation and AI can be gradually tapped, e.g. from a reduction of shift work and a higher flexibility to the associated strengthening of performance and competitiveness.

Due to the high dynamics of development of digitalisation and AI, it is advisable to establish and culturally anchor methods for continuous improvement (CIP) in the company on the one hand and to convey the basics of digitalisation and potentials for one's own company on the other hand, so that they can be incorporated into the CIP and applied.

6 Funding Note

This article contains results from the research and development project TransWork, which was funded by the Federal Ministry of Education and Research as part of the funding measure "Work in the Digitalized World" and supervised by the Project Management Agency Karlsruhe (PTKA) (funding code: 02L15A164). The responsibility for the content of this publication lies with the authors.

References

Frost, M., Jeske, T., & Terstegen, S. (2019). Die Zukunft der Arbeit mit Künstlicher Intelligenz gestalten. *ZWF Zeitschrift für wirtschaftlichen Fabrikbetrieb, 114*(6), 359–363.

Fraunhofer IAO. (2015). Innovationsnetzwerk Produktionsarbeit 4.0 des Fraunhofer-Instituts für Arbeitswirtschaft und Organisation IAO. https://www.engineering-produktion.iao.fraunhofer.de/de/innovationsnetzwerke/innovationsnetzwerk-produktionsarbeit-4-0.html. Accessed: 31. July 2021.

ifaa—Institut für angewandte Arbeitswissenschaft (2015). ifaa-Studie: Industrie 4.0 in der Metall- und Elektroindustrie. ifaa, Düsseldorf. www.arbeitswissenschaft.net/Studie_Digitalisierung_2015. Accessed: 15. June 2021

Jeske T, Eisele O, Würfels M, ifaa—Institut für angewandte Arbeitswissenschaft (Hrsg.) (2021). ifaa-Broschüre: Produktivität steigern—Erfolgreich mit Digitalisierung und Produktivitätsmanagement 4.0. ifaa, Düsseldorf. www.arbeitswissenschaft.net/Broschuere_PM40.

Jeske T, Würfels M, Frost M, Lennings F, ifaa—Institut für angewandte Arbeitswissenschaft (Hrsg.) (2020). ifaa-Studie: Produktivitätsstrategien im Wandel—Digitalisierung in der deutschen Wirtschaft. ifaa, Düsseldorf. www.arbeitswissenschaft.net/Studie_Digitalisierung_2019. Accessed: 15. June 2021.

Körtelt, B., Stahl, M., Utecht, J., & Gesamtmetall, (Hrsg.). (2020). *Zahlen 2020—Die Metall- und Elektro-Industrie in der Bundesrepublik Deutschland*. Gesamtmetall.

Terstegen, S., & Jeske, T. (2021). Digitalisierung und Künstliche Intelligenz nutzen—Chancen und Anforderungen der Arbeitsgestaltung. *ASU Arbeitsmed Sozialmed Umweltmed, 56*(1), 12–14.

Würfels, M., & Jeske, T. (2021). Auswirkungen der Digitalisierung auf Unternehmen und Beschäftigte—Analyse aktueller Entwicklungstendenzen. In: GfA (Hrsg.), *Arbeit HumAIne Gestalten. Bericht zum 67. Kongress der Gesellschaft für Arbeitswissenschaft vom 03.—05. März 2021. ISBN 978-3-936804-29-4,* GfA-Press, Dortmund, Beitrag B.12.12.

The Role of Humans in the Context of Sovereign Data Spaces

Data Marketplaces as Enablers of the World of Work 2030

Johannes Mayer, Thomas Bergs, Stefan Sander and Daniel Trauth

1 Introduction

The networking of 500 trillion IoT devices with the internet will lead to a global supply of 1.1666 yottabytes of data by 2030 (https://www.cisco.com/c/en/us/products/collateral/se/internet-of-things/at-a-glance-c45-731471.pdf?dtid=osscdc000283). The exact value of this data base is unknown, but the majority of German executives (75%) believe that the linking of data and analytics with artificial intelligence (AI) is gaining importance. Quantitatively, an increase in

J. Mayer (✉)
Werkzeugmaschinenlabor WZL, RWTH Aachen University, Aachen, Germany
e-mail: J.Mayer@wzl.rwth-aachen.de

T. Bergs
Werkzeugmaschinenlabor WZL, Fraunhofer-Institut für Produktionstechnologie IPT, RWTH Aachen University, Aachen, Germany
e-mail: t.bergs@wzl.rwth-aachen.de

S. Sander
IT-Recht, SDS Rechtsanwälte Sander Schöning PartG mbB, Duisburg, Deutschland
e-mail: sander@sds.ruhr

D. Trauth
senseering GmbH, Köln, Deutschland
e-mail: d.trauth@senseering.de

I. Knappertsbusch and K. Gondlach (eds.), *Work and AI 2030*,
https://doi.org/10.1007/978-3-658-40232-7_24

productivity, streamlining of processes and creation of customer experiences are expected, qualitatively, a market capitalization of about 5.53 trillion euros is forecasted (Betti et al., 2021).

To exploit the potentials of advanced data analysis using artificial intelligence (AI), a shift from the classical, repetitive monitoring and adjustment of a machine to the orchestration of a multitude of autonomously acting plants based on data analysis is required. The fusion of digital capabilities with industry-specific know-how of manufacturing technology will form the basis for decisions in the future and will become a success-critical competence of humans. Contrary to the general opinion that AI would predominantly cannibalize jobs, linking the expertise of data science and mechanical engineering strengthens human abilities, archives historical expert knowledge and increases process efficiency (Brown et al., 2018). Currently, there is a limited availability of this person-related data science expertise (Schürmann, 2017). The stock meets the demand at a maximum of 20–25% and without professional data preparation and analysis, no information can be extracted from collected manufacturing data. Based on current statistics, the shortage of data analysts in Germany is projected to reach 1.1 million professionals by 2030 (Strack et al., 2021).

So far, despite the recognition of the importance of linking data and AI, only 39% of companies in the manufacturing industry have been able to scale data-driven use cases beyond a single value stream. More demanding applications, such as predicting failures and their causes, are based on AI models that have been trained on a large amount of data. Effective training requires the exchange of data across company boundaries, as a single company can rarely provide such amounts of data. Collaboration and data exchange in hyper-connected alliances form the basis for the working world of 2030 and solve the challenge of lack of data and AI expertise within a so-called data economy (Trauth et al., 2020a).

2 Status Quo/Inventory: Data Markets in the Data Economy

The data economy describes the complete process from data collection and preparation to analysis and data use and marketing. Data exchange via an open infrastructure is the link between the individual process steps of the data economy and enables new digital business models. The networking between companies required for the data economy is currently still low. Only 49.3% of companies are digitally networked across companies, of which only 16% with at least two collaborating companies (Icks et al., 2017). If such networking exists, the required capacities, if available, are bound up to 80% of their time to prepare the data and ensure its quality (Bowne-Anderson, 2018).

To solve the challenge of lack of AI applications and insufficient networking, the World Economic Forum (WEF) proposes the implementation of an openly accessible platform (Betti et al., 2021). While maintaining data security, this platform is intended to dissolve internal data silos and release them for bidirectional exchange. Data markets are gaining increasing attention in this context for cross-industry trade in data and the networking of different actors (Trauth et al., 2020b). The main actors of a data market form a trust alliance and can be classified into platform operators, data providers and data users (Fig. 1).

A data market offers, on the one hand, today's companies without their own person-related AI expertise access to advanced analysis results. On the other hand, the employee of tomorrow (2030) gets access to a sufficiently large data base for their own AI analyses and effectively trained analysis models, which strengthen their abilities and archive their expertise.

A central aspect of the willingness to share IoT data and services is the trust in the mechanisms of secure data trade, which protect against the loss of know-how and competitive advantages. This challenge can be met by choosing a networked infrastructure that enables interoperability and portability of IoT data within and across domains, as well as guarantees data sovereignty and availability (Trauth et al., 2021). The GAIA-X initiative enables such a sovereign and self-determined data and service trade. The resulting digital ecosystem is characterized by openness and transparency, in which IoT data and services can be made available, merged, securely and trustfully shared and used.

An infrastructure supplemented by distributed ledger technologies (DLT) is predestined for applications where trust is important (Trauth et al., 2020a). The replacement of intermediaries enables tamper-proof digital transactions and the documentation of ownership rights. In combination with smart contracts, decentralization offers accelerated data exchange (on demand in real time) as well as process automation and optimization. The main features of DLT are the geographically decentralized distribution of data sets within the network. DLT create an identical data base among the actors and prevent unnoticed data manipulation by increasing transparency and traceability of transactions. For the scaling of an IoT data space for a data market, infrastructural and architectural issues need to be clarified and realized. For this reason, the target architecture of a market place is examined in more detail below.

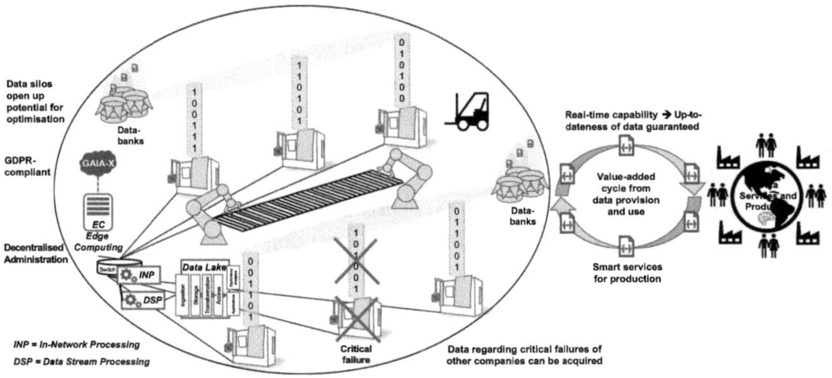

Fig. 1 Data marketplace in the manufacturing industry

3 Challenges and Solutions

3.1 Architecture of a Data Marketplace in an IoT Data Space

An IoT data space for a data marketplace should enforce a data management solution where each participant owns one or more nodes in a decentralized network (see Fig. 2). It is up to each participant whether their own nodes are executed locally on site (on premise) or in a cloud subscription. Data can be exchanged between these nodes with other participants. Each node is realized with a data management software for storing, visualizing and analyzing the data. The manager manages the data sources, which can connect to a data node with an IoT connector. A data exchange module, which is anchored in the node, enables a cross-node exchange of data.

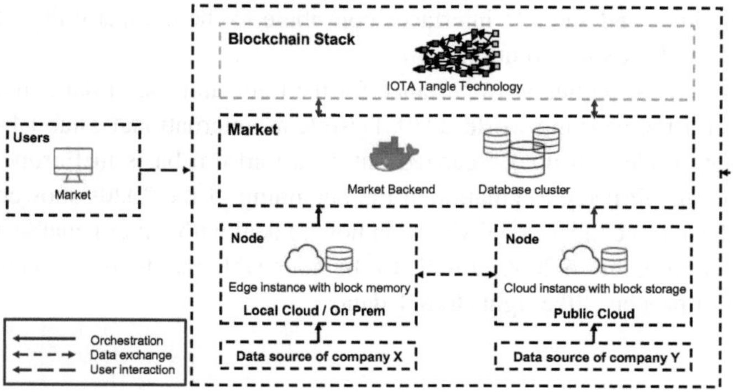

Fig. 2 Infrastructure of a decentralized data marketplace

The management task is taken over by a central cloud catalog. The nodes of a user are easily coordinated with this and flexibly combined to a meaningful conception. The basis for this is the registration of each node in the catalog and the transmission of the relevant metadata as well as the contained data sources and structure of the data. The data itself is thus never passed on to the cloud catalog for reasons of data sovereignty. Comparable to Google, each network participant can use a search mask in the catalog to locate external and internal data sources. The data of these sources are moved to one of their own nodes via a peer-to-peer connection between the nodes using the data exchange module. A decentralized data trade is maintained despite the central catalog instance, as it only mediates the connection between the nodes, but not the exchange of data. A blockchain stack creates the necessary trust regarding potential subsequent manipulations or deviations from offered data sets within the decentralized network. It lays the foundation to generate the critical mass of participants. Functionally, it guarantees via decentralization and digital identity the integrity of all IoT data points in the marketplace, offers a media-free payment instrument with the marketplace's own token and enables the automation of exchange, trade and service processes using smart contracts in real time.

For maximum scalability, it is crucial that a sovereign, functional IoT data space provides an open source pipeline to all data providers for the interoperable and secure data utilization along value creation networks. With different company sizes, the complexity of usage rights varies, especially in dealing with the data, so that a user, role and policy management is to be stored in the cloud catalog. Thus, any company or team structures with different access rights can be accurately digitally replicated in the data space. Integration possibilities for existing data exchange modules and suitable interfaces contribute to the compatibility of the IoT data space with existing solutions.

The legal context is of particular importance for the trade and use of data on a data marketplace for the working world 2030. Despite an international establishment by 2030, data trade will not be carried out on a uniform basis in Europe. The legislation on the ePrivacy Regulation at the beginning of the 2020s showed that many states have too controversial views on how to deal with data to enable a closed political action. Even in 2030, it will still be noticeable that there is a lack of an assignment of property-like rights to IoT data.

3.2 Approaches to Trading IoT Data and Services

The harmonized legal protection for trade secrets in Europe seemed at first glance to contradict the vision of data trading. Since the directive (EU) 2016/943 on the protection of confidential know-how, which was adopted for harmonization, had defined its scope of application in such a way that only information subject to actual protection measures was legally protected, information security management systems were introduced. Their basis is the classification of the existing information. Companies became aware that they have information that is not in need of protection and thus tradable in an IoT data space. The scope of application of Directive (EU) 2019/1024 even explicitly referred to access to information as a fundamental right enshrined in the EU Charter. The goal proclaimed by the EU Commission to promote "trustworthy data altruism" with so-called "data donations" created incentives to provide data from countless contexts without a predefined purpose, e.g. for AI.

Data trading in an IoT data space for data-driven value creation is subject to different design options with regard to the legal good traded and the time horizon of use. Marketplace participants could, for example, acquire either a) **the content-defined right of use for data,** b) the **data set itself** or c) a **copy of the data** for i) **permanent use,** ii) **permanent possession** or iii) a **time-limited use**. Depending on the design of the data trade, the legal bases to be considered vary and influence the legal relationships of a person in the working world 2030. The handling of contractual services and the compliance with contractual conditions, the possession of as well as the exploitation and usage rights to databases determine the legal framework.

From 2019 onwards, the legal science, as a result of the unlimited reproducibility of data, instead of "data ownership" pushed for "data possession" to map the access rights to data (Adam, 2020). However, this debate was overtaken by technical possibilities. An exclusive use of the acquired data is realized, for example, by limiting the usability of the data to the system environment of the marketplace without a possibility of download. Such techniques are used, for example, where it is important to balance data provision and use on the marketplace. If the acquired IoT data can only be used on the marketplace, this implies, conversely, that the user must integrate his own data into the marketplace if he wants to combine the data sets in an AI model. Depending on the configuration of the marketplace, the own IoT data can also be released for commercialization or only the model can be made available for download.

Providing a **copy of the IoT data** in the sense of a duplication as a legal good for trading will remain a relevant question or precondition for niche products. Data trading will establish itself to commercialize the procurement effort and scale revenues by being able to duplicate data almost free of charge (marginal cost theory). In the context of potentially acquirable **rights of use to IoT data,** whose implementation is ensured by technology (e.g. digital rights management (DRM)), the following must be clarified: The exclusive right of use would be suitable for the data marketplace and defines in the form of a contract the exclusive use of the IoT data within the defined content boundaries. If new types of data use occur subsequently, which were not foreseeable at the time of the contract agreement, the data provider should obtain the right to an appropriate remuneration. In the context of the still low maturity level of AI-supported data use, this protects the provider from under-calculated prices.

Despite all the euphoria regarding the potentials of AI, the data protection laws with their scope of application oriented towards personal data still have a braking effect in 2030. The working world in 2030 has become transparent in terms of the means of production due to omnipresent IoT devices. However, the employee is still a legal subject, protected by the legal order and, despite all digitalization, has not been degraded to a means of operation. Machines can act autonomously and make independent decisions such as buying data sets in the age of Industry 4.0, but with regard to the machine operator as a natural person, the following still applies in 2030: Data sets with timestamps can allow a conclusion about the machine operator and are therefore possibly subject to the restrictions of data protection law, so that their suitability for data trading is limited.

4 Outlook on the Role of Humans in the Context of Sovereign Data Spaces in 2030

A data marketplace can be described as an enabler of the working world 2030, as it revolutionizes the actions of every human being, regardless of the industry, in terms of dealing with AI. IoT data from various stakeholders of different industries are publicly available to everyone to enable, enrich and optimize AI models with specific IoT data. The implementation and use of a data marketplace already favor today the fast access to AI for companies without their own, person-bound expertise. The lack of a uniform, legal framework for data trading leads to dampened prices on data marketplaces. Higher prices would be achievable if a "poor quality" of the traded goods would lead to claims of the user against the provider of the data afterwards. However, since the found solution is limited to the factual moment of knowledge of the data and the normative framework of such transactions is missing, "rights of the user due to a defect" are only present where the respective contractual parties agree on such in their autonomy.

References

Adam, S. (2020). *Daten als Rechtsobjekte* (2063–2068). NJW 2020.

Betti, F., Bezamat, F., Bloempott, S., & Fendri, M. (2021). *Data Excellence: Transforming manufacturing and supply systems.* World Economic Forum.

Bowne-Anderson, H. (2018). What Data Scientists Really Do, According to 35 Data Scientists. https://hbr.org/2018/08/what-data-scientists-really-do-according-to-35-data-scientists. Accessed: 23. June 2021.

Brown, J., Gosling, T., Sethi, B., Sheppard, B., Stubbings, C., Sviokla, J., Williams, J., Zarubina, D., & Fisher, L. (2018). *Workforce of the future: The competing forces shaping 2030.* PricewaterhouseCoopers GmbH.

https://www.cisco.com/c/en/us/products/collateral/se/internet-of-things/at-a-glance-c45-731471.pdf?dtid=osscdc000283. Accessed: 23. June 2021.

Icks, A., Schröder, C., Brink, S., Dienes, C., & Schneck, S. (2017). Digitalisierungsprozesse von KMU im Produzierenden Gewerbe. https://www.econstor.eu/bitstream/10419/156246/1/882667238.pdf. Accessed: 23. June 2021.

Schürmann, H. (2017). Datenanalysten sind rar. https://www.vdi-nachrichten.com/karriere/datenanalysten-sind-rar/. Accessed: 23. June 2021.

Strack, R., Carrasco, M., Kolo, P., Nouri, N., Priddis, M., & George, R. (2021). *The Future of Jobs in the Era of AI.* Boston Consulting Group

Trauth, D., Bergs, T., Gülpen, C., Maaß, W., Mayer, J., Musa, H., Niemietz, P., Rohnfelder, A., Schaltegger, M., Seutter, S., Starke, J., Szych, E., & Unterberg, M. (2020). INTERNET OF PRODUCTION—TURNING DATA INTO VALUE—Monetarisierung von Fertigungsdaten, RWTH Aachen. https://doi.org/10.24406/ipt-n-589615.

Trauth, D., Niemietz, P., Mayer, J., Beckers, A., Prinz, W., Williams, R., & Bergs, T. (2020). *Distributed Ledger Technologien im Rheinischen Revier in Nordrhein-Westfalen.* RWTH Aachen. https://doi.org/10.31224/osf.io/5mdw6.

Trauth, D., Bergs, T., & Prinz, W. (2023). *The Monetization of Technical Data.* Springer.

AI in the Crafts

Opportunities and Challenges

Philipp Hartmann

1 Introduction

More than one in ten employees subject to social security contributions in Germany worked in a craft occupation[1] (Federal Statistical Office, 2020, p. 5). Crafts are one of the most important sectors of the German economy and occupy a fixed place in numerous economic cycles. In rural areas, crafts secure the employment and supply of the local population and form an essential pillar in vocational education: 28% of all trainees in Germany practice a craft occupation (Central Association of German Crafts, 2020a, b).

Accordingly, questions about the influence of artificial intelligence (AI) on the world of work in crafts are gaining in importance. Unlike in many areas of industrial production and services, there is hardly any discussion of this question so far. One reason for this may result from the heterogeneity of the sector "crafts". It comprises more than 130 different occupations; from carpenters to hairdressers, plasterers, bakers and organ builders. Therefore, a uniform view of the potential of AI in crafts is difficult. Nevertheless, crafts are largely united by the provision of an individual, craft service, the structure of the businesses and a number of challenges.

[1] Crafts are defined by the trades listed in the Crafts Code, which belong to the craft requiring a license or the craft not requiring a license (cf. Annex A HwO, Annex B HwO).

P. Hartmann (✉)
AI Strategy, UnternehmerTUM GmbH, München, Germany
e-mail: philipp.hartmann@unternehmertum.de

I. Knappertsbusch and K. Gondlach (eds.), *Work and AI 2030*,
https://doi.org/10.1007/978-3-658-40232-7_25

AI currently plays a subordinate role in crafts. In the future, however, the development of AI will influence many activities and processes in crafts—some more and sooner, others later. As electricity did more than 100 years ago, AI will also replace old work steps in crafts. To dress wood, carpenters today usually use an electric planing machine. Only in exceptional cases is the mechanical plane used. Similarly, in the future there will be work steps that are done by an AI. This results in opportunities for crafts on the one hand, and inevitable changes in the occupational profiles and processes on the other.

2 Status Quo: AI in Crafts 2021

The application of artificial intelligence in crafts still sounds like music of the future for many businesses. Specific AI applications in crafts are still waiting, especially in Germany.

And yet AI already influences the working world of most craftsmen—often without the craftsmen being aware of the use of AI: They plan their route through the evening traffic by app, dictate a message by voice recognition into their smartphone or have receipts automatically classified by the accounting programme. All these applications typically involve artificial intelligence.

Specific applications of artificial intelligence in crafts, on the other hand, rarely make it beyond the pilot phase:

- For example, a bakery reported on how it uses AI to calculate the sales forecast. This in turn leads to a more efficient use of resources, which is oriented towards the actual demand (Markel, 2020);
- a heating engineer, how he uses AI to handle repair orders (Guthardt, 2020).

In crafts, AI is not yet the technology of the masses. There are two structural reasons for this, among others: On the one hand, crafts are characterised by a large number of small businesses that can only invest limited resources in experiments with AI. On average, 9.6 people were employed in each craft business in 2018; of these, less than ten active people in four out of five businesses (79.96%), and less than five employees in 59.1%. This is also reflected in the turnover of the companies: 70% of the businesses generate an annual turnover of less than half a million euros. In contrast, 3.2% of the businesses generate a turnover of more than five million euros (Central Association of German Crafts, 2020c). As a consequence, crafts have less innovation capability compared to industry (Scholz, 2009).

On the other hand, craft businesses often lack competence in the field of IT and the digitalisation of processes. However, this is usually the prerequisite for the use of AI methods. Handwritten calculations on slips of paper are not suitable for training a machine learning model.

In many trades, start-ups have recognised the potential of using AI and are developing corresponding solutions, mostly outside Germany. Especially in the construction industry, there are a large number of start-ups that design solutions. Here are some examples:

- The Chinese start-up ViAct uses artificial intelligence to automate construction monitoring and improve safety on the construction site. ViAct automatically detects situations with risk of injury. The solution helps construction companies optimise their processes while increasing worker safety and compliance.
- The American start-up OpenSpace offers a photo- and video-based documentation tool. Photos are automatically taken via a helmet camera and independently linked to project plans. This makes it easier to monitor the construction progress and relieve the requirements for the documentation processes.
- Indus.ai provides an AI analysis platform for the construction industry that counts truck and material arrivals and measures equipment productivity in this way. For this purpose, Indus.ai uses live video stream cameras that are evaluated using AI methods.

But also in other areas, start-ups develop AI solutions for partial process steps:

- Enway develops autonomous industrial sweeping machines that allow building cleaners to clean large areas without personnel.
- FoodTracks offers bakeries an AI-based solution to get better order and sales forecasts.
- Setws develops robots that allow automatic handling of laundry in textile cleaning. So far, this work could only be done manually.

3 Challenges and Solutions

The development of artificial intelligence in the next ten years will have a significant impact on the working world of crafts. The crafts sector in Germany faces major challenges in many areas, regardless of the development of AI: First, most craft businesses face a shortage of skilled workers, which has worsened in recent years.

A study by the Competence Centre for Securing Skilled Workers at the Institute of German Economy estimates that 65,000 skilled workers are missing in the crafts sector and another 12,000 skilled workers in occupations with a craft component. To close this gap, many businesses increasingly rely on unskilled workers, often labour migrants from Eastern and South-eastern Europe (Hickmann et al., 2021).

Second, technological progress leads to increasing demands on the activities of most craft occupations. In many areas, the systems installed and maintained by craftsmen become increasingly digital and complex; software plays an increasingly important part. Examples include the development of vehicles, heating systems or the networking of home technology.

Third, the crafts sector is constantly under competitive pressure from industrial forms of production. This development began with the advent of industrialisation in the 19th century: While 100 years ago furniture was made by carpenters, individual production today only accounts for a small part in a high-priced segment. Accordingly, the focus of many craft activities shifted from production to assembly and maintenance of industrially produced goods. Manual production often takes place only in high-priced areas or where industrial production is not profitable due to individual requirements.

4 Outlook on AI in the Industrial Working World in 2030

The development of artificial intelligence influences these challenges and means both an opportunity and a challenge for the crafts sector:

1. *AI leads to an expansion of automatability and increases competitive pressure from industrial production.*

AI shifts the boundaries of automatability. Through new capabilities that arise from AI technology—perceptive, analytical, motor and generative skills—certain tasks can be automated that were previously reserved for humans. Accordingly, the activities of many craft occupations will change.

Similar to previous waves of technological development, automation through AI technology will lead to a shift towards an "industrial" production and a stronger competition for the crafts sector. By using AI, customer-specific requirements can be much more easily automatically translated into designs. Such tasks typically form the core of craft activities. New manufacturing methods such as additive manufacturing enable the economic production of very small quantities.

This also often eliminates process steps. For example, a house that is created by 3D printing only requires the presence of two technicians—significantly fewer workers than on traditional construction sites.

2. *Nevertheless, there are a variety of craft activities that cannot be (economically) automated using AI in the next ten years.*

Despite the technological development, AI will not be able to economically automate many activities in the crafts sector and will continue to require craft work. This is explained by the so-called *Moravec's paradox:* Methods of artificial intelligence can usually solve problems that are difficult for humans, such as playing Go or detecting tumours on X-rays. However, tasks that are easy for humans often pose a serious challenge for AIs[2]. Most craft activities fall into the second group of tasks and require the sensor-motor intelligence of humans. We hardly notice it in everyday life, but our environment consists of complex, chaotic and dynamic elements: Components deviate from the target dimensions, building materials behave differently than expected or unexpected obstacles arise. Craftsmen solve all these problems in their daily work, for example with a few targeted hammer blows or the use of construction foam. AI systems, however, face almost insurmountable challenges with such tasks.

In addition, the effort for automation in many areas is higher than the labour costs, despite foreseeable technological advances. For example, cleaning an office can be done by a low-skilled worker. The development of an automated system with AI and robotics, on the other hand, would probably be prohibitively expensive, if possible at all.

3. *With AI, craft businesses can increase their productivity—AI and humans will work hand in hand.*

Thus, AI is rather to be seen as an opportunity for the crafts sector, with which the challenges outlined above can be overcome, instead of an immediate threat to

[2]This was observed, among others, by Hans Moravec in the 1980s: "It is comparatively easy to make computers perform at adult level on intelligence tests or playing checkers, and difficult or impossible to give them the skills of a one-year-old in terms of perception and mobility", (Moravec 1988, p. 29). Although the performance of AI systems has increased radically in the last 30 years, this fundamental contradiction has not changed fundamentally.

the core of most activities. AI will take over various work steps as a new tool that are currently part of the craft. In particular, there is the possibility to increase the productivity of the craft activity by AI in the following areas:

- **Perform standardisable and repetitive activities**
  Individual tasks within a trade can be automated with AI methods in the future. These include mainly repetitive and standardised activities. For example, painting large areas can be done by painting robots and cleaning large areas by autonomous cleaning vehicles. The human takes care of complex areas as before.
- **Plan implementation and required parts**
  By using AI, the implementation can be planned and optimised automatically. Typically, most orders in the crafts sector are planned today based on experience by the master craftsman. AI-based systems can develop optimised assembly plans, make suggestions for the best cut and provide material forecasts independently of the competence of individual employees.
- **Monitor implementation/quality control**
  Artificial intelligence also simplifies the monitoring of implemented work. AI-based tools can detect possible collisions, delays and changes. On a construction site, for example, images taken by drones can be automatically analysed by an AI and thus detect possible errors or deviations from the original construction plan during the construction phase.
- **Administrative activities**
  Many tedious work steps such as creating quotes or parts of accounting can be done by AI-supported programmes. This reduces the workload of the employees in the long term.

4. *AI can help the crafts sector to cope with the shortage of skilled workers and the increasing technological requirements.*
   By AI taking over planning and monitoring activities, craft businesses can use unskilled workers much more flexibly. In this way, they can better instruct and initiate the work steps, and monitor the implementation automatically. Thus, AI can help to address the essential challenges in the crafts sector.
5. *AI leads to a further concentration in large businesses in the crafts sector.*

Although the core of many craft activities does not change by AI, its use will influence the structure of the industry. The systematic use of AI requires an active engagement with the technology and an investment in the AI tools as well as the necessary IT infrastructure. As described above, a large number of small businesses currently characterise the crafts sector in Germany. These cannot afford the investments necessary for progress for the most part. Accordingly, AI will reinforce the concentration in larger businesses that has long been underway (see Müller, 2013). Only larger businesses will be able to provide the necessary investments and resources for the use of AI. Also, an improved productivity by AI will give businesses a competitive advantage that can use it.

5 Summary and Practical Recommendations

The use of AI opens up the opportunity for the crafts sector to focus on the core of its activities, increase productivity and thus meet the challenges of the shortage of skilled workers with increasing requirements.

To exploit this potential, businesses first need an understanding of AI and the possible applications in the crafts sector. This prerequisite is already reflected on the part of the crafts sector: In a survey by the Central Association of German Crafts, 76% of the surveyed businesses expressed a wish for "increased education on AI" in order to realise a stronger use of AI (Central Association of German Crafts 2020a). Meanwhile, there are a variety of qualitative online courses that allow people without AI knowledge to get started in the topic.[3] In addition, AI should be part of the curriculum in every vocational training.

The craft businesses, in turn, should actively deal with AI solutions that can help them in their trade. They do not have to—and should not—develop and operate AI solutions independently in order to use them. Many start-ups develop solutions that businesses should be open to. In this way, AI can expand the possibilities as an additional tool, increase productivity and increase the attractiveness of certain activities.

The downsides should not be overlooked, however: AI methods will continue to automate craft activities and shift the balance in favour of fewer large businesses. This development cannot be stopped. However, if craft businesses actively use the opportunity, they not only prepare their businesses for the future, but also help to inspire newcomers to the crafts sector.

[3] For example, the free online course "The Elements of AI", which is offered jointly by the DIHK and appliedAI. To be found at https://www.elementsofai.de/.

References

Guthardt, S. (2020). Künstliche Intelligenz entlastet Handwerker. Deutsche Handwerks Zeitung. https://www.deutsche-handwerks-zeitung.de/kuenstliche-intelligenz-entlastet-handwerker-149270. Accessed: 24. Juni 2021.

Hickmann. H., Malin, L., Schirmer, S., & Werner, D. (2021) Fachkräfteengpässe in Unternehmen—Fachkräftemangel und Nachwuchsqualifizierung im Handwerk. Institut der deutschen Wirtschaft Köln e. V.: Köln. https://www.iwkoeln.de/fileadmin/user_upload/Studien/Kofa_kompakt/2021/KOFA_Studie_Handwerk_05_dk_komprimiert.pdf. Accessed: 24. June 2021; 29. July 2021.

Märkel, C. (2020) Künstliche Intelligenz: Wie Betriebe die schlauen Helfer für sich arbeiten lassen. handwerk magazin. https://www.handwerk-magazin.de/kuenstliche-intelligenz-wie-betriebe-die-schlauen-helfer-fuer-sich-arbeiten-lassen/150/4/407897. Accessed: 24. June 2021.

Moravec, H. (1988). *Mind Children*. Harvard University Press.

Müller, K. (2013). Strukturentwicklungen im Handwerk. In: Wirtschaftsdienst 93. Springer. pp. 636–642. https://doi.org/10.1007/s10273-013-1576-3. Accessed: 24. June 2021.

Scholz, J. (2009). Regionale Strukturpolitik am Beispiel Trier und Luxemburg: Entwicklung von Methoden, Instrumenten, Referenzprozessen und politischen Handlungsempfehlungen zur Förderung des Technologie- und Innovationstransfers im Handwerk. *Verwaltung & Management—Zeitschrift für allgemeine Verwaltung, 15*(3), 163–167.

Statistisches Bundesamt. (2020). *Produzierendes Gewerbe. Unternehmen, tätige Personen und Umsatz im Handwerk—Fachserie 4 Reihe 7.2.* Destatis.

Zentralverband Deutsches Handwerk. (2020a). Blitz-Umfrage: Künstliche Intelligenz und ihre Potenziale im Handwerk. https://www.zdh.de/fachbereiche/wirtschaft-energie-umwelt/konjunktur-umfragen/blitz-umfragen/blitz-umfrage-kuenstliche-intelligenz-und-ihre-potenziale-im-handwerk. Accessed: 24. June 2021.

Zentralverband Deutsches Handwerk. (2020b). Kennzahlen des Handwerks: Wirtschaftlicher Stellenwert des Handwerks 2020. https://www.zdh.de/daten-fakten/kennzahlen-des-handwerks. Accessed: 24. June 2021.

Zentralverband Deutsches Handwerk. (2020c). Handwerkszählung 2018. https://www.zdh.de/ueber-uns/fachbereich-wirtschaft-energie-umwelt/statistik/handwerkszaehlung/handwerkszaehlung-2018/. Accessed: 31. July 2021.

AI in the mobile world of work and logistics

Potentials in the Field of Mobility by Mathematical Methods of AI

Anita Schöbel, Henrike Stephani and Michael Burger

"What really counts is intuition" (Albert Einstein)

1 Introduction

AI is already widely used in mobility-related research. However, currently available methods are often not yet applied. For example, in the **quality control of vehicle parts**, there are a variety of automation approaches and image processing methods that could detect anomalies and poor quality, but their use often fails due to the assurance of satisfactory and adaptable image acquisition. **Transport connections and timetables** are still planned manually, as relationships and local peculiarities are not sufficiently taken into account in available standard software. Out of the multitude of possible solutions, it is usually impossible to select the best one and therefore one is satisfied if the timetable only meets the really necessary requirements. The same applies to the development and implementation of algorithms for **(dynamic) traffic control.** Traffic light schedules for complex

A. Schöbel (✉) · H. Stephani · M. Burger
Fraunhofer-Institut für Techno- und Wirtschaftsmathematik ITWM, TU Kaiserslautern, Kaiserslautern, Germany
e-mail: schoebel@mathematik.uni-kl.de

H. Stephani
e-mail: henrike.stephani@itwm.fraunhofer.de

M. Burger
e-mail: michael.burger@itwm.fraunhofer.de

I. Knappertsbusch and K. Gondlach (eds.), *Work and AI 2030*,
https://doi.org/10.1007/978-3-658-40232-7_26

intersections or large roundabouts are set up 'by hand' or are based on experience-based approaches. These, however, often cannot adapt to constantly changing traffic situations.

For this reason, repetitive tasks in the work environment are still widespread. People deal with the evaluation of plans and solutions or inspect components for verification. This type of task will become less. While human competence in responding to the unknown and in deciding novel questions is hard to replace by a computer, AI already outperforms humans in two other areas today: it performs repetitive processes, especially computationally intensive ones, better and faster. Moreover, it is able to establish relationships that are not or hardly visible to humans due to computing power and storage capacity. AI is therefore not intended to replace human intelligence, but to complement it.

In the following, we want to show what effects a further use of AI can have on the work environment.

2 Status Quo/Inventory and Case Studies

Due to the dynamics in the field of digitisation, the discrepancy in the degree of digitisation of different companies is large. At the same time, there is a huge, heterogeneous and sometimes confusing spectrum of AI applications of different maturity. Three application examples are intended to illustrate this spectrum and the resulting potentials.

Application 1: Quality Control in the Production of Vehicle Parts Within the transport equipment production, the safety and reliability requirements are particularly high in aircraft construction. Turbines, so-called "bladed disks" or BLISKs, which are installed in aircraft engines, must be reliably controlled. The components have different geometries and surface structures and low quantities. While human inspectors can proceed flexibly, a fully automatic inspection system must systematically ensure both the complete coverage of the component and adaptability to different types of errors. However, the human inspection process has serious disadvantages. First, it is time-consuming and costly. Above all, however, it is often not reliable, as the repeatability of human inspection cannot be ensured.

Application 2: Planning of Public Transport Services In the field of public transport, passengers should have fast and reliable connections for their daily needs. To design the service accordingly, various planning steps have to be coordinated with each other. For example, the timetable has an impact on the use of buses or trains and this in turn has an impact on the rosters. At the same time, the temporal and spatial wishes for stops, routes and timetables have to be taken into account. The operational implementation also requires great skill. How can one react to delays? Should a connection be held or not? Which train is allowed to enter the track first? For such decisions, experienced transport planners are supported by standard software in individual planning steps, but this is mainly about checking whether elaborately developed solutions are “admissible”, i.e. meet all conditions. The planning itself, however, is again a time-consuming process, in which different expertise produce different solutions.

Application 3: Intelligent Control and Regulation of Traffic Flows Intelligent traffic control systems and infrastructure as well as communication and cooperation between traffic participants and with infrastructure elements are equally important building blocks to create safe and efficient mobility solutions. AI-based algorithms provide the core of the control concepts that will be able to guide traffic flows more efficiently in the future. This will improve goods and passenger logistics and thus have an impact on many production and work processes. In first virtual studies, AI-based traffic light control could already be derived by techniques of *reinforcement learning*, which are superior to traditional, rule-based approaches in certain situations on a complex road network, such as the Opel roundabout in Kaiserslautern (Fig. 1), (Baumgart & Burger, 2021). Currently, however, the development of a control for a complex traffic light system is regularly set up by experts, although it is based on static rules.

In all three cases, the necessary work processes consist of a large number of always identical checks of quality criteria on the one hand or side conditions on the other hand. This type of repetitive work is still usually performed by humans and is error-prone, time-consuming and inferior to the work of computer systems.

Fig. 1 Opel roundabout in Kaiserslautern – map excerpt: OpenStreetMap ©

3 Challenges and Solutions

The desire to use methods of artificial intelligence to perform repetitive tasks by computer or to solve complex planning tasks is therefore great in various application areas.

Application 1: Quality Control in the Production of Vehicle Parts In the quality control of BLISKs, many process steps are already automatable. The verification of the coverage of the image acquisition is possible with the so-called *ray tracing,* (Glassner, 1989) (see Fig. 2). In addition, both the geometry of the component in digitised form and hybrid algorithms that model and train the error knowledge of the inspectors using image processing and AI methods, e.g. deep neural networks, are available (Stephani et al., 2017; Wang, et al., 2018). In Fig. 2 a computer-assisted recording setup is shown, which still has to be created or adapted manually for each geometry. For the creation of these algorithms, AI expertise is needed and for the recording planning a lot of manual interaction. This expertise is usually not available near production. External AI knowledge has to be combined with the production knowledge, which is essential for the correct operation and acceptance of the introduction of innovative solutions into the daily work routine.

Fig. 2 BLISK recording setup

Application 2: Planning of Public Transport Services Also in the field of traffic planning, there are AI-based partial solutions. Some individual planning steps can now be well executed by standard software, see (Borndörfer et al., 2018). These include the circulation planning of the vehicles and the service planning. However, the problem of timetable planning is mathematically difficult - therefore, mainly tools are available that visualise or check timetables for specific requirements. After two automatically generated timetables for the Dutch railway (Kroon et al., 2009) and the Berlin underground (Liebchen, 2008), the enthusiasm to design timetable planning completely automated in the future has been slowed down again due to the increasing demands and networks. In the case of line planning, there are so many different requirements that a uniform planning software is still far away. Also in the area of delay management, software solutions are not yet used due to technical and legal hurdles. The integration and coordination between the planning steps lead to a discrete optimisation problem of such significant size that a planning software for the coordination of the different plans is not to be expected in the coming years. First approaches to this are shown by Schöbel (Schöbel, 2017).

Application 3: Intelligent Control and Regulation of Traffic Flows Modern vehicles are already equipped with more and more assistance systems and autonomous driving functions. Partial or fully automated driving seems to be able to become reality in the near future. Cooperative driving to increase safety and efficiency or communication with infrastructure elements (e.g. traffic light, intersection etc.) and interaction with (active) traffic guidance systems is not yet sufficiently researched and developed today. In the field of communication, there are today, for example, very promising approaches with the 5G technology. Also in the field of AI-based methods, there are initial studies and concepts that on the one hand suggest great potential, but on the other hand also require research and development work until the real implementation with all the associated aspects (safety, legal, etc.) is achieved.

4 Outlook on AI Through Mathematical Methods in the Field of Mobility in 2030

For the future development of AI, two theses emerge, which we present, justify and substantiate with our case studies below.

1. *People will increasingly become creative decision-makers. Repetitive tasks, calculations and the merging of information will increasingly be taken over by AI algorithms.*

Computers are not meant to replace humans, but to support them in areas of tasks where they are clearly superior. These are especially the "memory", the playing through of different solutions and the linking of data according to certain patterns. Routine tasks will no longer be performed by humans in the future, but can be reliably done by computers. The resulting free space can be used to perform activities that require humans: Creative activities, the development of new ideas, the coordination between different planning areas, the modelling of relationships and the making of decisions, even with a measure of uncertainty.

In the future, breakthroughs will be achieved in many AI areas, but above all, the individual methods will be increasingly connected with each other. This leads to our second thesis.

2. *Future systems will become more reliable, secure and sustainable.*

By merging information and the consistent further development of the individual methods, human workers in various areas of mobility will be enabled to have a more comprehensive picture of certain issues. The impact of a decision on subsequent processes can be examined, different criteria can be considered and weighed against each other, different scenarios, even under uncertainty, can be simulated and compared with each other. This holistic view will improve systems, making them more reliable, secure and sustainable.

Application 1: Quality Control in the Production of Vehicle Parts In the next few years, methods for automatic, robot-assisted planning of inspections will be developed, which require less and less manual adjustment (Gospodnetic et al. est., 2021). AI-based automatic quality inspection will become the standard. Still existing disadvantages, namely the amount of data required and the traceability of the results, are currently being explored in a promising way (Došilović et al., 2018; Fend et al., 2021; Wang et al., 2018).

Inspection tasks that still have to be performed individually and repeatedly in the same way will be eliminated. Human inspectors will increasingly become "data curators" for the computer-assisted methods. On the other hand, they will be needed for the cases that do not have a clear inspection answer. AI systems can point out these special cases, support them with data and preliminary evaluations, and thus make the inspection decision more sound and objective.

Application 2: Planning of a Public Transport Service More and more traffic data will be available in digital form. With AI approaches, the robustness of a transport system can be evaluated and improved (Müller-Hannemann et al., 2021). In the future, planners will not have to create the required plans themselves, but define criteria and leave the system to create a representative selection. Using innovative visualisation, the advantages and disadvantages of generated plans are quickly grasped and the appropriate one can be selected from a holistic perspective. Planners can make creative and comprehensively informed decisions. At the same time, the transport systems will be better coordinated with each other and thus optimised in terms of their efficiency and reliability by playing through many more possibilities using artificial intelligence.

Application 3: Intelligent Control and Regulation of Traffic Flows In addition to other traffic data, the possibilities of capturing current traffic conditions, such as by airborne sensors, will increase. Likewise, real-time communication possibilities or robust exchange options of the mentioned data and information will be expanded (e.g. by 5G networks). All this will favour and stimulate the development of the previously outlined AI-based approaches to traffic flow control. This will make logistics processes and supply chains more reliable and sustainable, production more stable and efficient. Participation in traffic will become more pleasant and safe for the individual; time pressure and stress levels can be gradually reduced by skilful use of data and AI methods.

5 Summary and Practical Recommendations

From the perspective of AI experts, we give the following recommendations to align the working world with newly developed AI methods.

Develop AI for Mobility The goal is to allow a holistic view of problems, to avoid repetitive tasks and thus to support and inform creative human decisions. Therefore, mathematical methods of artificial intelligence should be further developed and applied to ever more challenging problems in all areas.

Easy Access to AI systems Since AI methods are often data-driven, the interface between human expertise and computers becomes more important. Where there used to be only one person with IT skills in a company, systems must be planned in the future so that AI is accessible in every work area. This means that computer systems must be understandable decision aids for humans. Here, visualisations and the consistent summarising of information are important building blocks that need to be worked on.

IT Education The interface to the computer, the correct classification and feedback will be part of the daily work of a large part of the population. A computer will make the same mistake as long as humans "explain" to it how to avoid it. Application experts must be trainers of data-driven approaches. Just as there is now a natural use of smartphones and social media in (almost) all population groups, there must be a natural use of AI methods in the working world in the future. For this, conditions must be created both in the area of development of AI methods and in the area of (further) education.

Use the Potential of Human Actors Already in school, but also later in university, the focus should be less on routine work, but on learning creativity techniques, modelling and decision techniques, so that the division of labour between AI and human actors can be designed advantageously.

References

Baumgart, U, & Burger, M. (2021). *A reinforcement learning approach for traffic control.* Proceedings of the 7th International conference on vehicle technology and intelligent transport systems (Vehits 2021).

Borndörfer, R., Klug, T., Lamorghese, L., Mannino, C., Reuther, M., & Schlechte, T. (2018). *Handbook of optimization in the railway industry*. Springer.

Došilović, F. K., Brčić, M., & Hlupić, N. (2018). Explainable artificial intelligence: A survey. International convention on information and communication technology, electronics, and microelectronics (MIPRO). *IEEE, 2018,* 0210–0215.

Fend, C., Moghiseh, A., Redenbach, C., & Schladitz, K. (2021). Reconstruction of highly porous structures from Fib-SEM using a dee neural network trained on synthetic images. *Journal of Microscopy, 281,* 16–27.

Glassner, A. S. (1989). *An introduction to ray tracing.* Morgan Kaufmann.

Gospodnetic, P., Mosbach, D., Rauhut, M., & Hagen, H. (2021). *Viewpoint Placement for Inspection Planning. Machine Vision and Applications – under minor revision.* Springer. https://doi.org/10.1007/s00138-021-01252-z.

Kroon, L. G., et al. (2009). The new Dutch timetable: The OR Revolution. *Interfaces, 39,* 6–17.

Liebchen, C. (2008). The First Optimized Railway Timetable in Practice. *Transportation Science, 42*(4), 420–435.

Müller-Hannemann, M., Rückert, R., Schiewe, A., & Schöbel, A. (2021). *Estimating the robustness of public transport systems Using machine learning.* arXiv preprint (arXiv:2106.08967), 2021: arXiv:2106.08967.

Schöbel, A. (2017). An Eigenmodel for Iterative Line Planning, Timetabling and Vehicle Scheduling in Public Transportation. *Transportation Research, C, 74,* 348–365.

Stephani, H., Weibel, T., & Moghiseh, A. (2017). Modellbasiertes Lernen in der Oberflächeninspektion. *at-Automatisierungstechnik, 65*(6), 406–415.

Wang, J., Ma, Y., Zhang, L., & Gao, R. X. (2018). Dazhong W (2018) Deep learning for smart manufacturing: Methods and applications. *Journal of Manufacturing Systems, 48,* 144–156.

Mobility in Urban Areas

How AI Enables Business Models and Creates Job Profiles

Verena Svensson

"Who is healthy and wants to work, has nothing to fear in this world."—G. E. Lessing

1 Introduction

Especially in urban areas, the question of efficient transport of people and goods from the origin to the destination is increasingly complex. Countless new options for routes and modes of transport emerge, accelerated by the increasing use of artificial intelligence. And this development is only at the beginning. According to a Delphi survey by the Munich Circle (2020), 92% of experts expect mobility to be strongly influenced by artificial intelligence—in no area is the expected impact higher.

2 Status Quo/Inventory and Case Studies

Various AI applications already play an important role in *mobility in urban areas* today (cf. e.g. Federal Ministry of Transport & Digital Infrastructure, 2018) in areas such as traffic control systems, repair and maintenance, autonomous driving, Mobility as a Service (MaaS).

V. Svensson (✉)
Stadtwerke Düsseldorf AG, Düsseldorf, Germany

I. Knappertsbusch and K. Gondlach (eds.), *Work and AI 2030*,
https://doi.org/10.1007/978-3-658-40232-7_27

Following the definition of different maturity levels of artificial intelligence by the Munich Circle (2020), technologies of type I (with purely reactive properties) and type II (with limited memory) are currently mainly used. Technology of type III (with understanding) will probably not be used until 2025. For type IV (with self-awareness), the question currently arises whether this will ever be used.

2.1 Traffic Control Systems

For years, a steady increase in traffic and vehicles on our roads has been observed. For individual transport (cars, motorcycles and mopeds, incl. taxi and car rental traffic), the Federal Ministry of Transport and Digital Infrastructure (2021) expects, for example, an average increase of 0.3% per year for the years 2020 to 2023.

To cope with this traffic volume, municipalities, federal and state governments rely on AI-based traffic control systems for road and rail transport. These help, for example, by intelligent traffic light switching, to avoid congestion. In rail transport, they enable increased frequencies by, for example, optimised routing and automated switch setting.

Traffic control systems are usually in municipal hands (e.g. city departments for traffic management) or in the sphere of influence of highly regulated operations (e.g. public transport companies (PTC)), whereby AI technologies are monitored.

2.2 Repair and Maintenance

A second area in which artificial intelligence has already gained a foothold is *repair and maintenance.* By combining historical data and enriching it with automatically obtained data (e.g. by means of optical sensors, cameras), the condition of vehicles can be reliably assessed. From this data analysis, necessary repairs and preventive maintenance measures can be derived. This approach is systematically used in rail transport, reduces failures and repair costs of the fleet and can increase the service life and availability of the fleet.

The framework conditions of this area are comparable to those of other industries in which machines are used. As e.g. Strötzel (2020) argues, further major progress can be expected here through AI applications within the framework of Industry 4.0.

2.3 Autonomous Driving

In the course of increasing driver assistance and automation of vehicles, artificial intelligence plays a key role. If one looks at the different levels of automation of driving in Tab. 1, one notices that technologies are already installed in vehicles that are approved today, which can be assigned to level 1 and 2.

It becomes clear that *autonomous driving* is not possible without artificial intelligence. For the different levels of automation, vehicles need corresponding systems that capture, analyse and evaluate a large amount of data, and then react according to the situation. Through the intelligence of the systems, they can learn from past situations, make predictions in new situations and gradually improve their decisions. Especially when autonomous vehicles are on the road with other road users in mixed traffic, the driving behaviour must be adapted to their decisions, which is only possible with artificial intelligence.

2.4 Mobility as a Service

Artificial intelligence enables new forms of mobility and related business models in the field of Mobility as a Service. Beckmann (2018) points out that *sharing providers* use AI algorithms to generate higher revenues. For example, the busi-

Tab. 1 Levels of automation of driving, own representation according to SAE International (2018)

Level	Name	Description
Driver pays attention to the environment		
0	**No automation**	The control is done manually by the driver
1	**Driver assistance**	The driver has the control over the vehicle
2	**Partial automation**	The vehicle has an advanced driver assistance or ADAS
Vehicle pays attention to the environment		
3	**Conditional automation**	The vehicle can recognise the environment
4	**High automation**	The vehicle can intervene and perform all driving tasks by itself
5	**Full automation**	The vehicle can drive autonomously, without any driver intervention

ness area is adapted to the demand and search behaviour of the users, depending on the parking time of the vehicles the price is automatically lowered—the optimal price point for customer incentivisation is learned here. In the dialogue with customers, *chat bots* and *voice bots* are used to ensure the availability of customer service around the clock, without waiting times and at low cost.

Mobility as a Service is used as an umbrella term for various services in the field of mobility that enable the customer to fulfil his or her transport request without using his or her own vehicle (cf. e.g. Kamargianni & Matyas, 2017, p. 4)

Ride Pooling providers can use AI applications to bundle transport requests from different customers in real time and determine the length of the detour that the customer still accepts to improve the utilisation of the vehicles, before the customer rates the service as bad and no longer uses it.

Artificial intelligence also enables *multimodal,* i.e. cross-modal *platforms,* that suggest to the customer the optimal combination of providers and means of transport for a desired route and let him or her book them across providers. Especially at the interface of providers, customers and service providers, AI applications come into play, which enable the optimal route composition and booking for each user based on learned preferences, availability of transport modes and environmental factors such as traffic volume, congestion information and weather.

3 Challenges and Solutions

The challenges of using artificial intelligence in mobility in urban areas are diverse and complex (cf. e.g. Cacilo et al., 2015). Among the most controversially discussed challenges are the handling of *data protection* and *ethical issues,* which will be addressed here.

3.1 Handling Data Protection

For the safe control of autonomous vehicles, large amounts of data from the environment are required. Vehicles of the company Waymo, for example, capture the surroundings in a radius of 200 m. This includes hand signals from cyclists, but also faces of pedestrians and license plates of other cars. The problem is that the affected road users are not aware of the capture and have not been asked for their consent.

A possible solution would be to develop an algorithm (and corresponding legal basis) that anonymises people and vehicles except in case of accidents. As the National Platform Future of Mobility (2020) and the Federal Ministry of Education and Research (2019) emphasise, a way must be found to use the benefits of technological progress without compromising data protection and data sovereignty of other road users.

3.2 Ethical Issues

By replacing human situation-specific decisions with technical algorithms, ethical dilemmas arise. Considering a case by Weber (2016), an autonomously controlled vehicle (level 4) that drives around a long curve in the mountains, where a fallen cyclist lies on the road, its options are to run over the cyclist and protect the occupants of the autonomous vehicle or to avoid the cyclist and endanger the occupants by falling into the valley. In the non-autonomous world, this decision is up to the driver. In autonomous driving from level 4 onwards, it depends on how the programmer has written the algorithm of the vehicle for such a situation. According to Graewe (2021), it can be assumed that the programming protects the occupants, as otherwise the sale of a known occupant-endangering autonomous vehicle would probably be difficult. However, such a fixed solution is illegal according to today's understanding.

Until the point in time when all road users are autonomously controlled and thus dangerous situations can be avoided, cases like the one described above are hardly solvable ethically and legally. A possible solution approach is the procedure of the federal government in 2021 to allow autonomous driving in controllable islands, then gradually expand them and connect them to a complete system (Graewe, 2021).

4 Outlook on AI in Urban Mobility in 2030

Due to the expected technological progress, it can be assumed that by 2030 significantly more AI applications will be affordable and used in mobility in urban areas. This development will also have an impact on the labour market.

Looking at the pure *number of jobs* in Germany until 2030 and beyond, studies such as Schade et al. (2020) conclude that there will be enough jobs available in the mobility sector. With regard to the *type of employment*, it can be expected that a massive upheaval will take place.

Thesis 1: AI Replaces Routine Work in the Repair and Maintenance Sector and Helps to Remedy the Shortage of Young Talent There

As described in section 1, AI solutions for repair and maintenance are already in use today and will gradually increase by 2030. Especially for fleets, such as those of public transport providers, the increase in the degree of automation is reflected in the elimination of activities that have been or were previously performed manually.

At first glance, this poses the risk of a lack of employment for trained specialists. However, a closer analysis of this field of employment reveals that it has been suffering from a shortage of young talent for years, which is exacerbated by the *demographic development* of the total population. Artificial intelligence can thus fill the gap that would otherwise be created by a lack of skilled workers (Schade et al., 2020).

Thesis 2: AI Reduces Vehicle Ownership and Employment in Automotive Production by Means of Autonomous Driving

Assuming that the introduction of autonomous driving on our roads will increase the proportion of shared vehicles, the number of vehicles in private ownership will probably decrease. This effect is expected to result in a decrease in the number of skilled workers required for *vehicle production*. This will in turn be reinforced by the increasing use of AI applications in the production environment (Strötzel, 2020). In addition, vehicles are nowadays almost exclusively designed as vehicles with alternative drive trains. In Germany alone, according to a survey by Bitkom (2017), of the approximately 840,000 employees in the automotive industry, about 300,000 are employed in the drive train, which is not necessary for vehicles with alternative drives.

Part of this overall projected decrease in resources required for automotive production will be absorbed in the purely quantitative analysis by the increased development resources for electric vehicles and AI applications for autonomous driving vehicles (Schade et al., 2020). It can therefore be assumed that artificial intelligence and autonomous driving will trigger an industrial transformation of a larger scale, which will completely redefine business models and value chains (Strötzel, 2020).

Thesis 3: AI Enables the Profitability of New Business Models and Creates New Jobs There

In section 2, it was shown that AI applications support *MaaS models*. In the future, their growth will be accelerated by the broader and more cost-effective deployment possibilities of AI and enable the profitability of the new business models, which are sometimes difficult to achieve, as Fischer (2019) points out. As a serious competitor, these offers will possibly take market shares away from other market participants, such as public transport, and cause a decline in the amount of work there.

At the same time, these new forms of urban mobility will create new employment opportunities. The focus of these activities lies in information technology and creative activities (Schade et al., 2020). The Münchner Kreis (2020) also welcomes the fact that the use of AI technologies allows resources to be aligned with the essential and real problems of the business model, as repetitive and monotonous tasks are automated by AI applications.

Thesis 4: AI Requires the Emergence of New Professions

Some of the challenges outlined in section 3 will intensify with the advancing use of AI. For example, the approval of autonomous driving of level 4 will require prior and permanent employment of ethicists. The position of the traffic ethicist is quite conceivable as a permanent part of development teams for AI technologies by 2030. Thus, ethical issues can be illuminated and expertly assessed already in the development phase. Nowadays, it is usually left to the IT developer himself how he programmes the algorithm (Fischer, 2019).

Furthermore, an increasing employment for lawyers can be expected. Approval and certification procedures as well as accident reports will have to be reviewed and processed anew for the use of AI technologies. Likewise, questions of how to deal with personal data will have to be examined and weighed. The profile of a traffic lawyer with a focus on autonomous driving is very likely by 2030.

Since machines will not be able and probably will not be able to show empathy and develop creative solutions by 2030 (survey Münchner Kreis, 2020), a larger proportion of employees than today will focus on these areas. The experts of the Münchner Kreis even expect that by 2035 the demand for cognitively demanding and creative activities will increase so much that there will be a real fight for talent.

5 Summary and Practical Recommendations

In summary, the advancement of the use of AI applications has a great influence on the structure of the employment situation in mobility in urban areas in 2030. Companies can prepare for the expected developments in various ways. They can consciously look for AI solutions for areas with a shortage of skilled workers, in order to counteract this. Furthermore, especially in the manufacturing industry, it is important to make early possible retraining and structural adjustments, in order not to fall behind in the upcoming upheavals caused by autonomous driving. To seize the opportunities of AI technologies, all companies in mobility in urban areas will expand their recruitment to include new professions and increasingly recruit specialists for AI developments. Talents who possess competencies and skills for this are well prepared for the requirements of the labour market of mobility in urban areas in 2030.

References

Beckmann, K. J. (2018). Digitalisierung und Mobilität. Chancen und Risiken für eine Verkehrswende. *Nachrichten der ARL, 2,* 12–16.

Bitkom (2017). Geschäftsmodelle in der Industrie 4.0. Chancen und Potenziale nutzen und aktiv mitgestalten. https://www.bitkom.org/sites/default/files/file/import/FirstSpirit-1496912702488170608-Faktenpapier-Geschaeftsmodelle-Industrie-40-Online.pdf. Accessed: 5. Juli 2021.

Bundesministerium für Bildung und Forschung. (2019). *Fortschrittsbericht zur Hightech-Strategie 2025*. BMBF.

Bundesministerium für Verkehr und digitale Infrastruktur. (2018). *Digitalisierung und Künstliche Intelligenz in der Mobilität Aktionsplan*. BMVI.

Bundesministerium für Verkehr und digitale Infrastruktur. (2021). *Gleitende Kurz- und Mittelfristprognose für den Güter- und für den Personenverkehr*. BMVI.

Cacilo, A., Schmidt, S., Wittlinger, P. et al. (2015). Hochautomatisiertes Fahren auf Autobahnen—Industriepolitische Schlussfolgerungen, Fraunhofer Institut für Arbeitswirtschaft und Organisation IAO. https://www.bmwi.de/Redaktion/DE/Downloads/H/hochautomatisiertes-fahren-auf-autobahnen.pdf. Accesssed: 22. Juni 2021.

Fischer, B. (2019). Das sind die Berufe der Zukunft. FAZ. 25.04.2019. https://www.faz.net/-gym-9mbhg. Accesssed: 22. Juni 2021.

Graewe, D. (2021). Autonomes Fahren: Traum der Ingenieure, Alptraum der Juristen. Beitrag zum Sammelband der 26. Interdisziplinären Wissenschaftlichen Konferenz Mittweida (IWKM) Online Tagungsband, https://monami.hs-mittweida.de/frontdoor/deliver/index/docId/12330/file/Graewe.pdf. Accesssed: 22. Juni 2021.

Kamargianni, M., & Matyas, M. (2017). *The business ecosystem of mobility as a service*. In: Proceedings of the 96th Transportation Research Board (TRB) Annual Meeting.

Münchner Kreis e. V. (Hrsg.) (2020). *Leben, Arbeit, Bildung 2035+. Zukunftsstudie Münchner Kreis Band VIII.* Bertelsmann Stiftung.

Nationale Plattform Zukunft der Mobilität. (2020). *Plattformbasierte intermodale Mobilität und Handlungsempfehlungen zu Daten und Sicherheit.* Dritter Zwischenbericht, Bundesministerium für Verkehr und digitale Infrastruktur.

Schade, W., Berthold, D., Doll, C., Grimm, A., Mader, S., Scherf, C., Sievers, L., & Wagner, U. (2020). *Synthese und Handlungsempfehlungen zu Beschäftigungseffekten nachhaltiger Mobilität. Arbeitspapier im Auftrag der Hans-Böckler-Stiftung.*

SAE International (2018) Taxonomy and definitions for terms related to driving automation systems for on-road motor Vehicles. https://www.sae.org/standards/content/j3016_201806/. Accesssed: 22. Juni 2021.

Strötzel, M. (2020). (Auto-)Mobilität zwischen Zwang und Teilhabe. Gewerkschaftliche Perspektiven auf die Probleme einer sozial-ökologischen Antriebs- und Verkehrswende. In A. Brunnengräber & T. Haas (Hrsg.), *Baustelle Elektromobilität. Edition Politik* (Vol. 95, pp. 383–408). Bielefeld: Transcript Verlag.

Weber, P. (2016). Dilemmasituationen beim autonomen Fahren. *NZV, 2016,* 249–254.

Industrial AI—Smart Factories and Team Robotics

Wolfgang H. Schulz, Vincent Geilenberg, Oliver Franck and Stanley Smolka

1 Introduction

Smart factories are characterised above all by their self-learning capabilities with implications for production, occupational safety, energy efficiency and resource utilisation. This also fits in perfectly with the Kaizen philosophy, which focuses among other things on process orientation and thus increases the innovation performance of companies (Imai, 2012). Production becomes more flexible on the one hand, but more robust on the other hand (VDA, 2007; Wolf et al., 2020). Production flexibility means the dynamic capabilities of the company to respond appropriately to data changes, so that both loss and insolvency risks are avoided or reduced. The past pandemic phase has impressively shown that production flexibility is not to be underestimated. A robust production process means that the products are error-free and the actual ordered quantity is delivered on time as agreed with the customer. Stochastic uncertainties such as occupational accidents,

W. H. Schulz (✉) · V. Geilenberg · O. Franck
Lehrstuhl für Mobilität, Handel und Logistik, Zeppelin Universität Friedrichshafen, Friedrichshafen, Germany
e-mail: wolfgang.schulz@zu.de

V. Geilenberg
e-mail: vincent.geilenberg@zu.de

O. Franck
e-mail: oliver.franck@zu.de

S. Smolka
Executive Education, Zeppelin Universität Friedrichshafen, Friedrichshafen, Germany
e-mail: stanley.smolka@zu.de

I. Knappertsbusch and K. Gondlach (eds.), *Work and AI 2030*,
https://doi.org/10.1007/978-3-658-40232-7_28

machine damage, disasters, bottlenecks at suppliers are absorbed by proactive actions within the framework of robust production management. Because robust production management partly conflicts with a cost-efficiency-oriented management thinking, it is rarely applied. It comes into play when managers pursue the goal of cost-effectiveness. An essential element for creating smart factories is the development of a capacity-oriented production architecture. In this respect, the smart factory approach is a further development of the holistic production systems concept, which dates back to the 1980s (Matuschek, 2016).

The reduction of costs, benefits especially companies that operate in a market that is under high adaptation pressure, has a high competitive intensity and/or is characterised by structural demand declines. That productivity can also be improved by a higher output is an opportunity for the industries that suffer from bottlenecks. These two characteristics apply in particular to the German automotive industry.

The reorganisation of the production process, the optimised value chain and the knowledge gains from the AI-based data analysis are the prerequisite for creating new value chains and developing and establishing new business models. Thus, industrial AI makes an important contribution to the essential entrepreneurial function of enabling arbitrage profits.

The automation of processes in companies leads to a decrease in the volume of work per head, which should economically lead to lower incomes. The lower labour costs lead to higher profits for the entrepreneurs, which are invested in further technical innovations, which in turn will reduce labour costs. This process of automation and digitisation in companies will go through two phases. In a first step, the workers will work together with the machines. The second step consists of a full automation of the entire tasks (Stengel et al., 2017). This process of digitisation takes place with an exponential growth, which is based on Moore's law, according to which the performance of computers is doubled every two to three years (Winkelhake, 2017). This effect has been observed for decades. Thus, it can be said that the automation of processes also runs exponentially and the digital development is faster than the creation of new tasks due to the extreme computing power of computers. Norbert Wiener, an American mathematician and the founder of cybernetics, had recognized the long-term consequences of the increasing digitization early on. Especially if no organised measures are taken by politics and society, digitisation could surprise society with its speed and lead to the biggest crisis of the labour market that humanity has ever experienced (Rifkin, 1995).

Against the background of the required system dynamic determination of employment effects, the development of team robotics must be examined in more detail. Team robotics comprises hybrid teams of workers and collaborative robots with different skills. Qualified human personnel work hand in hand with robots to solve complex manufacturing tasks in a team. The human-machine interaction is in focus. Through this complementary collaboration, new forms of work are created. It will certainly lead to structural unemployment in a first step. But team robotics offers new employment potentials with appropriate qualification.

2 Status Quo/Inventory and Case Studies

Science and research are mainly concerned with the possibilities of how smart factories will change the world of work. The work of Orellana and Torres is representative of the research literature that deals with the possibilities that can be implemented at the present time by AI-based digital applications. They examine the extent of efficiency improvement of old-fashioned machines by digitising them using the exemplary use case of the mining company Sage Mills. Old-fashioned means that the machines are fully depreciated economically and thus the Marx-Engels effect cannot be further exploited. They distinguish four phases of Industry 4.0. In the first phase, the company has isolated digital applications with a limited analytical capacity. In the second phase, the company is able to fully digitise the machines and enable them to exchange data with each other. In the third phase, the suppliers and partner companies are also digitally integrated. In the fourth phase, the operation is 100% digitised (Orellana & Torres, 2019).

The authors succeed in proving that a company with old-fashioned machines can achieve a cost-effective transition from phase 1 to phase 2 by digitally optimising these machines instead of buying new ones. This contribution is important because it means for the world of work that a longer economic service life of machines reduces the qualification pressure for the workers. Qualification and conversion costs are postponed to the future. For the company, such an effect is favourable because the capital accumulation due to the Marx-Engels effect (or capacity expansion effect) is stronger.

The partial improvements of machine utilisation are a characteristic of the first wave of digitisation linked to a weak artificial intelligence. As the AI algorithms improve and approach the artificial general intelligence, there are possibilities to redesign and reorganise the entire production process. Most of the current research works deal with these possibilities. The central characteristics of a smart factory are described by Shi, Z., Xie, Y., Xue, W., Chen, Y., Fu, L., & Xu, X. (2020). Overall, a smart factory is characterised by three different features. The sensors are able to self-organise by receiving, processing and meaningfully interpreting all the information from their environment. In addition, all machines in the smart factory are strongly integrated with each other by robot vision and artificial intelligence technologies. Moreover, the objects communicate with the employees through virtual reality applications. In this respect, this paper establishes a connection to the changed requirements for the world of work due to artificial intelligence.

Wang et al. undertake the project to describe all the essential characteristics of the smart factory (Wang et al., 2016).

Prinz et al. link the requirements for the workforce in the course of the AI-based development with the smart factory (Prinz et al., 2016).

Herrmann examines the main challenges and risks regarding the implementation of a smart factory (Herrman, 2018).

Veza et al. develop the concept of the virtual enterprise, which is a consortium of several existing companies that form a common interoperable production network, but do not create a new legal entity (Veza et al., 2015).

Stocker, Brandl, Michalczuk and Rosenberger put the question of the role of the human being in the context of the production process of Industry 4.0 at the centre (Stocker et al., 2014).

3 Challenges and Solutions

The German automotive industry can be used as a leading industry for the implementation of industrial AI.

The competitiveness of the German automotive industry is currently threatened by a number of developments: the diesel scandal, the lack of test benches for the new vehicle cycles to determine the actual emissions are just examples of self-inflicted problems of the German automotive industry. In addition, the supply-side market structure has changed with new providers from other countries such as Tesla from the USA and BYD from China. Furthermore, these newcomers represent new technological fields: electromobility and autonomous driving. In addition, the experiences that the own automotive production is significantly limited by bottlenecks in the semiconductor industry. Therefore, the willingness to use artificial intelligence is particularly high in this industry. With the project "AI platform concept for storing and processing learning and test data" in the years 2018 and 2019, the automotive industry has set the starting point to integrate artificial intelligence more strongly. In the KIP concept project, the requirements for such a platform are to be defined and the platform itself is to be specified. Also, the legal issues associated with such a platform are to be clarified in this project. Finally, a valid operating concept is to be developed that also describes the necessary processes and tools for secure data exchange. This project is part of the so-called AI family (https://ki-familie.vdali.de), which was initiated and developed from the VDA lead initiative autonomous and connected driving. 80 partners from science and industry are involved in the projects. The projects of the AI family receive funding from the BMWi. The AI family creates the basis for the German automotive industry to quickly catch up with its competitors from the global internet economy and to maintain and expand its international competitive leadership in the field of autonomous driving. In addition, this platform can contribute significantly to the sustainable promotion of application-oriented AI research in Germany, as in addition to the practice-relevant multisensory data sets, infrastructures are also created that research cooperations can access in the future.

The interests pursued with the establishment of the Catena-X Automotive Network platform are different, because here a networking between vehicle manufacturers and suppliers is intended to make production more efficient and environmentally friendly. These activities represent a first step in the direction of smart factories and team robotics, because Catena-X offers the opportunity for a transparent value chain, so that production can be optimised.

In addition to the antitrust issues, the AI efforts of the German automotive industry show that beyond the technical realisation of AI platforms, cooperation models are needed that enable collaboration and data exchange in the long term. As a cooperation model especially for industrial AI, the theory of institutional role models has proven itself (Luhmann, 1984; Schulz & Wieker, 2016; Schulz et al., 2019). The theory of institutional role models comprises five phases:

1. Identification phase: Identification of institutional, technical and economic roles for the system architecture of AI-based applications or platforms.
2. Initialisation phase: Initialisation of an Open Distributed Ecosystem (ODE) to fulfil massive decentralisation.
3. System architecture process: Based on the initialisation and identification phases, a secure and powerful system architecture is designed.
4. Governance: Design and parameterisation of a rule system (compliance) based on the institutional role model (IRM).
5. Autopoiesis: Future viability of the AI ecosystem through evaluation and feedback processes.

4 Outlook on AI in Smart Factories and Team Robotics in 2030

The realisation of smart factories and team robotics will be possible from 2030 onwards, because the following developments have prevailed:

- Artificial Narrow Intelligence will be replaced by Artificial General Intelligence.
- In 5G campus networks, edge devices can be interconnected with the high bandwidth and guaranteed low latency of 5G to form a local edge cloud that can then meet the real-time requirements of the factory.
- Data infrastructures meet the industry's requirements for data sovereignty, decentralisation in heterogeneous multi-cloud systems and edge support.
- Enabling a sustainable economy. Sovereignty as self-determination at all levels. In a networked economy, self-determination means above all the freedom to determine the technology of choice or the location of choice.

5 Summary and Practical Recommendations

Artificial intelligence with smart factories and team robotics will fundamentally change the global economy in the near future, which is why it is urgent to scientifically capture and evaluate the changes or rather disruptions triggered by artificial intelligence. In addition, data-driven business cooperations are becoming increasingly relevant in this digital environment, which is why novel solutions are needed for how these can be structured and implemented successfully for all involved companies. One such solution approach consists in the form of the theory of institutional role models (IRM).

The automation of processes in companies leads to a decrease in the volume of work per head, which should economically lead to lower incomes. The lower labour costs lead to higher profits for the entrepreneurs, which are invested in further technical innovations, which in turn will reduce the labour costs. This process of automation and digitisation in companies will go through two specific phases. In a first step, the workers will work together with the machines. The second step consists of a full automation of the entire range of tasks (Stengel et al., 2017). This process of digitisation takes place with an exponential growth, which is based on Moore's law, according to which the performance of computers doubles every two to three years (Winkelhake, 2017). This effect has been observed for decades. Thus, it can be said that the automation of processes also runs exponentially and the digital development is faster than the creation of new fields of tasks due to the extreme computing power of computers. Norbert Wiener, an American mathematician and the founder of cybernetics, had early recognised the long-term consequences of the increasing digitisation. Especially if no organised measures are taken by politics and society, digitisation could surprise society with its speed and lead to the biggest crisis of the labour market that humanity has ever experienced (Rifkin, 1995).

Against the background of the required system-dynamic determination of employment effects, the development of team robotics must be examined more closely. Team robotics comprises hybrid teams of workers and collaborative robots with different skills. Qualified human personnel work hand in hand with robots to solve complex manufacturing tasks in a team. The human-machine interaction is in focus. Through this complementary cooperation, new forms of work places are created. It will certainly lead to structural unemployment in a first step. But team robotics offers new employment potentials with appropriate qualification.

References

Herrmann, F. (2018). The smart factory and its risks. *Systems, 6*(4), 38.

Imai, M. (2012). *Gemba Kaizen: A Commonsense Approach to a Continuous Improvement Strategy*. McGraw-Hill Professional. 2nd Ed.

Luhmann, N. (1984). *Soziale Systeme: Grundriss einer allgemeinen Theorie*. Suhrkamp.

Matuschek, I. (2016). *Industrie 4.0, Arbeit 4.0–Gesellschaft 4.0. Eine Literaturstudie*. RLS-Studien.

Orellana, F., & Torres, R. (2019). From legacy-based factories to smart factories level 2 according to the industry 4.0. *International Journal of Computer Integrated Manufacturing, 32*(4–5) 441–451.

Prinz, C., Morlock, F., Freith, S., Kreggenfeld, N., Kreimeier, D., & Kuhlenkötter, B. (2016). Learning factory modules for smart factories in industrie 4.0. *Procedia CiRp, 54,* 113–118.

Rifkin, J. (1995). *The end of work: The decline of the global labor force and the dawn of the post-market era*. Putnam's Sons.

Schulz, W. H., & Wieker, H. (2016). *Co-operative Intelligent Transport Systems: Neue Marktchancen durch den Systemverbund aus Automobil-und Telekommunikationsindustrie. Future Telco III –Powerplay für Kommunikationsunternehmen,* (pp. 138–150). Institute of Electrical and Electronics Engineers.

Schulz, W. H., Joisten, N., & Arnegger, B. (2019). Development of the institutional role model as a contribution to the implementation of co-operative transport systems. https://ssrn.com/abstract=3421107 or http://dx.doi.org/10.2139/ssrn.3421107.

Shi, Z., Xie, Y., Xue, W., Chen, Y., Fu, L., & Xu, X. (2020). Smart factory in Industry 4.0. *Systems Research and Behavioral Science, 37*(4), 607–617.

Stengel, O., Van Looy, A., & Wallaschkowksi, S. (2017). *Digitalzeitalter—Digitalgesellschaft: Das Ende des Industriezeitalters und der Beginn einer neuen Epoche*. Springer.

Stocker, A., Brandl, P., Michalczuk, R., & Rosenberger, M. (2014). Mensch-zentrierte IKT-Lösungen in einer Smart Factory. *e & i Elektrotechnik und Informationstechnik, 131*(7), 207–211.

VDA (2007). *Das gemeinsame Qualitätsmanagement in der Lieferkette. Produktherstellung und Lieferung—Robuster Produktionsprozess.*

Veza, I., Mladineo, M., & Gjeldum, N. (2015). Managing innovative production network of smart factories. *IFAC-PapersOnLine, 48*(3), 555–560.

Wang, S., Wan, J., Li, D., & Zhang, C. (2016). Implementing smart factory of industrie 4.0: an outlook. *International journal of distributed sensor networks, 12*(1), 3159805.

Winkelhake, U. (2017). *Die digitale Transformation der Automobilindustrie: Treiber—Roadmap—Praxis*. Springer.

Wolf, S., Jordan, M. M., Seifert, I., Evertz, M., & Korzynietz, R. (2020). Wie Industrieproduktion nachhaltig gestaltet werden kann. In *Klima* (pp. 164–179). Springer Vieweg.

AI in the Automotive Industry

How AI is Changing the Automotive World

Peter Schlicht

1 Introduction

Mobility as a basic requirement of modern life is subject to constant change—just like social life as a whole. This affects the demand ("When can I move where and how?"), the type ("How do I get from A to B?"), and the experience of mobility ("How do I spend the time in motion?"). Many such changes take place slowly; e.g. the increase of technical assistance in the vehicle. Other changes, however,—so-called disruptions—change mobility behaviour abruptly. Examples of this are the introduction of the railway or the mass production of automobiles. These disruptions are consequences of essential technical innovations (steam engine, assembly line production), which found an immediate application to the mobility products and thus fundamentally changed the demand, type and experience of mobility.

Artificial intelligence, especially deep learning, is another essential technical innovation that has found its way into everyday products due to various causes:

- Modern chips are able to execute complex neural networks efficiently (i.e. fast and energy-saving).
- Due to the strong networking (mobile communications, Internet of Things etc.) huge amounts of data are available to train deep neural networks and offer customer functions.

P. Schlicht (✉)
Safe Artificial Intelligence, CARIAD SE, Wolfsburg, Germany
e-mail: Peter.Schlicht@cariad.Technology
URL: https://www.linkedin.com/in/peter-schlicht-298660104/; https://scholar.google.de/citations?user=Wg1dFekAAAAJ&hl=de

I. Knappertsbusch and K. Gondlach (eds.), *Work and AI 2030*,
https://doi.org/10.1007/978-3-658-40232-7_29

- The great commercial interest further ensures that the AI field is actively researched.
- The necessary hardware becomes cheaper.

These factors ensure a high diffusion of AI functions for example in smartphones, household appliances and data centres.

The academic and industrial research has made astonishing progress in recent years favoured by this hype: by now neural networks are able to solve tasks previously considered unsolvable [(Grechishnikova, 2021), (Silver et al., 2018)].

Also driver assistance functions even including automatic driving seem possible based on AI-driven perception functions—just like developing completely new mobility products.

Whether AI triggers the next disruption for the world of mobility—possibly together with the spread of electromobility—can only be judged in retrospect. But the combination of neural networks and data-driven software development with automotive software products undoubtedly enables essential changes in the type and experience of mobility of the future.

2 Status Quo: An Industry in Transition

The automotive industry is undergoing profound change worldwide: The switch from the combustion engine to electromobility, from the automobile manufacturer to the mobility service provider and from manual driving to highly assisted "being moved" by software-dominated vehicles takes place simultaneously with far-reaching social changes such as digitalisation and the strongly focused pursuit of global sustainability. These changes affect the entire corporate landscape from customer needs to business models, from product development to production and from corporate culture to responsibility as a "big player" in society.

AI methods are a catalyst for these changes and are particularly evident on the product side: Modern driver assistance systems use neural networks for the interpretation of manifold sensor data (cameras, lidars, radars, GPS, Car2X communication, map, control unit data) and "understand" the current driving situation. Emergency braking, parking or driving assistance systems are based on this understanding. Together with information about the interior, this understanding can also be used to optimise and individualise comfort and infotainment functions for the vehicle occupants. At the same time, all vehicles can build up a central knowledge in the cloud and thus optimise traffic flows or mobility offers or pass on fleet knowledge to third parties.

Such learning software connects numerous sensors and actuators in the vehicle and therefore entails another technical change in modern vehicles: the introduction of highly networked and performant vehicle architectures (the basic framework of control units, sensors, communication and supply systems) and a strong and secure connection of the vehicles with the cloud. Of course, such a change in the basic topology of the vehicles also entails a fundamentally new way of developing vehicle software. Whereas "traditional" software development had the goal of implementing a function of a control unit, now all software components have to interact in a highly complex "system of systems". This is especially true for data-driven development, i.e. when the specification of the function results from its use.

3 Challenges and Solutions

The entry of highly complex software and learning functions into modern vehicles described above poses a number of challenges to the technical development of these vehicles.

Data-Driven Development Processes

As a safety-critical system, vehicles are subject to regulatory requirements in terms of the development process and the resulting quality (safety, robustness, etc.). These standards [(SIG, 2015), (ISO 26262–1:2018, 2018)] generally envisage three specific phases in the development of vehicle software:

1. Requirement Elicitation: from the definition of the customer function, a description of the operational environment and possible safety risks are derived. After that, the function is specified: SW architecture, requirements for each component, etc.
2. Implementation of the function, i.e. the translation of the requirements into executable code.
3. Integration and testing of the function at different integration levels and test locations—here, the fulfilment of the specification of the function is checked and documented.

On the other hand, AI technologies such as deep neural networks are mainly used for the implementation of such functional components that cannot be fully specified. There are various reasons for this—two examples are given here:

- the absence of exact physical or chemical models, e.g. with regard to acoustics, temperature profiles or the charging characteristics of vehicle batteries.

- the inherent uncertainty due to the functional application in the constantly changing world. This is especially a challenge for the specification of video-based perception functions.

For such functions, it is hardly possible to formulate a complete set of requirements whose fulfilment would guarantee compliance with all standards. In the data-driven development of neural networks, requirements are therefore formulated in terms of data, runtime and safety properties and evaluated on the basis of statistical analysis of the resulting neural networks. From this evaluation, new requirements arise for both the training and the data. The data-driven development process is inherently more iterative. As a result, requirements and test results become dynamically changing artefacts of the development process. This iterative process as well as normative requirements for the data sets or statistical evaluation processes to be used are only inadequately reflected in most existing standards (M. Gharib, 2018).

In order to bring AI development in line with the standard development processes, the iterativity of data-driven development must be reconciled with the strict traceability in the classical development process and extended by mechanisms for data sets and their quality. For this purpose, there are a variety of technological approaches and work processes that are summarised under the keyword "DevOps" or "MLOps".

Another challenge for the data-driven development of automotive software lies in the high dynamics of academic research in the field. Especially in the area of "safe AI", the state-of-the-art changes at a high frequency. In order to keep up with the methodological progress, a high permeability of emerging AI technologies into the development chain is necessary.

Functional Safety

Functional safety is a central requirement and relevant acceptance criterion for novel safety-critical vehicle functions. This is especially true when the vehicle functions take over partial responsibility for the driving function. For driver-assistance systems, the assurance is described by [(ISO 26262–1:2018, 2018)]. With regard to the assurance of automated vehicles, this standard is complemented by a multitude of further approaches [(21448:2019, 2019) and additionally (ISO/DIS 21448 2021—Preview), (ISO/TR 4804:2020, 2020)]. The use of AI functions in complex safety-critical software systems brings additional challenges [e.g. (Gesina Schwalbe, 2020)].

The dependency of the functional quality of neural networks on the quality of the data (diversity, labelling quality, etc.) as well as the already mentioned under-specification of the corresponding functional components has the consequence that a proof of complete safety seems unrealistic. Rather, it seems necessary to support the safety argument by statistical measurement methods and thus make an acceptable residual risk plausible. Subsequently, a continuous monitoring of the safety in connection with an efficient error mitigation process can document and ensure the safety in a sustainable manner [(Apollo, 2019), (Edge Case Research, 2019)]. "Safety" and "quality" thus become accompanying processes of development.

The world in which AI functions are rolled out changes continuously, for example due to ageing processes on the vehicle or the emergence of new, previously unseen objects. Consequently, the functional quality and performance of data-driven functions potentially decrease, which can have a negative impact on the overall safety. The sensitivity of neural networks to small changes [(Christian Szegedy, 2014)] can lead to a significant change in performance even with small environmental changes.

It is therefore likely that safety-critical AI functions will be updated. Vehicles will thus run different software versions throughout their life cycle. Moreover, they will become test sites in a continuous safety and development process. This has implications for the vehicle architecture: re-programmability, accessibility and scalability of automotive control units become enablers for safe AI functions. In a traditionally cost-sensitive industry, this implies that further business models need to be developed that exploit the resulting extra efforts.

The handling of the remaining residual risk is a concern for society. Cross-industry research consortia such as conference workshops (e.g. **SAIAD**) and research consortia (e.g. **KI-Absicherung**) help to successfully push forward the state-of-the-art for standardisable safety strategies.

Networking in Vehicle and Organisation

It is in the nature of data-driven development that AI benefits from the availability of as many and as good data as possible. Consequently, the potentials for using AI are particularly high when both the data sources in vehicle and backend as well as the various function-developing teams are highly networked.

Nowadays, the software components in vehicles are distributed across numerous control units. In the corresponding vehicle architectures, cost efficiency can be increased on the one hand by distributing data streams small and control unit-specific. AI functions that access many sensor streams or replace individual components, on the other hand, benefit from a centralisation of computing resources and data streams. Vehicle architectures resulting from this paradigm allow for greater flexibility in the design of the control units. In addition, the complexity of the software and middleware running on the control unit increases.

Similarly, the development of AI functions is favoured by a high degree of organisational networking. In this way, the competencies required for AI development (data scientists, statisticians, data engineers) can be optimally deployed, AI potentials identified and leveraged. This networking can be supported by agile working methods and the creation of high technical transparency in the company. The formation of competence centres increases the exchange between function teams and enables the identification of potentials for the implementation of AI functions.

The incremental development of functions and the possibility to change the functionality even afterwards lead to a convergence of traditionally separated business areas such as design, development, safety, strategy and aftersales, in order to ensure fast product cycles and customer orientation "by design".

AI-based tools can substantially accelerate and partially automate the development process. The identification and development of such tools, for example for test automation or simulation, represents a significant potential for increasing efficiency in the development process. To identify and implement AI-based development workflows, specific *labs* or cooperation initiatives are helpful.

Adding Necessary Competencies

The proportion of AI functions in the vehicle as well as the use of AI tools in the development process are expected to increase significantly [e.g. (FutureBridge, 2020)]. This entails an organisational challenge in the shift of necessary competencies.

- The necessity of statistical rigor in the quality and safety evaluation of neural networks or the analysis of the large amounts of data generated both in the vehicle and in the development process create a high demand for statistical expertise.
- The data-driven development requires massive and scalable computing clusters along with development tools running on them ("MLOPS")—this creates an increased demand for data scientists and DevOps engineers.
- The development of AI tools and the continuous further development of already deployed vehicle software create an increased demand for safety managers as well as experts for CI/CD and transfer learning.
- At the same time, the function-specific expertise remains important for the model selection and the identification of potentials for AI functions.

With the increasing importance of AI functions and tools, it can become a competitive advantage to implement efficient internal training formats in addition to attractiveness in the labour market.

4 Outlook on AI in Automotive in 2030

Vehicles in 2030

More than half of the vehicle software is now supported or fully implemented with AI methods. The computationally intensive AI models are now almost exclusively executed energy-efficiently on dedicated AI-processors. This makes it possible to process the high-resolution modern sensor data streams in near real time. Vehicle data is used live in the automotive backend to keep automatic driving functions at the highest level of safety and to react to traffic events across the fleet. The high availability of mobile data communication has also massively changed the mobility space: traffic lights, separate traffic areas and parking spaces in the city centre are a thing of the past in many places.

Just as in the early 2000s in the mobile phone sector, after an initial diversity, three essential operating systems for automotive software platforms have now established themselves and are available as a licensing model to all vehicle manufacturers. The operating systems were developed by automotive manufacturers, technology corporations and industry consortia and have a number of basic mechanisms for safety and security and the rollout of AI functions to the vehicle platforms.

Technical Development in 2030

Thanks to the development of automotive operating systems with standardised interfaces and numerous AI-based tools, a large part of the actual software development and its monitoring in the field runs highly automatically. The identification of new products is based on a sophisticated expert analysis of the data from the mobility fleets. New functions are developed independently of new hardware platforms and vehicle models and brought to market at high frequency. By now, a large part of the revenue of automotive manufacturers is generated after the vehicle has left the sales structures, or comes from other business areas such as data marketing, financial and energy services.

The work of the developers has also fundamentally changed: In small, function-specific teams that make use of the numerous competence centres of the technical development, new products (hardware and software) are constantly developed and first made available to a small customer base. If the functions prove to be worthwhile, are rolled out at large scale.

5 Summary and Practical Recommendations

AI methods are making their way into the products and development process of the automotive industry. This provides significantly improved functions, but also poses significant challenges for the industry: Automotive manufacturers become software companies with highly complex and AI-based software products.

Both the development process and the organisational structure of it will take these changes into account and the companies will transform into dynamic and agile mobility service providers of the future.

For the introduction of AI methods into safety-critical automotive software and software development products, technical, organisational and process-related challenges must be overcome. These will lead to increased networking and centralisation on many levels: vehicle architectures, development tools and processes, and organisational structures.

References

21448:2019, ISO/PAS. 01 2019.

Apollo. (2019). A whitepaper on automated driving safety. https://apollo.auto/platform/whitepaper.html. Accessed: 29. Juli 2021.

Edge Case Research. (2019). Key Ideas: UL 4600 Safety Standard for Autonomous Vehicles.https://edge-case-research.com/ul4600/. Accessed: 29. Juli 2021.

FutureBridge (2020). Artificial Intelligence reshaping the automotive industry. https://www.futurebridge.com/industry/perspectives-mobility/artificial-intelligence-reshaping-the-automotive-industry. Accessed: 29. Juli 2021.

Grechishnikova, D. (2021). Transformer neural network for protein-specific de novo drug generation as a machine translation problem. *Science and Reports, 11,* 2021.

Gharib, M., Lollini, P., Botta, M., Amparore, E., Donatelli, S., & Bondavalli, A. (2018). *On the safety of automotive systems incorporating machine learning based components: A Position Paper.* 48th Annual IEEE/IFIP International Conference on Dependable Systems and Networks Workshops (DSN-W), 271–274.

ISO 26262–1:2018. 12 2018.

ISO/DIS 21448. 2021—Preview.

ISO/TR 4804:2020. 1. 12 2020.

Schwalbe, G., Knie, B., Sämann, T., Dobberphul, T., Gauerhof, L., Raafatnia, S., & Rocco, V. (2020). *Structuring the Safety Argumentation for Deep Neural Network Based Perception in Automotive Applications*. Bd. vol 12235, in Computer Safety, Reliability, and Security. SAFECOMP 2020 Workshops. SAFECOMP 2020. Lecture Notes in Computer Science, von Ortmeier F, Schoitsch E, Bitsch F, Ferreira P (Hrsg.), Casimiro A Cham: Springer.

SIG, VDA QMC Working Group 13/Automotive (2015). Automotive SPICE Process Assessment/Reference Model. 3.0.

Silver, D., et al. (2018). *A general reinforcement learning algorithm that masters chess, shogi, and Go through self-play* (p. 362). Science.

Szegedy, C., Zaremba, W., Sutskever, I., Bruna, J., Erhan, D., Goodfellow, I., & Fergus, R. (2014). *Intriguing properties of neural networks. nd International Conference on Learning Representations.* ICLR.

AI in the Rail Sector

Emerging Use Cases and Potential Impact on Employment

Konrad Scheuermann, Ingo Kucz and Sabina Jeschke

1 Introduction

Climate neutrality by 2050—this goal is now legally anchored in the German Federal Climate Protection Act. The transport sector plays a decisive role in its realisation. In 2019, its share of national greenhouse gas emissions was 20.3%, higher than ever before. The sector-specific greenhouse gas emissions increased by 7.9% during this period.[1]

The federal government has set itself the goal of reversing this trend and has assigned a central role to rail transport. The coalition agreement from 2017 sets the goal of doubling the number of passengers by 2030 and shifting more freight to rail. The implementation of this goal is reflected in numerous initiatives, in particular the Master Plan for Rail Transport[2] as well as the Deutsche Bahn corporate strategy "Strong Rail".[3]

[1] Figures from UBA (2021).

[2] cf. BMVI (2020a).

[3] cf. Deutsche Bahn (n. d.).

K. Scheuermann (✉)
Digital Products and Projects, DB Regio AG, Berlin, Germany
e-mail: Konrad.Scheuermann@deutschebahn.com

I. Kucz
White Octopus GmbH, Berlin, Germany

S. Jeschke
Deloitte Deutschland & RWTH Aachen, Aachen, Germany

I. Knappertsbusch and K. Gondlach (eds.), *Work and AI 2030*,
https://doi.org/10.1007/978-3-658-40232-7_30

2 Status Quo

The challenges are manifold. The biggest obstacle for a short-term massive increase are capacity bottlenecks in the German rail network. They already cause quality deficits today, which hinder the switch to rail. These bottlenecks can only be resolved in the medium to long term by building new train railroads or expanding existing ones, apart from punctual measures. In addition, the rail system differs significantly from other transport systems in its complexity. It combines technology from 120 years ago with the latest digital technology and is also still largely characterised by national regulation and a lack of international/European standardisation. This complexity also increases the requirements for the employees who have to understand and operate the interfaces between the different technical worlds. Finally, the worsening shortage of skilled workers acts as a further challenge. Deutsche Bahn (DB) currently recruits about 20,000 people per year, many of them in training occupations.[4] With a decline in school leavers and increasing numbers of students, it will become increasingly difficult to find the necessary staff to operate the rail system.

At the same time, opportunities are opening up that could work to the lasting advantage of the rail sector. The awareness of the challenges posed by climate change has increased significantly. To some extent, this also applies to the political willingness to take ambitious measures to contain it. This is expressed in particular in the significantly increased contributions from the federal budget for the expansion of the rail infrastructure.[5] In parallel, we are experiencing a fundamental technological change, specifically the advancement of the 4th Industrial Revolution. This is characterised by the increasing intelligence of technical systems, which is driven by the use of artificial intelligence (AI).

AI in the rail sector is not science fiction. In addition to specific, already implemented methods and use cases, many areas of application are in the development stage. In the following, we would like to illustrate this using the example of the rail system.

[4] cf. Fockenbrock (2019).

[5] cf. BMVI (2020b).

3 Challenges and Solutions

3.1 AI in Rail Operations: More Capacity and Higher Quality

Building new train railroads to solve the capacity problem is time-consuming and costly. With AI, on the other hand, the existing infrastructure can be used more efficiently. Today, the track-side signalling and safety technology determines the number and length of the so-called block distances and thus the distance between two trains. In the future, trains could run closer together, because the rear one registers in real time when the front one brakes. The distance between two trains could thus be reduced to the safety-necessary distance, without any additional investments in track-side signal infrastructure. Estimates assume that this could increase the capacity in the rail network by up to 35%.[6]

The fuller the rail network, the more challenging it becomes to organise the traffic and maintain the level of quality. This is especially true as very different trains use the same infrastructure, these include fast ICEs, frequently stopping regional trains and the slow freight trains. Even today, a disproportionate share of the delays at Deutsche Bahn (German rail system) occur on overloaded train lines and in nodes, where many train lines meet, such as between Cologne and the Ruhr area. With artificial intelligence, the disposition of traffic can be significantly facilitated. The technology allows to virtually play through disposition scenarios in quasi real time and thus to consider the respective effects in a disposition decision. For example, the decision to let an already delayed train wait a little longer may be unpleasant for the passengers, but sensible for the overall system, in order to keep the impact on other trains low. In initial tests at the S-Bahn Stuttgart, punctuality in the event of a disruption could be increased by 3%.[7,8]

Keeping the impact of disruptions low is one thing—even better would be if disruptions did not occur at all. In an open, complex and technically driven system, this is a great challenge. Nevertheless, data analysis and AI also offer solutions here. The key lies in the constant monitoring of the system and the timely prevention of disruptions by predictive maintenance[9]. For example, it is possible

[6] cf. Deutsche Bahn (2019).

[7] cf. Deutsche Bahn (2021).

[8] cf. Koenen (2021).

[9] cf. Weißhaupt (2020).

to determine from the current curve of a switch drive whether it will fail soon.[10] Such an intelligent switch can then send a message and order the repair on its own. This is possible because an AI was trained with historical data to recognise patterns that lead to a failure of the drive. This principle can be transferred to all components of the system. Camera images are particularly suitable as sensor data. If trains pass through a camera gate before entering the workshop, which identifies external damage and irregularities, the staff in the workshop can reduce the inspection and focus on the repair of the defects.[11] The trains are then back in service more rapidly and better maintained.

3.2 AI for the Customer: Better Informed and Easier to Reach the Destination

AI not only ensures a smoother operation behind the scenes, but also offers visible and tangible improvements for customers. One example is customer information:[12] Is my train on time? Will I catch my connection? A consistent and reliable information in real time is one of the central customer expectations. Here, Deutsche Bahn has been using big data analyses to predict delays and arrival times since 2018.[13] For this purpose, historical and real-time data from different data sources are used to identify influencing factors for punctuality. These include calendar information such as month, weekday, time, but also train type, current delays, the scheduled travel time, circulation information and train conflicts. In contrast to the previous linear extrapolation of delays, today, based on the historical data, prediction models are trained using methods of machine learning and big data technologies. Because the error propagation is usually not linear—under certain circumstances, delays can be *made up* again, in other cases they multiply. To make a prediction for a specific ICE, the trained prediction models are combined with the real-time data. The resulting forecast is then fed into the information systems on the train and at the station as well as into the DB Navigator app.

However, looking at the train journey is not enough. Customers are travelling from door to door and rarely live directly at the long-distance station. The travel

[10] cf. Thomas (2017).

[11] cf. Jeschke (2021).

[12] cf. Kucz et al. (2020).

[13] cf. Jeschke (2020).

chain usually links several modes of transport, e.g. bus, ICE, suburban train, underground, etc. The choice of possibilities has rather increased than decreased with rental bikes, car sharing and e-scooters—at least in the city. With a multitude of possibilities, however, it becomes increasingly difficult for us humans to make the right choice with complex preferences. It is not always about speed: Maybe I prefer to ride a bike in good weather, but take a taxi in bad weather? And what if the ICE is late and the connection is in danger? AI can pre-structure this complex decision for the human and give him a suitable recommendation in real time—provided the computer knows the users, i.e. has enough data.

Finally, ticket purchase is an area where AI technology enables solutions that make life easier for customers. One such solution is the automated travel billing in the background instead of buying a ticket. The latter is often associated with a confrontation of the local tariff system for occasional and visiting customers, in order to select the right ticket. An alternative are so-called check-in/check-out systems, where customers register and deregister with their smartphone at the beginning and end of the journey and the system automatically calculates the cheapest tariff for the travelled distance. A step further are so-called be-in/be-out systems. With these, no action by the customers is required, as the smartphone automatically registers in the vehicle (bus, underground). The main challenge here is the unambiguous assignment of a person to a vehicle, which is especially a technical challenge for several crossing or parallel running modes and lines of transport. For this purpose, it is necessary to combine several location and sensor data. The corresponding models are based on AI algorithms. Instead of pure be-in/be-out models, however, mixed models such as check-in/be-out are currently still in use, which are introduced by the transport associations.[14]

4 Outlook on AI in the Mobile Work Environment in 2030

These examples show: AI is already in use today, without customers necessarily actively perceiving it. The examples also show, however, that relevant AI applications are still in the development stage, as the example of the capacity increase of up to 35% that is important for the transport transition shows. The vast majority of AI applications are still to be developed.

[14] cf. Castrillo and Binder (2020).

Since AI applications are currently only at the beginning of their development, their effects on work in rail operations or in the rail sector as a whole are largely unclear or even (necessarily) speculative.

4.1 Changed Job Profiles & Composition of the Workforce

Rail operations, as a real-time service, depend to a large extent on operational staff—visible to customers in the service and largely invisible in the diverse tasks in the background. This starts with vehicle maintenance and includes the control centre to driving staff, a large proportion of the employees in the rail sector are skilled workers. However, the majority of these job profiles have so far had no significant relation to AI.

For the workforce, an increasing application landscape of AI applications has two essential, direct effects:

- The demand for highly qualified AI experts in the rail sector is increasing. Many applications are not developed by the operators themselves, such as Deutsche Bahn or other rail transport or infrastructure companies. But even the definition, description, procurement, rollout and application of AI requires a substantial staff. In companies in the rail sector that are typically dominated by direct work areas, the proportion of indirect employees with highly sought-after and well-paid profiles is growing.
- With increasing penetration of AI applications, the training and further education needs of their users increase. Application, interpretation and comprehension skills for recommendations or interpretation-required results of AI applications will trigger an immense learning and need for trust: How can I be sure that a counter-intuitive train sequence is good for the overall system? How can I trust results whose calculation I cannot comprehend? How do I resolve the conflict of objectives between AI-based values and my tactile intelligence in vehicle inspection?

4.2 Growing Workforce and More Service for the Transport Transition

As shown above, the significant immense increase in infrastructure capacity is a key lever of AI technology in rail operations. The AI-induced capacity increase enables significantly more services in regional and long-distance transport and lays the foundation for the modal shift. AI technology can thus become a driver of the transport transition. Morerail traffic, however, also means more staff in all non-automatable areas. With the strong growth in services, the coordination effort in the overall system increases, in order to ensure a smooth customer experience in the rail system for travellers. As a first approximation, a positive employment effect is likely to result even with a technology push.

An increasing demand for staff is contrasted at the same time by the already prevailing shortage of skilled workers. If the overall passenger volume increases as forecasted and politically desired, staffed services in trains, stations and other service facilities will reach their limits. A flanking of the growth by technology use, especially AI, is thus necessary to secure the transport transition. At the same time, the use of AI enables a technology-based service offensive for an improved customer experience, without further straining the scarce resource of staff.

The use of AI and other technologies is, in our view, a way of resolving the tension between the ambitious goals of the transport transition and the existing shortage of skilled workers.

5 Summary and Recommendations

Growth in rail traffic and the increase in service quality is not an automatism, but largely shaped by mainly politically set framework conditions. Through the financing of the infrastructure and the public procurement of regional transport, the state and political decision makers have the levers for a technological leap themselves in their hands. The potentials and possible effects of AI on the workforce structures and sizes outlined above seem large. An immense capacity increase or a significantly improved customer experience through AI create a space of possibilities that needs to be understood and above all entered by courageous political decisions:

- AI applications and their employment effects are an immense lever to achieve ambitious climate and transport transition goals. But: The pace in the development and rollout of these applications is still too slow. The climate crisis needs the expected efficiency and growth boosts from AI in the rail system as soon as possible. The promotion of AI-based technologies is to be initiated and expanded on a large scale.
- The current efficiency-driven tendering practice in regional transport often still hinders technological innovations. Here it is important to set the right incentives to underpin the transport transition technologically.

- Finally, the technological renewal and its design together with the employees that is necessary for the transport transition is a task for the entire industry. Deutsche Bahn, as a central player with a state-owned owner, is one of the most important drivers of the transport transition, but will also not be able to realise it on its own.

References

BMVI. (2020a). Masterplan Schienenverkehr. https://www.bmvi.de/SharedDocs/DE/Anlage/E/masterplan-schienenverkehr.pdf?__blob=publicationFile. Accessed: 28. Juli 2021.

BMVI (2020b): Bund und DB unterzeichnen größtes Modernisierungsprogramm für das Schienennetz. https://www.bmvi.de/SharedDocs/DE/Pressemitteilungen/2020/001-scheuer-starke-schiene-unterzeichnung-lufv.html. Accessed: 28. Juli 2021

Castrillo, J. L., & Binder, M. (2020). Vertriebsinnovationen auf Basis von elektronischen Fahrterfassungssystemen In Verband Deutscher Verkehrsunternehmen e. V. (VDV) (Hrsg.): *Digitale Transformation des ÖPNV. Chancen, Lösungen und Herausforderungen für die Branche* (pp. 296–300).

Deutsche Bahn. (2019). Digitale Schiene Deutschland. Die Zukunft der Eisenbahn. https://digitale-schiene-deutschland.de/Downloads/Broschüre_DigitaleSchiene_2019.pdf. Accessed: 28. Juli 2021.

Deutsche Bahn. (2021). Künstliche Intelligenz macht die Bahn pünktlicher und zuverlässiger. https://www.deutschebahn.com/de/presse/pressestart_zentrales_uebersicht/Kuenstliche-Intelligenz-macht-die-Bahn-puenktlicher-und-zuverlaessiger-6201662. Accessed: 18. Juli 2021.

Bahn, D. (o. J.). Unsere Strategie Starke Schiene. https://ir.deutschebahn.com/de/db-konzern/strategie/unsere-strategie-starke-schiene. Accessed: 28. Juli 2021.

Fockenbrock, D. (2019). Lok sucht Lokführer: Die Personaloffensive der Bahn ist in vollem Gange. https://www.handelsblatt.com/unternehmen/handel-konsumgueter/fachkraeftemangel-lok-sucht-lokfuehrer-die-personaloffensive-der-bahn-ist-in-vollem-gange/25314628.html?ticket=ST-944563-M2ZceyGNoZhlPFwFU6Q1-ap2. Accessed: 28. Juli 2021.

Jeschke, S. (2020). Data Lakes und „very big data“ für intelligente Reisendeninformation. https://de.linkedin.com/pulse/data-lakes-und-very-big-f%C3%BCr-intelligente-prof-dr-sabina-jeschke-1f. Accessed: 20.Juli 2021.

Jeschke, S. (2021). KI macht die Bahn pünktlicher und zuverlässiger. https://de.linkedin.com/pulse/ki-macht-die-bahn-p%C3%BCnktlicher-und-zuverl%C3%A4ssiger-jeschke. Accessed: 20. Juli 2021.

Koenen, J. (2021). Bahn bringt Künstliche Intelligenz aufs Gleis. https://www.handelsblatt.com/unternehmen/handel-konsumgueter/schienenverkehr-bahn-bringt-kuenstliche-intelligenz-aufs-gleis/27222608.html. Accessed: 18. Juli 2021.

Kucz, I., Naji, S., Ritzer, P., & Ypma L. (2020). Wer spricht? Gedanken zur digitalen Transformation in der Kundenansprache, in: Verband Deutscher Verkehrsunternehmen e.V.

(VDV) (Hrsg.), *Digitale Transformation des ÖPNV. Chancen, Lösungen und Herausforderungen für die Branche* (pp. 339–346).

Thomas, Peter (2017): Doktor Diana lauscht am Gleis. https://www.faz.net/aktuell/technik-motor/technik/virtuelle-diagnoseplattform-doktor-diana-lauscht-am-gleis-14727367.html. Accessed: 20. Juli 2021.

UBA (2021): Vorjahreschätzung der deutschen Treibhausgas-Emissionen für das Jahr 2020. https://www.umweltbundesamt.de/sites/default/files/medien/361/dokumente/2021_03_10_trendtabellen_thg_nach_sektoren_v1.0.xlsx. Accessed: 18. Juli 2021.

Weißhaupt, B. (2020). Predictive Maintenance als zusätzlicher Baustein in der Instandhaltungsstrategie, in: Verband Deutscher Verkehrsunternehmen e. V. (VDV) (Hrsg.), Digitale *Transformation des ÖPNV. Chancen, Lösungen und Herausforderungen für die Branche* (pp. 260–267).

AI as an Opportunity for the Future Airline Business

Present and future solutions

Susan Wegner and Didem Uzun

Before joining Deloitte Digital Dr. Susan Wegner was responsible as Vice President for the business unit Artificial Intelligence&Data Analytics at Lufthansa Industry Solutions. The article was written during this time. Many thanks to the whole team for the inspiring case studies and the support for the outlook on 2030.

1 Introduction

As in other industries, the technological, ecological and social trends in aviation are developing rapidly and leading, in some cases, to a completely changing ecosystem. The Covid pandemic has accelerated this development, as it has shown that much less business travel will be necessary in the future. Video technology has advanced enormously in a short time and holograms (i.e. 3D representations of people) potentially will be used more often in the future. The small airlines that have survived the pandemic have merged. Electric Vertical Takeoff & Landing or flying taxis are pushing into the market and short-haul flights are competing with ever faster rail connec-

D.-I. Wegner (✉)
Deloitte Digital, Managing Director, Berlin, Germany
e-mail: swegner@deloitte.de

D. Uzun
Lufthansa Industry Solutions, Business Development AI & Data Analytics, Norderstedt, Germany
e-mail: didem.uzun@lhind.dlh.de

I. Knappertsbusch and K. Gondlach (eds.), *Work and AI 2030*,
https://doi.org/10.1007/978-3-658-40232-7_31

tions. Sustainability is more in focus and reinforces the demand for climate-neutral flying, enabled for example by using synthetic fuels. Passengers want a more personal travel experience from individualised airport processes, automatic handling, more self-service, but also increased security, e through biometric digital check-in for instance.

It has also became clear how important digitisation or automation is for a more flexible and dynamic flight operation. AI enables the automation of intelligent behaviour and is already used for many routine tasks to increase efficiency and speed. Today, and even more so in the future, AI applications enable new possibilities to find solutions to problems that are yet too complex to solve for humans alone. Thus, AI transforms the entire aviation industry and with it the working environments of aviation personnel.

2 Current Status and Case Studies

2.1 Relief and Acceleration of Tedious Text Tasks by Natural Language Processing

Especially the Corona year 2020 with many flight changes has clearly shown that an efficient and timely processing of customer inquiries is crucial. Depending on the situation, customer inquiries come in different forms, e.g. verbally during a flight, by e-mail or through a call-centre call. Processing these different types and kinds of comments is often monotonous and time-consuming. AI enablesaccelerating and improving such evaluations. It is already able to understand natural language—whether written or spoken—and to sort incoming e-mails and automatically forward them to the responsible department or employee. (Fig. 1)

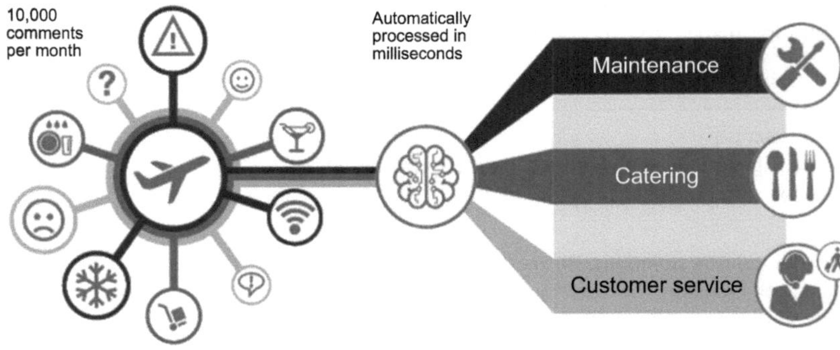

Fig. 1 Automatic analysis of customer comments (Lufthansa Industry Solutions, 2021)

In order for such an application to evaluate the customer inquiries, it must first learn the meaning of the respective texts based on historical data, such as existing and previously mostly manually evaluated passenger surveys. AI is then able to assign a comment, an e-mail or a text passage to a topic. Even incomplete or incorrectly written sentences and words can be analysed and interpretations can be performed, such as that "no sufficient cooling" simply means: It was too warm.

AI is also used for text analysis in other areas, such as cash flow analysis. AI already enables an automatic classification of all transactions to planning categories in seconds. The application is based on a hybrid approach of human classification, fixed rule set and self-learning AI, which can achieve an accuracy of 90%. This is a use case and problem AI offers a solution to which was not possible due to the complexity before, and thus also provides employees with new tasks. At the beginning, the participation of the experts is crucial, as they have to check the assignment and make corrections if necessary. Since the AI learns continuously, the need for correction during the implementation process becomes less over time, however, a random sample check will remain necessary for the time being.

Additional, already existing areas of AI application are automatic e-mail replies when querying the aviation freight status. For employees, the "flood" of e-mails that they have to process is significantly reduced, so that they have more capacity for more complex and unusual e-mail processing. Also, the automatic recognition of structured information, for example from maintenance logs, enables a prediction of future damage in the sense of predictive maintenance and thus to provide assistance to the employees when certain parts need to be overhauled or replaced.

2.2 Acceleration of Work Steps by Image Processing

Especially in the field of image processing, massive progress has been made in recent years, which has led to the use of this AI technology in very different areas of the aviation industry and has already changed the working world significantly.

For the maintenance of aircrafts, any damage that occurs is photographed for documentation purposes and the repair or the required new components are described. This resulted in an extensive data collection, which represents an essential knowledge base for maintenance and repair. For any new damage, which technicians were not familiar with, they oriented themselves based on the damage image and turned to their colleagues with the question "Have you ever seen this damage before?" When there was no answer available a tedious textual search in the database followed. Today, the technicians have the possibility to search for similar damages based on the current damage image. So a kind of "Google image search" specifically for aircraft damage. This makes the work of the technicians much easier and saves several working hours per year.

One of the essential optimisation topics of aviation is the minimisation of the time that an aircraft is on the ground. The previous manual processing of data often led to erroneous data amounts and incomplete conclusions. By combining camera live streams at airports and other system data, the data is analysed in real time (Microsoft, 2021). This helps the overall coordinators to track and manage the ground times to the second and changes their previous work envirnment from manual to automatic. The platform even warns in real time about delays, e.g. about a late arriving l fuel truck.

2.3 Change of Work Processes by Forecasting Systems

In order to not only reactively address the minimisation of ground times, predictive planning is essential. Especially unscheduled repairs of aircrafts generate significantly longer ground times or flight cancellations. Modern aircrafts produce a large amount of data, which enables the prediction of demand-oriented maintenance (Technik, 2021). For example, sensor data, such as the temperature of brakes and engines as well as various error messages from the aircraft, in connection with the flight plan, can help to plan the maintenance in a timely manner. By adjusting the maintenance, the ground times of the aircraft are reduced, resulting in more revenue and less costs (e.g. because certain maintenance can be brought forward or postponed). The entire planning process for the maintenance of aircrafts changes due to corresponding recommendations of the AI and thus also the work steps of the responsible employees. Thus, the collection, analysis, and evaluation of the data leads to the fact that aircraft fleets can be maintained better in the long term and aircrafts simply fail less often.

Nevertheless, also crew failures and misplanning can potentially lead to flight cancellations, too. l. Here as well, the deployment planning can be improved and optimised by AI recommendations. A prediction by employee seniority and aircraft type a prediction up to seven days in advance is possible already. This facilitates the prediction for the planning of the standby personnel, reduces costs, but also increases the satisfaction of the employees by reducing standby times.

A forecast of when an aircraft will land can already be determined by an AI with a fairly high accuracy, while the aircraft is still in the air. This enables the early information of the passengers about delays, the optimisation of the passenger transit for connecting flights and a timely replenishment of water, fuel etc. to minimise the airplane's ground times. This consequently also means a change of the work processes of the employees involved.

3 Challenges and Solutions

3.1 AI as a Service

AI is also considered a flexible, but difficult to use tool in the aviation industry. This perception is not entirely wrong. AI projects are among the more demanding tasks that an IT team can undertake. The biggest hurdles are complexity, lack of data and lack of experts. So far, AI development has been more like individual programming, for which experts often had to be found or trained. Especially in aviation, there are currently rather few AI experts employed. This makes the projects lengthy and expensive or they fail due to the necessary expertise. In addition, only very large and financially strong companies have been able to handle such projects. With a cloud service, however, AI has come within the reach of all companies, regardless the company size. AI as a service lowers the challenges, because many subproblems have already been solved by the provider in advance. It is already used successfully in aviation today, e.g. for the analysis of passenger comments. *AI as a Service* simplifies, accelerates and makes the use of AI much more cost-efficient (Lufthansa Industry Solutions, 2021).

4 Outlook on AI as an Opportunity for the Future Airline Business in 2030

The above examples show that some routine tasks in aviation can already be taken over by AI today. To avoid errors and to identify them quickly in case of emergency, the processes are subject to particularly strict safety guidelines and are therefore highly standardised. This opens up multiple possibilities for AI use cases in the future to make the processes more efficient and less error-prone. AI supports humans in identifying larger contexts. In the following, let us look at some of the typical stages of a flight journey.

4.1 Check-In

The fast lane known as “online check-in” is already familiar to many travellers and a popular feature. It not only saves time, but can also contribute to safety in the future, through a biometric, digital check-in or through digital vaccination certificates. Also, at some airports baggage check-in now can be done by passengers themselves. If we imagine that at check-in and baggage drop an assignment of the seat, the baggage dimensions and the weight is done, it becomes clear that this can offer great advantages for the baggage loading process.

4.2 Baggage Loading

In the baggage loading process, it is important that the weight is evenly distributed in the plane, in order to simplify the settings of the take-off and landing parameters, as well as to optimise them from a sustainability perspective. An AI that takes the temporal component of the process steps seat, baggage space and connecting flights into account, can offer enormous time savings and also achieve optimisation of passenger waiting times.

4.3 Security Check

Advanced image processing has great potential to cover the security check risks even more intelligently. For example, suspicious objects can be identified faster through intelligent video analysis. An AI can help to screen objects in the security check continuously and give a risk assessment. At certain thresholds, the AI can support the security personnel in the search by pointing out suspicious persons. However, image processing requires strict GDPR compliance, which is already ensured today with modern anonymisation methods. Additionally, robots or drones will probably also be used in the departure hall, which are able to analyse the current situation. They can detect potential hazards, prioritise them and give recommendations to optimise the entire security check. Also, the annoying repetition of the sentence “Have you remembered to take your keys out of your pocket?” by the security staff will in the future certainly be taken over by individualised versions said by robots. For example, based on passenger reactions or emotions at the security check, robots will automatically explain what steps need to be done. By analysing the emotions of the passengers via cameras, suspicious passengers can also be checked more closely or appropriate work steps can be recommended to the security personnel.

4.4 Boarding

With the collected information about the passengers from check-in to boarding, a further contribution to the optimal distribution of the weight of the aircraft load can now be checked and verified. The baggage items, baggage loading spaces, passenger weights and passenger seats are known. Using the Tetris principle, the AI checks one last time before boarding whether there were any changes to the seats and passes them on to the passengers at the gate when entering the plane. This not only ensures an optimal weight distribution in the plane, but also an optimised passenger handling with lower waiting times.

4.5 Service on Board

For the service on board, there will probably be less cabin crew and possibly also less cockpit staff. The prerequisite is an easy-to-use on-board system for passengers and crew.

The emotion analysis from the security check can also improve the customer experience on board in the future. AI can recognise emotions of passengers and react specifically to them. It will be able to recognise whether a passenger is thirsty or hungry. It will probably also recognise which drink the passenger is most likely to order. But apart from an individualised service, it will also be able to monitor the health status of each individual passenger at any time. Heart attacks account for the largest share of serious incidents on board, with half of these cases being silent heart attacks. Health conditions can be monitored by an AI using sensors and the right staff can be informed in case of emergency. This way, emergency chains can gain decisive speed by using data transmission to save lives. Such an AI can also be used to support passengers with fear of flying. For example, the entertainment system can show relaxing content to the passenger depending on the pulse level. This is also transferable to the cabin crew to increase flight safety.

These application examples are already possible in practice today. Whether we can use these solutions in Europe soon will depend on smart solutions for data protection compliance.

4.6 Sustainability Enabled by AI

As in other industries, many airlines are concerned with the topic of sustainability. AI applications can help to develop promising solutions that only start with synthetic fuels. Also, the *Corporate Digital Responsibility Initiative* (2021) has identified climate and resource protection as fields of action, but also the application of AI, especially the transparent handling of it or the data. The pressure on the aviation industry is growing and pushing the entire industry towards sustainability. This has become evident in the establishment of sustainability managers or the development of such departments.

For example, already today CO_2-dashboards as infographics provide information on emission values. With AI, up-to-date dashboards with local and global CO_2 consumption and causes can be created, and the CO_2 certificate trade can be restructured. Combined with CO_2 compensation offers, customers can be actively encouraged to travel more consciously and sustainably.

Thus, with AI airlines can anchor sustainability in their product portfolio more efficiently. Passengers can be offered different products that create incentives for sustainable travel, for example by offering price advantages or more flexible flight rates for every kilogram CO2 saved by a passenger. On board, passengers can be offered options that not only offer the choice of vegetarian, but also sustainable menus. These two examples, supported by an AI, mean not only CO_2-, but also cost and waste reduction at the process level for an airline.

5 Summary and Practical Recommendations

The briefly outlined current and future AI applications in the aviation industry show that AI systems can relieve us humans of daily tasks, so that we can use our valuable time more efficiently. They are particularly suitable for processing monotonous tasks that tire and bore humans, but also for tasks that are too complex for humans to solve alone. For the working environments in the aviation industry, this means an active collaboration with AI, for example the integration of recommendation engines into the work processes, the collaboration with robots, but also with novel technologies such as holograms. It is crucial that the transparency of the algorithmic recommendations proves compliance with ethics. This essentially means: decisions that affect people should ultimately still be made by people. The use of AI in the aviation industry will increase the speed, efficiency and safety in the future, as well as enable more complex technologies such as autonomous, vision-based navigation and data ecosystems.

References

Microsoft (2021). Lufthansa CityLine streamlines aircraft turnaround, reduces flight delays with Azure Video Analyzer, https://customers.microsoft.com/en-au/story/1369447387472683550-lufthansa-zerog-travel-transportation-azure-video-analyzer. Accessed: 6. Juli 2021.

Lufthansa Technik (2021). AVIATAR The Digital Operations Suite, https://www.lufthansa-technik.com/aviatar. Accessed: 6. Juli 2021.

Lufthansa Industry Solutions (2021). Whitepaper: Artificial Intelligence as a Service (AIaaS), https://www.lufthansa-industry-solutions.com/de-de/studien/whitepaper-artificial-intelligence-as-a-service-aiaas. Accessed: 6. Juli 2021.

Corporate Digital Responsibility Initiative (2021). Gemeinsam Verantwortung übernehmen—Die CDR-Initiative, https://cdr-initiative.de/initiative. Accessed: 6. Juli 2021.

AI in Intralogistics

How the Use of AI Will Change the Organisation of Work in Intralogistics

Norbert Bach and Sven Lindig

1 Introduction

The term *intralogistics* has been defined as the organisation, control, execution and optimisation of the internal material flow and the associated information flows (Arnold, 2006, p. 1; Günther, 2006, p. 6). The related business challenges concern the alignment of quantities in relation to locations and the calculation of optimal unit sizes and transport routes. In the past, operations research methods with given objective functions (e.g. linear optimisation, combinatorial optimisation) were used for this purpose. In contrast, AI systems can recognise patterns in large amounts of data in real time and compare them with historical data, thus predicting probabilities of occurrence that result from ongoing operations. The future of AI in intralogistics considered here thus relates to both access to material flow-related data in real time and evaluation of individual data and pattern recognition in comparison to information on reference objects collected in the cloud. For these reasons, we address the following questions:

N. Bach (✉)
Fachgebiet Unternehmensführung/Organisation, Technische Universität Ilmenau, Ilmenau, Germany
e-mail: norbert.bach@tu-ilmenau.de

S. Lindig
Lindig Fördertechnik GmbH, Krauthausen, Germany
e-mail: sven.lindig@lindig.com

I. Knappertsbusch and K. Gondlach (eds.), *Work and AI 2030*,
https://doi.org/10.1007/978-3-658-40232-7_32

- How is AI already used in intralogistics today?
- What specific challenges of further development of AI does intralogistics face?
- What future scenarios could be considered for the year 2030?

2 Status Quo: Use of AI in Intralogistics Today

The individual identification of goods and materials is currently still mostly done with the help of barcodes, QR codes and RFID chips (Günthner & Hompel, 2010; Hippenmeyer & Moosmann, 2016). Operations research methods for evaluating these data have been in use in intralogistics since the 1960s. The prediction of demands based on AI enables a higher customer orientation with regard to external and internal customers—customer orientation is the biggest driver of AI in intralogistics. Data evaluation is either done in a central computer or—in swarm-based approaches—in distributed computers.

The *prediction of needs* particularly affects the design and stocking of storage systems, sorting and picking processes. For example, it has proven successful in online retailing to place pre-orders based on predicted incoming orders. This enables next-day deliveries, which reduces the return rate. Also well-known is the slotting of goods, i.e. storage in close proximity. AI determines probabilities for frequently simultaneously requested goods based on pattern recognition. This reduces transport routes, increases efficiency and conserves resources (Poll, 2019).

In addition, AI enables the ideal packaging and assembly of goods on load carriers. Which order with which items, volume and weight is assigned to which employee and which transport means when? This also creates transparency about the material flow chain as a basis for optimisation in other areas, e.g. personnel planning. Likewise, by specifically using low-load times, capacities can be smoothed and demand peaks cushioned.

Depending on the use case, *swarm-based AI systems* are also used, e.g. for driverless transport systems (DTS). Swarms of DTS that can perceive their environment independently of each other navigate autonomously to their destinations. The allocation of transport orders and the routes of the individual DTS are negotiated by the DTS among themselves. If obstacles or disturbances occur, the swarm reacts in real time. Load carrier utilisation is also optimised based on swarm intelligence, including the detection of weak points in operation, e.g. leaving a temperature window or the occurrence of vibrations (Heistermann, n.d.).

Furthermore, AI is used for maintenance, repair and deployment planning of machines and plants. Predictive maintenance (preventive maintenance) serves to detect the impending failure of an operating resource and to initiate preventive measures.

Finally, AI in intralogistics eases the burden on *employees*. In pick-by-voice picking, AI-based speech recognition enables error-free communication with the system and confirmation by voice input: The picker always has free hands. Image recognition in combination with VR glasses enables the capture and control of goods—more accurate and reliable than the human eye. If, in addition to image recognition, a comparison with a digital image is made, AI is also able to detect faulty goods and sort them out (Poll, 2019).

3 Current Challenges for the Further Development of AI in Intralogistics

The AI-based optimisation of intralogistics requires the capture of large amounts of data and, for the recognition of patterns, access to reference data sets. This raises the question of who owns the data and who is allowed to access it. For example, a manufacturer of industrial trucks must have an interest in obtaining and evaluating all available data on the operating conditions of the vehicles they build. The customer benefits from the evaluation of the data, but at the same time grants a deep insight into their operation. So far, too few device operators are aware of the criticality of the issues regulated in the fine print of the contract. Customers will demand a higher level of anonymisation of their data or agree to a data transfer only in aggregated form. This issue will also have an impact on the development of standardised data formats that would enable evaluation across manufacturers and systems.

Although intralogistics is limited to internal material flows, the materials do not always belong to the owner of the operation. Marketplaces like Amazon also ship on behalf of third parties and shop-in-shop concepts use retail space. This raises the question of whose interests an optimisation of the location-based intralogistics serves and who participates to what extent in the savings. This affects legal issues as well as the objective function that is given to the agents for negotiation processes in swarm-based AI.

The safety of people in the workplace is also regulated by law. To protect people, spatial areas for DTS and robots are currently separated as closed areas or the people working in this area are protected by grids and cages. In the future, humans and robots will use a common space. How the safety of humans can still be ensured and what legal regulations have to be complied with remains an open question.

Data capture via barcodes and QR codes will increasingly be replaced by pattern recognition. This only works if reference data is available in digital formats and algorithms allow reliable recognition in real time. This is not yet widely available.

4 Outlook on AI in Intralogistics in 2030

Despite digitisation and additive manufacturing, intralogistics is growing
For many of the legal challenges described, there will be standardised solutions by 2030, as will the issues of data security and standards. Legal corporate boundaries will therefore play an increasingly minor role in logistics, so that intralogistics and supply chain management will merge seamlessly. The intralogistics issues that have so far been related to individual locations can therefore be extended to larger units of analysis, which enables the exploitation of further potential and also new business models. Additive manufacturing methods such as 3D printing will also not lead to a shrinking of the industry. Although some tasks in the inbound logistics area will be eliminated, due to the trend towards individualisation and the possibilities of flexible production, the outbound logistics will become more complex and diverse.

Increasing quality of data leads to higher forecasting quality
By 2030, the human factor, which is still often used in data entry, will largely be replaced by sensors or AI-based image, speech and pattern recognition in comparison to digital images and reference values. This includes that faulty sensor values are also detected and compensated by statistical methods. In the future, neither faulty data nor errors caused by sensors will lead to incorrect forecasts, as the AI detects, compensates and takes into account errors in the forecast. New methods will be available from statistics on how to deal with missing data and data gaps. The errors in pattern recognition and forecasting that can still be observed today will therefore become less and less.

Intralogistics becomes three-dimensional
Currently, the optimisation of paths is usually done in two dimensions—vehicles are located on the ground. In order to use existing spaces better, vehicles will be able to move three-dimensionally in the future, either in guided systems or by using flying drones. Flying drones are already used today for automated inventory, and by 2030 drones will be a common means of transport. This affects both indoor and outdoor use. The flying objects will have higher payloads and the control systems will master difficult environmental conditions. The extension of the solution spectrum into the three-dimensional dimension entails even higher requirements for AI.

Optimisation in the digital twin, implementation in the real system

By 2030, it will be possible to create a digital twin of real or planned logistics systems, despite their complexity. In addition to the pattern recognition in big data that is already used today, the digital twin allows a combination of simulating the future and evaluating historical big data. This further increases the forecasting quality and is probably also necessary due to the expansion of the forecasting to entire supply chains. For example, the picking process in a new warehouse can be checked for its practical suitability, even though the building is still in the planning stage. Virtual reality makes it possible to demonstrate what the future warehouse will look like already during its development. This also includes sample processes that can be rehearsed in an augmented reality environment.

From reaction to anticipation

The simulation of demand developments that is also possible with the help of digital twins has an impact on the entire supply chain and all factors required for it. This affects both personnel planning and the deployment planning of conveying equipment. With regard to the conveying equipment, including the drones, predictive maintenance becomes prescriptive maintenance. Based on a forecast of the next failure, the maintenance time, the corresponding personnel and ideally also the spare parts provision are planned in advance and in a low-load time. Possibly, this planning goes so far as to influence the parameters of the equipment concerned. Damage-relevant movements are no longer performed with full power, but as gently as possible with optimal use of the infrastructure, in order not to provoke a failure before the planned maintenance date. Also in the material planning, a much better control of the required goods is possible by anticipatory logistics.

Co-robots ease the burden on the logistics worker

The diverse manual activities that are still common in intralogistics today will be largely done by robots by 2030. This particularly affects sorting and picking processes as well as inspections. Networked and AI-controlled robots ensure an almost simultaneous and error-free communication between the machines used and thus contribute to increased productivity. The solutions already existing today for intelligent containers and intelligent vehicles benefit again from the digital twin, as simulations in real time can also be used in the AI-based negotiations for optimal utilisation and paths. Quality control in 2030 will be carried out almost exclusively by robots and in comparison with a digital image.

The human being remains the decisive factor in intralogistics

Even in 2030, computers will only execute what humans order and what results from human-programmed learning algorithms. However, the level of demand for the remaining intralogistics activities shifts. In fact, the use of AI eases the everyday life of a logistics worker. In addition, the AI-controlled personnel planning leads to higher planning reliability through networking with other systems. Through smartwatches, wearables and exoskeletons, humans are also networked and become more and more a fixed part of a networked logistics system. Humans are no longer just recipients of optimisation results, their real-time data flows into the optimisation and protects humans from damage by robots or AGVs. This also has an impact on the design of workplaces, as users can virtually try out their future work environment in augmented reality environments.

In view of the scenarios described, the future will require even more precise thinking, creativity and adaptability of the logistical processes. This will require computer scientists, engineers and business economists, but also skilled workers who apply the new digital tools. Overall, areas with a high level of demand will be more in demand, while helper activities will lose importance. But digitisation is not meant to replace people! No one who is willing to adapt to the changing conditions already today, to learn and to develop further has to worry. Nevertheless, the overall number of employees will hardly change in the next ten years. Because what is often overlookedis the fact that digitisation also creates numerous jobs. In addition, there is a continuing high demand for young talent and a large number of people who will retire in the next decade.

5 Summary and Practical Recommendations

The tasks of intralogistics were optimised with algorithms in the past. By using AI and pattern recognition, large savings potentials have already been realised today and, in particular, a higher customer orientation has been achieved. In addition, the reduction of transport routes and inventory levels also contributes to environmental protection.

By 2030, tasks of intralogistics will be thought of in three dimensions by using drones. Today's common pattern recognition will be complemented by simulations in the digital twin. The anticipatory logistics possible on the basis of AI no longer affects spatially delimited tasks of intralogistics.After clarifying legal issues, the goal is to optimise cross-cutting supply chains. Humans will become integral parts of logistics systems through exoskeletons and wearables and will be able to work safely with robots and transport systems.

Companies can prepare for these challenges. Similar to Industry 4.0, it is necessary to create the conditions for networking within one's own company, both in terms of technology, transport containers and conveying equipment, as well as the early adoption of innovations in legal standards and data standards into one's own administrative processes. Finally, companies should recognise early on which tasks can be taken over by robots and drones in the future and for which new tasks the employees should be trained.

References

Arnold, D. (2006). Intralogistik—Die späte Taufe einer längst erwachsenen Branche. In D. Arnold (Eds.), *Intralogistik—Potentiale, Perspektiven, Prognosen* (pp. 1–3). Springer.

Günther, P. (2006). Intralogistik—eine starke Branche stellt sich vor. In D. Arnold (Eds.), *Intralogistik* (pp. 5–16). Springer.

Günthner, W. A., & Hompel, M. (2010). *Internet der Dinge in der Intralogistik*. Springer.

Heistermann, F. (n. d.) *Die Logistik der Zukunft—Digitalisierung, Robotics und Künstliche Intelligenz. Vom Gabel- zum Datenstapler?* Themenheft der Initiative Die Wirtschaftsmacher—eine Initiative deutscher Logistiker.

Hippenmeyer, H, & Moosmann, T. (2016). *Automatische Identifikation für Industrie 4.0*.

Poll, D. (2019). *Warum KI in der Intralogistik nicht mehr wegzudenken ist*. In: Produktion. Technik und Wirtschaft für die Deutsche Industrie.

AI in the medical and pharmaceutical world of work

AI Makes Medicine More Efficient, Individual and Preventive

Joachim Hornegger

1 Introduction

The history of medicine is marked by fundamental, sometimes disruptive innovations. Such milestones tend to go hand in hand with technological development, especially since the second half of the nineteenth century: The discovery of viruses and bacteria revolutionised the fight against infectious diseases, the introduction of anaesthesia alleviated the suffering and pain of patients, the use of X-rays marked the beginning of medical imaging, which opened up more and more applications in the course of digitalisation.

Today, medicine faces a new disruptive change: Artificial intelligence (AI) and machine learning (ML) have the potential to take prevention, diagnosis and therapy to a new level. With it, we cannot only store and retrieve individual and structural health data as needed, but also make them usable for automated and accelerated analysis and decision processes. AI and ML are fundamentally new because they can recognise patterns—similarities, deviations, parallels, repetitions, correlations, clusters, classes—in an unmanageable amount of digital information. They establish a separate discipline in computer science that goes far beyond the simple processing of complex measurement and survey data.

The spectrum of applications for pattern recognition in medicine and the entire health care system is broad. This article will focus on three key areas of

J. Hornegger (✉)
Friedrich-Alexander Universität Erlangen-Nürnberg (FAU), Erlangen, Germany
e-mail: praesident@fau.de

I. Knappertsbusch and K. Gondlach (eds.), *Work and AI 2030*,
https://doi.org/10.1007/978-3-658-40232-7_33

application: 1) the detection of anomalies in medical image data, 2) the processing and use of sensor data and 3) the integration of health data for data mining.

2 Status Quo and Case Studies

2.1 Detecting Anomalies: AI and the Evaluation of Medical Image Data

X-ray, ultrasound (US), computed (CT) and magnetic resonance tomography (MRI)—imaging techniques have revolutionised medicine because they allow a view into the living organism. Estimates suggest that imaging provides about 90% of the total stock of medical data and is thus best suited for self-learning systems. Machine learning can support diagnostics particularly well when medical information is already available digitally. Especially in oncology or dermatology, AI-supported diagnosis programs are expected to provide valuable decision support because the characteristic image features of the different types of cancer are sometimes difficult to recognise and require intense training phase for the physician.

Initial demonstrations show how powerful AI is even today in the evaluation of images. The University of Heidelberg, for example, had more than 150 dermatologists compete against a computer algorithm in 2019, which had been trained provides a large number of annotated open-source images. The task was to classify suspicious lesions on 100 images and to distinguish between moles and black skin cancer. As a result, the algorithm outperformed 136 of the 157 dermatologists across all levels of experience in terms of average specificity and sensitivity (Brinker, et al., 2019). The largest study to date on automated skin cancer diagnosis, led by the Medical University of Vienna, also from 2019, reached similar results (Lancet, 2019).

The central question here is not necessarily whether the computer is better than the doctor. In many cases, a computer program is also superior to humans because it has constant performance 24/7. Monotonous and lengthy evaluations increase the risk of human errors, both in terms of false-positive results and possible micro-metastases being overlooked.

Intelligent algorithms can help here to drastically reduce the workload, for example by separating complicated from simple cases which can then be manually examined in a targeted manner. Especially in the screening of breast cancer, studies suggest that a combination of AI and experienced physician delivers the best diagnostic results (Scinexx, 2018). Similarly positive effects are promised by the interaction between physician and AI in the evaluation of very complex examinations, such as whole-body CTs.

Self-learning algorithms will more than revolutionise the classic imaging by X-ray, US, CT and MRI. The close collaboration between medicine and computer science will increasingly lead to completely new technologies. An example of this is the project “4D + nanoSCOPE”, in which FAU is involved as the lead partner. To better understand the bone structure and anatomy of humans and to detect, for example, damage caused by stress, X-ray microscopyis to be made possible for the first time on the living organism. The layer image calculation could also be used beyond medical research—for example, to examine microfractures or corrosion processes in natural and synthetic materials.

Another example of automated image analysis beyond CT and MRI is a method that is being researched at the Max Planck Center for Physics and Medicine in Erlangen. Here, blood is to be examined for certain diseases without having to send it to the laboratory. During analysis, the a blood sample flows through a narrow transparent channel, while the individual cells are captured by a high-speed camera. Up to 4000 images per second are searched by an intelligent algorithm for features that indicate diseases. Recently, the researchers have achieved a much-acclaimed discovery: A change in the red and white blood platelets can lead to “Long Covid”. A finding that could also pave the way for a possible therapy. The method is expected to find its way to clinical routine in the next two years and could become the AI-supported standard procedure even in general practice in the medium term (Fast & efficient diagnoses by artificial intelligence, 2021).

2.2 Using Sensor Data: From Lifestyle Wearable to Health Monitoring

In the monitoring of vital functions and motion data using embedded systems or microcomputers, medicine can benefit from an already established innovation from the lifestyle sector—so-called wearables. These are already successfully used to match information of certain biomarkers with those of our everyday behaviour and to draw conclusions about our health status. The latest apps turn electronic bracelets or watches into diagnostic devices for diabetes, arrhythmia or sleep apnoea. Wearables are based onhighly sensitive sensors that provide reliable information about breathing rhythm, heart rate or movement patterns, and require powerful algorithms that evaluate these data. Self-learning systems, which are fed with a sufficiently large amount of data, are later able to analyse deviating patterns and decide, for example, whether an increased pulse is the result of physical activity or should be interpreted as an alarm signal for a health disorder.

In clinical research, there are promising approaches to integrate sensors for monitoring lifestyles and movement patterns into practical everyday objects. For example, a pair of glasses was developed at FAU that look very similar to normal glasses but can capture up to 100 different biomarkers—from vital functions to food intake. The sensors in the temple measure, for example, muscle contractions and analyse chewing sounds. In this way, they provide reliable information about individual eating behaviour and even about the type of food. From such data, the system can derive behavioural recommendations, for example, for diabetics and also serve the doctor as a support for the therapy.

Another project on which neurologists and computer scientists are also researching in the Nuremberg metropolitan region is the monitoring of Parkinson's patients. The aim is to depict the everyday life of the patients as authentically as possible and to assess the disease progression more objectively. So far, the project has developed a sensor-equipped shoe that delivers over 700 measured values for gait analysis, including speed, stride length and evenness. Numerous clinical studies have confirmed that the parameters recorded correlate very well with the symptoms of the disease. Just as important as the construction of the shoe was the development of self-learning algorithms that can both understand the pattern of the individual gait and evaluate changes. In addition, a complex analysis of context factors is required to be able to assess, for example, whether a slowed step indicates a deteriorated health condition or whether the patient is just walking with his elderly mother. The intended goal is to generate objective scores from the sensor data, on the basis of which the appropriate therapy will be decided in the future.

Initial research projects are already dealing with sensors that are not placed on, but in the body. Smart implants being designed to support the healing of bone fractures. They do not only collect data, but should be able to specifically stimulate bone growth (Medica, 2019).

2.3 Information Systems: Data Mining for New Medical Standards

AI can advance medicine wherever electronic data is available. The more data is digitally recorded, the more information is accessible to systematic evaluation and thus also to learning methods and algorithms. Hospital information systems and electronic patient records, for example, are a true treasure for medical research, for the organisation of individual treatments or for the further development of telemedical services.

According to estimates, currently only about 50% of all patient information in Germany is digitised. Increasing this percentage gradually is the prerequisite for organizing medical care more efficiently. This is the only way for medical facilities to know, for example, immediately what previous illnesses the patient has, whether X-rays exist or which medications have already been prescribed. This can avoid multiple examinations, sometimes even unnecessary operations.

However, with the help of AI applications, medical informatics can not only optimise organisational processes, but also offer targeted decision support for diagnosis, therapy and follow-up care. An example illustrates this: A patient comes to his family doctor with non-specific complaints. He uses an AI-supported program that not only has access to the patient's anamnesis data, but can also compare the symptoms with a database of millions of disease cases. At the same time, the program searches the available literature for possible clues to the complaints.

Such a scenario can only become reality if the actors of the health care system—family doctors, hospitals, care facilities—are networked with each other on the one hand and use a homogeneous or compatible data infrastructure on the other. We are still far from both today, but there are approaches to change that. Since 2018, the BMBF has been funding several data integration centres in Germany as part of the Medical Informatics Initiative, one of them at the Universitätsklinikum Erlangen. Their goal is to pool the data of regional studies and health service providers and make them accessible for structured analysis (https://www.medizininformatik-initiative.de/de/konsortien/datenintegrationszentren).

3 Challenges and Solutions: The Handling of Data must be Regulated

The application areas already described show the gigantic potential of AI-supported medical informatics. Of course, as with any new technology, there are also reservations and—not to be left unmentioned—specific risks.

The convincing successes of self-learning algorithms in the evaluation of medical image data have triggered more than just storms of enthusiasm. . Occasionally, the—certainly not entirely serious—demand was raised to stop investing in the training of radiologists. The assumption that AI could one day replace medical professionals is unfounded and just as absurd as the idea that the calculator could make mathematicians unemployed. AI is not a competition, but a valuable decision support system that can significantly accelerate and improve processes. The final decision must be made by the doctor, the traditional examination is not made obsolete by AI.

However, more serious is the danger that commercial diagnostic programs—in a similar way to wearables—are used by patients without medical supervision. Anyone who relies solely on their smartphone for skin cancer screening or determines their heart rhythm exclusively with their own smart watch is taking high risks.

At the same time, however, there must also be an awareness in doctors' offices and clinics not to rely blindly on advanced technologies, because even intelligent algorithms can produce errors. This also applies to medical research in the evaluation of causalities. Identifying spurious correlations between different variables of a data set is ultimately not the responsibility of the machine, but of the human.

The most serious challenge in establishing self-learning systems in medicine is certainly data protection. Self-learning algorithms depend on extensive information—if this is not available or access to it is denied, AI cannot exploit its potential. On the other hand, individuals must be reliably protected from becoming transparent patients against their will. The solution to this dilemma, according to many experts, lies in voluntary data donation. In order to create or increase the willingness for this, individuals must be informed about the benefit of their donation and the use of personal data must be made transparent. Also, the possibility to revoke consent at any time can contribute to more acceptance and trust.

4 Outlook on AI in the Medical and Pharmaceutical Work Environment in 2030

Whether glasses, shoes or other wearables, sensor-based mobile computing in combination with AI algorithms holds enormous potential for the medical sector. —they range from promoting a healthy lifestyle to early diagnosis and monitoring of chronic diseases to support in acute emergencies. The monitoring of health status and even medication could increasingly be shifted from doctors' offices and hospitals to the home enviroment—with expected positive effects for cost reduction in the health care system and for the psychological well-being of patients. Overall, AI-supported analysis of sensor data can help to make medicine more personalised and to focus more on preventive measures.

In addition, data mining offers considerable potential. The ambitious and rewarding goal is to pool the growing data treasures in a national infrastructure. On this basis, gigantic amounts of medical information could be meaningfully evaluated and empirical knowledge systematically generated and made usable by means of AI algorithms. Data mining programs are able to recognise, learn and interpret certain patterns in organised data sets. The detection of health risks depending on various factors—genetic makeup, lifestyle, demographic variables up to regional environmental influences—can not only improve the treatment of acute cases, but also revolutionise medical research, for example on the origin of diseases. It is to be expected that data mining algorithms in combination with exploratory multivariate statistics will reveal new correlations and causal relationships that have remained hidden to science so far.

We can also relate this back to imaging; by using large image databases, deep learning systems can do more than just easily detect known anomalies such as tumours. They can also help to decipher symptoms that can develop into serious diseases in later stages. An example of this are structural changes of the optic nerve head in the human eye and their influence on a later glaucoma disease (Diener et al., 2021).

5 Summary

We can have high expectations of what AI can achieve in medicine; the leap from research labs to widespread application can succeed. In many areas of application, intensive research is being conducted—but no one can predict what will have become established in medical practice in the next ten years. The example of autonomous driving shows that ideas sometimes take decades to work safely in everyday life. And some ambitious goals, such as deciphering the song of whales, could prove to be permanently unattainable. Artificial intelligence will change many things in medicine, but not everything.

References

Brinker, T J et al. (2019). Deep learning outperformed 136 of 157 dermatologists in a head-to-head dermoscopic melanoma image classification task. https://www.sciencedirect.com/science/article/pii/S0959804919302217. Accessed: 6. Juli 2021.

The Lancet (2019). Comparison of the accuracy of human readers versus machine-learning algorithms for pigmented skin lesion classification: an open, web-based, international, diagnostic study. https://www.thelancet.com/journals/lanonc/article/PIIS1470-2045(19)30333-X/fulltext. Accessed: 6. Juli 2021.

Scinexx (2018). KI erkennt Brustkrebs. https://www.scinexx.de/news/technik/ki-erkennt-brustkrebs/. Accessed: 6. Juli 2021.

Max-Planck-Zentrum für Physik und Medizin (2021). Schnelle und effiziente Diagnosen durch künstliche Intelligenz. http://www.mpzpm.de/de/news-events/news-detail/article/News/detail/schnelle-und-effiziente-diagnosen-durch-kuenstliche-intelligenz/. Accessed: 6. Juli 2021.

Medica (2019). Smarte Implantate: Verbesserung der Heilung von Knochen. https://www.medica.de/de/News/Archiv/Smarte_Implantate_Verbesserung_der_Heilung_von_Knochen. Accessed: 6. Juli 2021.

https://www.medizininformatik-initiative.de/de/konsortien/datenintegrationszentren.

Diener, R., Treder, M., & Eter, N. (2021). *Diagnostik von Erkrankungen des Sehnervenkopfes in Zeiten von künstlicher Intelligenz und Big Data.* https://doi.org/10.1007/s00347-021-01385-6.

AI in the Clinical Treatment Path

Potentials and Challenges for Health Care Providers—Opportunities for Patients

Thomas Hummel and Monika Rimmele

1 Introduction

The development of modern medicine is a remarkable success story. Scientific research in a broad field of disciplines and their implementation in medical technology and pharmaceutical applications as well as the organisation of an efficient health care system have enabled the prevention, detection and treatment of diseases at a very high-quality level. Many diseases, for which there was hardly any chance of recovery a few years ago, can now be successfully treated or at least influenced in their development so that the life span and quality of life of the patients are sustainably improved.[1] Although challenges and solutions differ in industrialised and emerging countries[2]; medical care is continuously improving worldwide.

This success also creates new challenges. In a longer living and on average older population, the demand for health services increases. This is contrasted by

[1] (Ortiz-Ospina & Roser, 2016).
[2] (WHO 2020).

T. Hummel (✉)
Strategy and Innovation, Siemens Healthineers, Erlangen, Germany
e-mail: thomas.hummel@siemens-healthineers.com

M. Rimmele
DiGa Factory, Berlin, Germany

I. Knappertsbusch and K. Gondlach (eds.), *Work and AI 2030*,
https://doi.org/10.1007/978-3-658-40232-7_34

a growing shortage of skilled workers in medical professions.[3] At the same time, medical innovations lead to a flood of publications, therapies, drugs and guidelines. The containment of costs in the health system as well as new regulatory requirements also increase the administrative burden for the medical staff. This situation is no longer sustainable and requires new solutions.[4]

The digitisation of the health care system offers enormous opportunities. Great potentials are attributed especially to the use of AI methods in diagnosis, therapy decisions, the monitoring of treatment success, as well as the non-stationary, long-term care of chronic diseases.

2 Status Quo—Inventory and Case Studies

The interest in digital medicine has been high for years.[5] The Covid-19 pandemic has additionally shown how vulnerable modern society is and how critical the responsiveness of medicine is.[6]

Digital solutions enable the simplification, standardisation and automation of repetitive and administrative processes. They can support the medical staff by providing relevant data and AI-based analysis results in the context of specific interventions, which allows the relief of scarce resources and re-focusing on the interaction with the patient. Similarly, in remote care management, potential problems can be detected early and targeted reactions can be initiated by using AI. The availability of accurate data from digital biomarkers and the insights derived from them by means of AI can further improve the quality of diagnosis, therapy decision, monitoring of treatment success and care of chronically ill patients.

Digital solutions are not necessarily new. Telemedical applications are implemented today in different degrees of maturity, also AI applications have been available for some years. However, due to the complexity of diagnosis and treatment especially in multimorbid disease patterns, feasibility studies and prototypes still prevail at the moment, although some AI applications are currently being brought into clinical routine or have already been successfully brought into

[3] (Institute of Medicine (US) National Cancer Policy Forum, 2009).

[4] (Global Burden of Disease Health Financing Collaborator Network, 2019).

[5] (Elmer, 2019).

[6] (Mumm & Rodler et al., 2021).

clinical routine.[7] In the current pandemic, for example, new algorithms for the diagnosis of COVID-induced pneumonia were provided in a short time in computer tomography based on existing AI solutions and brought into application in accelerated approval procedures.[8]

Accordingly, there is a certain euphoria regarding the potentials of digitisation and AI in the health care sector.[9] Estimates of market sizes as an indicator for the expected penetration of the health care system with these solutions should be treated with caution. The market is unclear, besides startups and companies from the medical technology, biotechnology and pharmaceutical industries, also technology companies, health insurance companies and state health institutions and foundations are active. AI applications have a very high innovation speed and there is no generally accepted, clear definition of different AI markets in the health sector.

3 Challenges and Solutions

Given the considerable potential, the question arises as to what is slowing down the implementation of the potential of AI. These challenges fall into four thematic complexes:

Sociological challenges arise from reservations and doubts due to the lack of familiarity with AI technologies in the medical environment[10] as well as the resistance to changes in socio-technical systems.[11] The concern of being replaced by AI, as well as doubts about the validity of the results, are partly based on knowledge gaps.[12] Such difficulties can be addressed by involving the affected parties early on.[13]

Open financing issues lead to *economic challenges*. For most AI-based solutions, there is no reimbursement by the payers of the health systems.[14] Since high

[7] (Briganti & Le Moine, 2020)

[8] (Greenspan & San José Estépar, et al., 2020).

[9] (Bohr & Memarzadeh, 2020)

[10] (Gaube et al., 2021).

[11] (Elish & Watkins, 2020).

[12] (Allen et al., 2020).

[13] (Huisman et al., 2021).

[14] (Chen et al., 2021).

quality does not necessarily lead to concrete efficiency gains, the use of AI is not always economically feasible.[15] The increasing focus of medicine on quality rather than quantity will require more measurable and traceable, specific results here. In particular, the explainability of the results of AI algorithms is of great importance.[16]

Regulatory, legal and ethical challenges arise, among other things, from data protection[17], whereby regional and local differences in framework conditions have to be taken into account, such as HIPAA in the USA and GDPR in Europe. Further topics are the question of the classification of an AI algorithm as a medical device and the resulting legal requirements, liability issues, as well as the efficient testing and approval of the quality of algorithms that are created in the short cycles of software development rather than the longer cycles of traditional medical technology.[18]

Technological challenges lie on the one hand in the availability of high-quality and quantitatively sufficient data for the training of AI algorithms, and on the other hand also in the necessary expertise and the high development speed of AI applications.[19] Although AI technologies are not the core competence of health service providers, they have access to relevant data. Partnerships and the provision of AI services by companies offer a high potential for solving these challenges.

4 Outlook on AI in the Clinical Care Pathway in 2030

Any outlook over a decade into the future, especially in a highly dynamic field such as artificial intelligence in healthcare, is a glimpse into the crystal ball. A specific scenario for 2030 can be sketched out based on four theses:

First, the place where healthcare takes place shifts. While today the patient mostly comes to the hospital or the doctor for medical care, care and nursing will come to the patient in the future. Data will be collected by digital biomarkers and

[15] (Sorace, 2020).

[16] (Amann et al., 2020).

[17] (Gerke et al., 2020).

[18] (Wu et al., 2021).

[19] (Kelly et al., 2019).

continuously evaluated by AI applications and translated into corresponding recommendations for the medical staff. This will support patients in their daily life with the disease and achieve a targeted prioritisation of treatments, such as hospitalisation or an intervention by specialists. Interventions that are centralised today (such as checking the healing or disease progression) will be decentralised, while activities that are decentralised today (such as consulting experts) will be centralised by the interplay of AI solutions and telemedicine applications. This spatial shift of health services will be necessary to cope with the increasing number of chronically ill or long-term patients with the same or increasing quality, without overloading the health system.

Second, diagnosis and therapy decisions will continue to be made by the doctor, but routinely with the comprehensive involvement of AI systems. AI will not replace the doctor; the potential of AI lies rather in simplifying and supporting various activities. Examples are triage by urgency, preparation of diagnosis and therapy recommendation by in-depth data analysis and guidelines, or automatic documentation of the patient conversation by speech recognition. Such technologies are partly already in use today. The AI support of the medical staff will be widely accepted standard by 2030, similar to how the smartphone is a matter of course today. Ultimately, the staff shortage in the medical field will not be solved otherwise.

Third, by the ability to quickly process large amounts of data, AI solutions will enable the early and accurate diagnosis of diseases. This will lead to a shift from costly and often strenuous therapies to prevention, early diagnosis and accordingly less severe treatments. For example, a liquid biopsy can provide faster and less invasive more accurate information about the patient's tumour disease; the course of certain chronic diseases can be slowed down by targeted treatment. This is made possible by the AI-based recognition of complex patterns in large amounts of data. This shift to prevention, early detection and focused therapy is inevitable in view of the high costs of modern therapies.

Fourth, AI applications will enable the transition from quantity, i.e. billing according to procedures performed, to quality, billing according to treatment success. Today, this is hindered not only by institutional barriers and specific interests, but also by the difficulty of measuring treatment success in a comprehensible and meaningful way. By the better availability of more accurate data and their analysis by AI algorithms, quality can be increased both at the level of a single treatment (more precise treatments through more precise diagnosis) and at the population level (recognition of patterns in specific disease patterns and their treatments).

In general, AI will permeate the treatment of patients at all stages, support the medical staff and relieve them of administrative and automatable routine tasks, and ensure cost-effective and high-quality medical care.

It should be noted that the digitisation of healthcare is also associated with risks. A dehumanised and anonymised medicine, the misuse of data, the danger of faulty recommendations by algorithms trained with biased data, or the uncritical belief in AI recommendations are frequently expressed fears. It is beyond question that these issues need to be actively addressed—by legal and regulatory frameworks as well as by education and targeted design of solutions.

5 Summary and Practical Recommendations

AI use offers extensive potential to increase the efficiency and quality of medical care. The realisation of these potentials requires the solution of many challenges.

For healthcare providers, pilot projects are a good approach to test the use of AI not only, but also to actively shape it and find solutions for the challenges presented in sect. 3. The goal is to build up real experience with the technology and find acceptable solutions as well as overcome doubts and reservations. Since AI or digital solutions are not the core competence of healthcare providers, partnerships with technology companies to jointly shape the digital future are advisable. It is crucial that these pilot projects are transferred to concrete clinical routine as soon as possible, regular reimbursement is enabled and thus real value in healthcare is generated.

From the perspective of state health institutions and regulatory authorities, not only clarity regarding the legal framework, but especially regarding the establishment of cross-cutting infrastructures for data exchange and the billing of AI use in medicine is necessary. With an innovative legislation, such as the enabling of digital health applications (DiGA) by the Digital Care Act (DVG), Germany has taken some good steps in recent years to implement the digitisation of health care in a forward-looking way. Based on this foundation, further steps are needed to economically promote AI applications.

Of central importance for the acceptance and further development of AI are the patients, especially where they themselves are to use the solutions, for example in the area of prevention or care of chronically ill. At this point, the healthcare providers and health insurers as well as state agencies can contribute considerably by educating. The private sector, i.e. IT, medical technology and pharmaceutical companies, must do their part by being transparent about their business models and providing real value for the patient to support the success of AI in medicine.

References

Allen, B., Agarwal, S., Coombs, L., Wald, C., & Dreyer, K. (2020). 2020 ACR Data Science Institute Artificial Intelligence Survey. https://doi.org/10.1016/j.jacr.2021.04.002.

Amann, J., Blasimme, A., & Vayena, E., et al. (2020). Explainability for artificial intelligence in healthcare: A multidisciplinary perspective. *BMC Medical Informatics and Decision Making,20*, 310. https://doi.org/10.1186/s12911-020-01332-6

Bohr, A., & Memarzadeh, K. (2020). The rise of artificial intelligence in healthcare applications. *Artificial Intelligence in Healthcare,* 25–60. https://doi.org/10.1016/B978-0-12-818438-7.00002-2.

Briganti, G., & Le Moine, O. (2020). Artificial intelligence in medicine: Today and tomorrow. *Frontiers in Medicine,7*, 27. https://doi.org/10.3389/fmed.2020.00027

Chen, M. M., Golding, L. P., & Nicola, G. N. (2021). Who will pay for AI? Radiology: *Artificial Intelligence 2021; 3*(3), e210030. https://doi.org/10.1148/ryai.2021210030.

Elish, M. C., & Watkins E. A. (2020). Repairing innovation: A study of integrating AI in Clinical Care, Data & Society Research Institute, https://datasociety.net/pubs/repairing-innovation.pdf.

Elmer, A. (2019). Die Digitalisierung des Gesundheitswesens—Handlungsempfehlungen für Politik und Akteure, GGW 2017, Jg. 17, Heft 3 (Juli), 23–30, https://www.wido.de/fileadmin/Dateien/Dokumente/Publikationen_Produkte/GGW/wido_ggw_0317_elmer.pdf, Accessed: 12. Juli 2021.

Gaube, S., Suresh, H., & Raue, M. et al. (2021). Do as AI say: susceptibility in deployment of clinical decision-aids. npj Digit. Med. 4, 31. https://doi.org/10.1038/s41746-021-00385-9.

Gerke, S., Minssen, T., & Cohen, G. (2020). Ethical and legal challenges of artificial intelligence-driven healthcare. *Artificial Intelligence in Healthcare, 295–336.* doi: https://doi.org/10.1016/B978-0-12-818438-7.00012-5.

Global Burden of Disease Health Financing Collaborator Network. (2019). Past, present, and future of global health financing: a review of development assistance, government, out-of-pocket, and other private spending on health for 195 countries, 1995–2050. *The Lancet, 393*(10187), 2233–2260, June 01, 2019, DOI: https://doi.org/10.1016/S0140-6736(19)30841-4.

Greenspan, H., San José Estépar, R., Niessen, W. J., Siegel, E., & Nielsen, M. (2020). Position paper on COVID-19 imaging and AI: From the clinical needs and technological challenges to initial AI solutions at the lab and national level towards a new era for AI in healthcare. *Medical image analysis, 66*, 101800. https://doi.org/10.1016/j.media.2020.101800.

Huisman, M., Ranschaert, E. R., & Parker, W., et al. (2021). An international survey on AI in radiology in 1041 radiologists and radiology residents, part 2: Expectations, hurdles to implementation, and education. *European Radiology*. https://doi.org/10.1007/s00330-021-07782-4

Institute of Medicine (US) National Cancer Policy Forum. (2009). Supply and demand in the health care workforce. In: Ensuring quality cancer care through the oncology workforce: Sustaining care in the 21st Century: Workshop summary. National Academies Press (US). https://www.ncbi.nlm.nih.gov/books/NBK215247/. Accessed: 12. Juli 2021.

Kelly, C. J., Karthikesalingam, A., & Suleyman, M., et al. (2019). Key challenges for delivering clinical impact with artificial intelligence. *BMC Medicine,17*, 195. https://doi.org/10.1186/s12916-019-1426-2

Mumm, J. N., & Rodler, S., et al. (2021). (2021) Digitale Innovation in der Medizin—die COVID-19-Pandemie als Akzelerator von „digital health". *Urol J. Urogynäkol. AT,28*, 1–5. https://doi.org/10.1007/s41972-020-00126-2

Ortiz-Ospina, E., & Roser, M. (2016). Global health. https://ourworldindata.org/health-meta. Accessed: 12. Juli 2021.

Sorace, J. (2020). Payment Reform in the Era of Advanced Diagnostics, Artificial Intelligence, and Machine Learning. *Journal of pathology informatics,11*, 6. https://doi.org/10.4103/jpi.jpi_63_19

WHO. (2020). Global spending on health: Weathering the storm. https://www.who.int/publications/i/item/9789240017788. Accessed: 12. Juli 2021.

Wu, E., Wu, K., & Daneshjou, R., et al. (2021). How medical AI devices are evaluated: Limitations and recommendations from an analysis of FDA approvals. *Nature Medicine,27*, 582–584. https://doi.org/10.1038/s41591-021-01312-x

To Make Medicine That No One Has Ever Seen Before

The path from individual scenarios to comprehensive use of AI

Thorsten Gressling

Dr. Monika Lessl, Senior Vice President,
Head of Corporate R&D and Social Innovation.
and Director of Bayer Foundation.

"The use of artificial intelligence is indispensable for the development of new therapies. AI approaches are applied along the entire value chain—from the discovery and optimizsation of new active ingredients, to the conduct of clinical trials, to pharmacovigilance, to name just a few examples."

1 Introduction: From Isolated Solutions to Transformation

Today, the use of AI to solve specific questions for standard processes is a possibility of automation. By using AI, there are already punctual efficiency gains in traditional functions such as accounting, purchasing and human resources. In

T. Gressling (✉)
Distruptive Technology, Bayer AG, Wuppertal, Germany
e-mail: thorsten.gressling@bayer.com

I. Knappertsbusch and K. Gondlach (eds.), *Work and AI 2030*,
https://doi.org/10.1007/978-3-658-40232-7_35

addition to these functions, there are also adjacent areas (e.g. logistics, supply chains) that have specific use scenarios for the pharmaceutical-chemical industry that go beyond simple efficiency gains and can only be mapped with the use of machine learning.

The introduction of deep learning in the pharmaceutical industry, on the other hand, represents a disruptive step (Bayer, 2020) due to the complexity of the research object (molecule or gene) and the principle of process engineering.

Finally, the fourth generation of AI will be discussed and how methods, systems and people are already being prepared for this next step (Meakin et al., 2021).

2 Status Quo/Stocktaking and Case Studies

Less than 5% of organisations in the healthcare sector currently use or invest in AI technologies (Cesljar, 2019). However, the introduction via basic processes is conceivable simple.

a) Use in the traditional functions.
In the area of *"supportive systems"* there are—as in other industries—traditional application scenarios that are already delivered by the suppliers of the systems and can thus be installed almost incidentally.

Purchased AI. These systems play an important role as building blocks for the fourth generation of AI. In general, this "bread and butter" AI can be found, for example, in the financial systems, in tax and in the ERP—as it is already laid out in many modules in SAP, for example (Cesljar, 2019). The use in education is also taking place: With the help of AI, the employees can be better supported qualitatively and the training contents can be offered more precisely. Here, Bayer uses the system *Degreed*, for example (Degreed, 2021).

Own models. The use of machine learning (ML) in sales and marketing is actually still traditional data science: AI serves to provide the commercial management with measurable variables for strategy and tactics, for key products in key markets (Commercial Business Insights, Integrated Multi-Channel Marketing) and to advise and support the business (Bayer Career Germany, 2021a). In addition to the 'purchased models' described in the previous paragraph, own ML models are also developed in different contexts (sales, patient data, communication and social data) that also add to the stock of AI models (Bayer Career Germany, 2021a).

AI platform. With the "*Precision AI*" project, Bayer is developing a system that includes all data, platform reporting and AI services as well as serves all commercial activities of Bayer worldwide (Bayer Career Germany, 2021b). This project is another fundamental building block on the way to the overarching AI structure, in addition to the purchased models.

b) Special fields of application of the pharmaceutical-chemical industry.
The pharmaceutical industry also has very specific process steps in which AI is being newly integrated in a special way. Here, AI begins to make a more far-reaching contribution than the introduction of traditional-technical innovations in the *Molecule to Market* (Myshko & R, 2019).

Assistance in the digitised laboratory. An exciting use case within the value chain of the pharmaceutical industry is the *Lab of the future.* In the laboratory, AI is implemented in the form of *cognitive systems.* This means the use of language, the recognition of images and the understanding of contents and actions. The digital twin ("*Smart Lab*") not only provides the laboratory technicians with information about the direct workplace, but also seamlessly integrates adjacent IT systems and thus creates the *flow* in the daily work process. Bayer is implementing *Smart Labs* with different techniques and manufacturers. However, it will take many years before AI becomes widespread in this area. Besides the technology, it is especially the change management that requires a lot of care and a clear approach. Exciting new perspectives are upcoming with the use of language models like ChatGPT.

Assistance systems for the engineer. All groups of employees are affected by the introduction of AI, including laboratory managers and chemists. Because the introduction of AI-based assistance systems is also in full swing in the intellectual space. An example is the prediction of synthesis routes for the synthesis of active substances. The *'intuition',* which was previously reserved for academics, thus migrates into learning systems.

Here, the employee rights are particularly affected. This concerns both the risks in dealing with strong biometric data, e.g. by bots (voice recognition) and the opportunities that arise from the experience in dealing with assistance systems in the laboratory environment.

In the development of an active substance into a drug the identification of critical process parameters can be improved by adding ML to the mechanistic modelling. As in other industrial sectors, the support of production by data scientists is normal. However, it should be noted that today mostly traditional statistics are used and only in rare cases AI (third generation in the sense of deep learning) is used. A further reuse of these trained models is currently not given. To use the knowledge that is available in this way in the mathematical models in the company in a sustainable way will be a next step in the connection between production and R&D.

Use in production. Compared to the innovations in laboratory and experiment, the use in production is relatively traditional. Today, systems in combination of big data and data science are an integral part of quality assurance—not only at Bayer. Interesting in this context is the discussion to what extent models that are designed in production can be played back into development. Together with large partners such as Siemens (Siemens, 2018), Bayer is working on these issues.

Assistance systems for the patient: ChatGPT, Google Assistant and Alexa in Healthcare. The use of AI in the form of cognitive systems also makes its way into contact with customers. Examples are interactions via voice systems: AMI (*"Ask Med by Bayer"*) works as a Google Assistant voice action that is activated by asking the voice assistant to talk to "Bayer Pharmaceuticals" (Schwartz, 2021). The development of such systems is not yet complete. For example, a system based on Amazon Alexa in the consumer health area is being tested (Bulik, 2021). The next level is using this as an assistance system in clinical trials. Here, however, the hurdle is much higher, as here systems like ChatGPT has to meet strict requirements as a medical product.

AI in Clinical Trials: Maintaining relationships and understanding huge amounts of data. If one follows the sequence of the development of a drug, the next difference to other industries is the test of efficacy in three so-called "clinical phases". In the development of a drug, there is no way around product safety through these studies. Here, there are two points where the use of modern technologies brings about a change (ReportLinker, 2021):

On the one hand, with the help of AI, the *measurement* (that is, the data acquisition) is possible as close as possible to the subject. By 2030, much more accurate information on efficacy and side effects will be available through better developed *data ingestion* technologies.

On the other hand, better communication (again in the form of cognitive systems) and a more intensive *relationship with the subject* can be established. Another advantage is logistics: A demand-based management of test drugs is possible.

And finally, these new algorithms can help to find the *right patients* for the clinical trials. Here, too, the same applies: Secure access to medical data is the highest maxim, especially and particularly for applications with AI. Since we are in a regulated environment, the AI strategy of the federal government comes in handy here: AI regulation & AI process (intelligence and (AI) is a key to the world of tomorrow, 2021) support the use and development of these sensitive systems.

c) Disruptive possibilities.
The use of AI can also create disruptive possibilities that go beyond the functions described so far.

Research: Drug design and AI. The acceleration of drug discovery and reduction of development costs is achieved by the use of AI in *molecule design* for many years. New, however, is the creative height of the latest generation. More than 230 startups offer AI in connection with drug discovery (Smith, 2017). This includes:

- *Screening* of small-molecule libraries to identify new drug candidates,
- Innovations in AI/ML for *target identification,* validation and prioritisation,
- *Design* of new drug candidates and *repurposing* of drugs.

Also companies that have been in the field of in silico for a long time are bringing AI into their products. For example, the company Schrödinger and Bayer have formed a technology alliance to develop a de novo development solution to accelerate the discovery of innovative, high-quality drugs (BusinessWire, 2020). The technology can specify and evaluate billions of synthetically producible, virtual compounds. Machine learning is combined with Bayer's own in silico models to predict absorption, distribution, metabolism, excretion and toxicity (ADMET). In addition, the use of machine learning can reduce the number of necessary animal experiments.

Computer Vision and Medical Imaging. Bayer and Merck have recently received the status "breakthrough designation" from the FDA for a software with AI for pattern recognition. This new software is used via the Radimetrics™ platform from Bayer to detect a rare but severe disease and help save and improve lives (Oelrich, 2020).

Fast drug development. Once a molecule is found, it has to be developed into a product. Here we expect that the quality of the AI models will increase significantly in the next few years, so that the material and resource use will also be positively influenced. Specifically, four application fields have been identified, in which ML-based systems are being developed or already used:

- Prediction of **material properties** and the formulation of recipies,
- Optimisation of **processes** and workflows using process mining and AI-enriched adapters,
- Evaluation of natural **language,** such as the entries in the laboratory journals or patents using *Natural Language Understanding (NLU)*,
- *Additional* use of AI in the existing mechanistic **simulation.** Here, the use of learning systems can significantly expand the limits of the simulations.

New business models. Using the example of disease prediction, a whole new connection between drug and patient emerges. *Technology-Driven Disease Prediction to Advance Patient Care* is an example from Bayer for the use of AI on data (Devoy, 2020). But also in the area of *cognitive AI* disruptive business models can emerge. Here, Amazon has filed many exciting patents around *Alexa in Healthcare (*Arsene, 2020). For example, Alexa can recognise from the voice pattern whether a user has a cold. The system helps patients with diabetes to monitor their condition (as a validated application in the sense of a medical device), and supports older people in adhering to their medication intake.

3 Challenges and Solutions: Every Epoch Dreams of the Next

Balance between lean value and disruption. Although artificial intelligence is producing more and more groundbreaking innovations, a balance between reality and future must be maintained in everyday life (Meakin et al., 2021). In most cases, investments must generate a ROI. This evolutionary approach is decided from the respective specific case. At the moment, relatively independent frontline teams each bear the responsibility for 'their' implementation. However, Bayer also takes a systematic, overarching perspective and is on the way to understand and prepare the next step towards integration of AI.

Putting it all together. How does the construction of a holistic model work in the next decade? From the previous chapters it becomes apparent that work is being done in all areas on the introduction of AI in addition to existing systems.

This is according to the maturity model the normal first step in the adaptation of new technologies. The subsequent step is always a beginning centralisation. Bayer maintains a register, in which all 'deep learning' applications and their model types are registered. This is the second step for the jump into the holistic view: What do we have at all and on which domains does this run concretely?

4 Outlook on AI in the Medical and Pharmaceutical Work Environment in 2030

A very crucial point is: The merging of deep learning on a subsymbolic level is not implemented yet (2021) and subject of basic research. And just as in the 1990s, enterprise-wide data models and catalogues were postulated as central integration, shouldn't an integration of AI also run over such a symbolic layer?

Symbolic integration. However, due to the immense implementation and maintenance effort, such an enterprise-wide repository has not been established by any company. But well prepared, the growth of the connective is at least no longer a surprise. Here, the way via a data mesh and enterprise-wide information structures is a good step, which Bayer is also pursuing.

Subsymbolic integration. A valid scenario for this can be the communication of the AI instances among each other using their own language, as described several times in sensational observations (Coldewey, 2016; Nieva, 2017): The AI islands start to coordinate themselves independently and added value is created. But what can emerge from this?

When "the company" starts to think. For the year 2040, the singularity is predicted. This is the point in time when systems based on deep learning start to design and develop themselves exponentially. What sounds like distant science fiction can currently (2023) be shown in simple *proof-of-concepts*. Thus, it is also vital for organisations to prepare for this point. 2030 is a good milestone to check whether a company is on the right track. Because the singularity will not only bring us as individuals into completely new questions, but also companies as a whole. Why should the singularity be limited to simple structures?

References

Arsene, C. (2020). Alexa in healthcare: 17 real use cases you should know about, https://www.digitalauthority.me/resources/alexa-in-healthcare/. Accessed: 1. Mai 2021.

Bayer, A. G. (2020) AI in pharma, https://pharma.bayer.com/ai-pharma. Accessed: 1. Mai 2021.

Bayer Career Germany (2021a). Commercial data scientist (m/f/d), https://karriere.bayer.de/en/job/commercial-data-scientist-m-f-d--SF357428_en_US. Accessed: 3. Juli 2021.

Bayer Career Germany (2021b), Precision AI solution operations (m/f/d), https://karriere.bayer.de/en/job/precision-ai-solution-operations-m-f-d--SF357441_en_US. Accessed: 3. Juli 2021.

Bulik, B. S. (2021). How might pharma buddy up with Alexa? Bayer Consumer's first interactive voice ad is food for thought, https://www.fiercepharma.com/marketing/bayer-consumer-streams-first-alexa-smart-speaker-interactive-ad-for-otc-product. Accessed: 3. Juli 2021.

Bundesministerium für Bildung und Forschung (2021). Künstliche Intelligenz (KI) ist ein Schlüssel zur Welt von morgen, https://www.ki-strategie-deutschland.de/home.html Accessed: 3. Juli 2021.

BusinessWire (2020). Schrödinger und Bayer entwickeln gemeinsam de novo-Entwicklungstechnologie zur Beschleunigung der Arzneimittelforschung. In: ZDNet.de, https://www.zdnet.de/press-release/schroedinger-und-bayer-entwickeln-gemeinsam-de-novo-entwicklungstechnologie-zur-beschleunigung-der-arzneimittelforschung/. Accessed: 4. Juli 2021.

Cesljar, D. (2019). SAP AI, https://customer-first-cloud.de/knowhow/sap-ai/. Accessed: 3. Juli 2021.

Coldewey, D. (2016) Google's AI translation tool seems to have invented its own secret internal language. In: TechCrunch, https://techcrunch.com/2016/11/22/googles-ai-translation-tool-seems-to-have-invented-its-own-secret-internal-language/. Accessed: 6. Juli 2021.

Degreed (2021). Bayer inspires employees to learn every day with degreed, https://www.globenewswire.com/en/news-release/2021/05/25/2235077/0/en/Bayer-Inspires-Employees-to-Learn-Every-Day-with-Degreed.html. Accessed: 3. Juli 2021.

Devoy, M. (2020). Artificial Intelligence—Technology-driven disease prediction to advance patient care, https://pharma.bayer.com/artificial-intelligence-technology-driven-disease-prediction-advance-patient-care. Accessed: 1. Mai 2021.

Meakin, T., Palmer, J., Sartori, V., & Vickers, J. (2021). Winning with AI is a state of mind, https://www.mckinsey.com/business-functions/mckinsey-analytics/our-insights/winning-with-ai-is-a-state-of-mind. Zugegriffe: 3. Juli 2021

Myshko, D., & Robinson, R. (2019). Artificial Intelligence: Molecule to market, in: PharmaVOICE, https://www.pharmavoice.com/article/2019-01-pharma-ai/. Accessed: 1. Mai 2021.

Nieva, R. (2017). Facebook put cork in chatbots that created a secret language. In: CNet, https://www.cnet.com/news/what-happens-when-ai-bots-invent-their-own-language/. Accessed: 4. Juli 2021.

Oelrich, S. (2020). Artificial Intelligence—When we suddenly know what we don't know, https://pharma.bayer.com/artificial-intelligence-when-we-suddenly-know-what-we-dont-know. Accessed: 3. Juli 2021.
ReportLinker. (2021). AI in pharma global market report 2021: COVID-19 Growth And Change, https://www.globenewswire.com/news-release/2021/03/08/2188622/0/en/AI-In-Pharma-Global-Market-Report-2021-COVID-19-Growth-And-Change.html. Accessed: 1. Mai 2021.
Schwartz, E. H. (2021). Bayer launches AMI voice assistant for doctors on google assistant, https://voicebot.ai/2021/04/19/bayer-launches-ami-voice-assistant-for-doctors-on-google-assistant. Accessed: 3. Juli 2021.
Siemens, A. G. (2018). Siemens und PSE kooperieren bei modellbasierten Lösungen, https://press.siemens.com/global/de/pressemitteilung/siemens-und-pse-kooperieren-bei-modellbasierten-loesungen. Accessed: 3. Juli 2021.
Smith, S. (2017). 230 Startups using Artificial Intelligence in drug discovery, https://blog.benchsci.com/startups-using-artificial-intelligence-in-drug-discovery. Accessed: 4. Juli 2021.

AI in the Health Market

Opportunities and challenges in a customer-centred health market

Stefan Knupfer and Stefan Weigert

1 Introduction: Health Market in Transition

For decades, the health insurance funds in Germany have focused mainly on reimbursing the costs for insured persons who become ill. But the realisation is gaining ground that it is a fallacy to consider this first health market solely as a repair operation. The role of the health insurance funds is already developing away from being mere cost bearers to being health partners. It may seem paradoxical that even a mass system like the statutory health insurance, which nine out of ten people in Germany entrust their health to, wants to offer each and every individual tailor-made services for health maintenance, health recovery or living with a disease. But the health insurance funds are able to do that. They have no choice, their insured persons expect from their health partner the same digital, individual, value-added and uncomplicated services that they are already used to in other areas of life—and rightly so. Accordingly, health insurance funds are redefining their purpose and business model, to achieve stronger customer centricity.

When aligning with the needs of their customers, there is no way around artificial intelligence for the health insurance landscape. It enables the organisations to analyse the large amount of data that is already collected in connection with the

S. Knupfer (✉) · S. Weigert
AOK PLUS, Dresden, Germany
e-mail: VORSTAND@plus.aok.de

S. Weigert
e-mail: stefan.weigert@plus.aok.de

I. Knappertsbusch and K. Gondlach (eds.), *Work and AI 2030*,
https://doi.org/10.1007/978-3-658-40232-7_36

health of the citizens and to use it for the benefit of the individual's health. Where artificial intelligence is already in use and what tasks it can foreseeably take over in the coming yearswill be illuminated in this article.

2 Inventory and Case Studies

Whether health data from wearables, results of imaging procedures, medication plans or medical diagnoses: There is no lack of fodder for artificial intelligence in the health sector (Wennker, 2020, p. 6). Accordingly, the possibilities of use and the handling of the sensitive health data by the actors require strong AI ethics.

Potential lies in the internal processes of the health insurance landscape and in the products and services for customers. A care tailored to the personal needs of the insured is not a waste, on the contrary. It makes sense to prevent diseases. It relieves the stakeholders of the statutory health insurance funds not only financially[1], but also creates the opportunity to make health a more conscious part of their everyday life and to provide the insured with the necessary resources to achieve the optimal personal health level for their individual circumstance.

2.1 Health Apps and Chatbots

One area where both health insurance funds and companies that are pushing into the health market from other sectors are gaining a foothold with varying success are health apps. Their sheer number is confusing and constantly growing. Starting from fitness apps that accompany experienced users during strength training or help beginners learn first yoga exercises, the market ranges from health apps that help their users manage their disease with specific diagnoses, to applications certified as medical devices (Jorzig, 2020, p. 116 ff.). Numerous health insurance funds create customer service apps that offer access to many of their services around the clock. Only some of these examples are AI-supported, but there are

[1] The expenditures for drugs, medical treatments and hospital stays amounted to more than 168 billion EUR in 2020 (GKV-Spitzenverband, 2020).

positive use cases. From avatars of customer advisors who answer simple customer inquiries on the website of health insurance funds and thus save waiting time on the phone or ways to branches, to chatbots that help patients classify disease symptoms outside of typical office hours[2] and thus also can also reduce the number of unnecessary emergency room visits. Artificial intelligence creates low-threshold contact points for clarifying health questions and can presumably contribute to a real relief for the health care system.

2.2 Customer Advice and Customer Feedback

Artificial intelligence also ensures more speed and efficiency in the workflow within the health insurance funds, which benefits the customers. The processing of inventory data is one such field where AI takes over tedious bureaucratic tasks. It processes submitted forms with relevant insurance data, e.g. for family insurance, by training an image recognition algorithm that detects gaps and errors and initiates necessary steps. When evaluating customer feedback, artificial intelligence analyses a large number of information in a structured way. Such data comes e.g. from the Net Promoter Score (NPS), which gives companies valuable clues about the satisfaction of their customers. Using natural language processing (NLP) and sentiment analysis, it is possible to evaluate the feedbacks both in terms of content and the mood of the customers. Instead of leaving optimisation potentials unused, processes and facts that are rated particularly positively can be further developed and concrete measures to improve customer satisfaction can be derived from negative experiences.

2.3 Care Pathway Analysis

Care pathway analysis provides information on whether treatments have taken place according to guidelines, at the right time and with the right doctor. For this purpose, the billing data of doctors, hospitals and pharmacies are translated into parts that are specific for the respective disease and its treatment. For example, certain diagnoses indicate the severity of the disease or certain treatments. The

[2] Ada Health is an example for such a service.

process is comparable to a public transport map: Each patient stops at different stations during his or her illness, whether it is a treatment, an improvement or a deterioration of the disease. However, since each treatment can be very different individually, the many paths on the map are completely mixed up and resemble more a ball of wool than a clear map. But the paths of several tens of thousands of patients with the same disease cannot be evaluated manually. This is where artificial intelligence can identifiy and cluster similar paths. This creates an overview of how typical treatment paths run or where—sticking with the public transport analogy—there are diversions or the destination is not reached. For health insurance funds, this results in important conclusions, among other things, in which areas contracts need to be further developed in favour of a better care of the insured.

2.4 Billing Audit

That the vast majority of service providers in the health care system bill correctly with the health insurance funds is beyond question. However, there are black sheep who cause economic damage to the community of insured persons in the millions by fraud—be it by means of forged prescriptions, invented treatments or manipulated bills. An AI-based radar enables health insurance funds to find anomalies in billing behaviour in large amounts of data and initiate an appropriate audit. This way, fraud prevention and the recovery of undue payments can be done even more efficiently and is less dependent on information from individuals.

3 Challenges and Solutions

3.1 Transparency and Trust

The use of artificial intelligence in the health market creates the need to establish a basis of trust between informed customers and employees and the algorithms and learning systems they face, which they do not yet know or understand. It is crucial that artificial intelligence is used for support, but never to overrule decisions made by people. To create trust in the predictions and recommendations made by AI, it is essential to make the decision comprehensible by Explainable AI, by e.g. presenting the data that was decisive for the AI together with the recommendations (Wennker, 2020, p. 67).

This is preceded by the fact that the systems used must be trained with a large amount of good data. Especially in a sensitive, confidential realm such as personal health, it must not happen that a lack of diversity in the training sets leads to people being discriminated against or wrongly advised.

3.2 Data Use and Data Protection

While this article was being written, there were still too many broken patient journeys and too few networked actors in the health care system. Health insurance companies are therefore in a pioneering role: they have long been collecting considerable information about examinations, treatments and diagnoses. A desirable future vision is that they can also evaluate this data—of course with the consent of the insured—in such a way that they can give their insured even more individual recommendations for achieving their optimal personal health level.

However, the preventive, event-independent data analysis is currently impossible. The Social Code does not develop at the same speed as the health market does. Even if § 68b SGB V offers new approaches for data analysis and individual recommendations, the statutory health insurance companies are still in a rather rigid corset, while globally operating corporations such as Google or Amazon already deal with health data in a very different way—with the difference that the SHI does not pursue profit interests. The problem at the present time is therefore not artificial intelligence itself, but the disagreement on how health data should be used.

3.3 Digital Health Literacy

An increasingly digitised health care system can only fully realise its potential if the citizens can also use the digital offers correctly. In 2020, the AOK conducted the first nationwide representative study on digital health literacy (Kolpatzik et al., 2020). The result was a wake-up call: More than half of the people in Germany today have only limited health literacy—across all population groups. This means that the majority find it difficult to find, critically evaluate and apply the health-related information that is relevant for them in their personal life situation. The study also highlighted that chronically ill people have a lower digital health literacy than people without chronic diseases.

To increase equal opportunities and enable all people to have equally good access to digital health offers, the health insurance companies were obliged by the law for better care through digitisation and innovation of December 2019[3] to promote the digital health literacy of their insured. This is not only about improving how people deal with health information, apps and data security. The increased self-efficacy of the insured will also relieve the health care system, e.g. by reducing doctor contacts or avoiding visits to the emergency room. The health insurance companies fullfill this task, e.g. by reimbursing or offering corresponding learning opportunities. However, health literacy would be strengthened even more if health in general—following the principle of "health in all policies"—received more room to manoeuvre in all areas of life and politics.

4 Outlook on AI in the Health Market in 2030

4.1 Prevention and the Health Insurance as a Health Partner

The traditional roles in the health market will continue to change in the coming years. The focus will shift towards prevention, instead of continuing to concentrate almost all resources and means on curing diseases. This includes the health insurance companies: In view of the increasing digitalisation and automation of billing processes, their function as cost bearers, who distribute the contributed funds to the service providers, is increasingly receding into the background. The future-oriented health insurance company will provide more value to its customers as a health partner—and not only of those who are ill, but also of those who feel healthy. It does not only intervene where there is something to heal. Many chronic diseases are strongly influenced by the long-term lifestyle and habits of the patients and predictive systems will enable interventions by detecting risk factors before medical treatment is necessary.

4.2 Empathy

The above-described possibilities for using artificial intelligence for administrative routine tasks have the potential to relieve the actors in the health market

[3] BGBl. I p. 2562.

first by accelerating processes and second by creating time for empathic communication between service providers and their patients or between employees of health insurance companies and their customers (Juffernbruch, 2020, p. 254 ff.). Experiential empathy is indispensable in a customer-focused health market, whose actors often accompany people through very stressful situations. However, the claim that empathic communication is reserved for humans can no longer be maintained: Chatbots are already capable of incorporating emotions—such as facial expressions—into dialogues with users (ibid.).

In this context, those people who are not familiar with dealing with digital, intelligent systems should not be forgotten. To build bridges for curious people and especially patients who can benefit from AI-supported health services in everyday life, health insurance companies can draw on their regional roots. Thus, the increasing efficiency in processing customer requests and the shift of large parts of customer services to the digital space offers the opportunity to reinvent the still widespread local branches and create places for exchange and learning opportunities to build digital competencies with low-threshold offers.

4.3 Handling Data

In order for AI systems to really unfold their potential benefit for prevention and health maintenance, the provision of medical data is indispensable. Nevertheless, the definition of data protection in the health context must always ensure that individuals retain control over their data and the decision who may access it at all times. Health insurance companies want to proactively approach their customers individually and many customers already expect this. In addition, the evaluation of comparative data and the preventive addressing of insured persons for the purpose of disease avoidance represents a possibility to also reach the people who do not yet have the resources to positively shape their personal health.

From the potential of personalised health care, a new form of the solidarity principle emerges, in which insured people will be willing to share their personal medical data anonymously, so that offers and treatments for them and the community can be further developed. Statutory health insurers as the first health partners also represent an important counterweight to large companies in the future, who collect sensitive data from their customers for economic motives. It is up to the health insurance landscape to carefully select their external partners, who often penetrate into the health sector from other economic sectors and provide important impulses with their innovations—always guided by the question of which products and services offer real added value for their insured. Many health insurance companies have long recognised the potential and are involved as mentors, be it in accelerator programmes or in their own hubs. They share their long-standing experience in the health sector with the founders and open doors for exchange with traditional actors, to jointly create suitable offers for patients.

5 Summary and Practical Recommendations

Originally developed to enable broad access to medical care in a socially just way, the health insurance companies in Germany are now attributed a very different role by their insured persons. The development is driven by the increasing digitalisation in other areas of life and business, where the process started much earlier than in the health sector. The health insurance companies have accepted the mandate and continuously adapt their processes and offers to the needs of the people.

This change is always closely linked to innovations in prevention, diagnosis and therapies. This results in the AI-supported disease avoidance moving further into the centre of the relationship between health insurance companies and insured persons. To fill this vision with life, a fundamental change is involved: the development towards an agile organisation. This goes hand in hand with a comprehensive cultural change process. Orientation is provided by the purpose—the question of why and what for of the whole organisation, each team and each individual.

The mindset, however, does not yet eliminate the structural hurdles that the health sector faces. For example, there is still a lack of a framework for a nationwide, secure use of artificial intelligence. The federal government must initiate this. The handling of data, which already exist, but are not converted into valuable information, must also change[4]. In addition to the data protection debate, more speed is needed with which the data in the health sector are exchanged. Only if service providers send billing data to the health insurance companies in a timely manner—preferably on a daily basis—without months of delay, can the useful individual health offers for patients be generated (Jorzig, 2020, p. 116 ff.)[5].

[4] Zimmermann (2021).

[5] Ada Health is an example of this.

References

GKV-Spitzenverband. (2020). GKV-Kennzahlen Vergleich 1.–4. Quartal 2020 zum 1.–4. Quartal 2019. https://www.gkv-spitzenverband.de/service/zahlen_und_grafiken/gkv_kennzahlen/gkv_kennzahlen.jsp. Accessed: 11. Juni 2021.

Jorzig, A. (2020). *Digitalisierung im Gesundheitswesen: Ein kompakter Streifzug durch Recht, Technik und Ethik*. Springer.

Juffernbruch, K. (2020). KI als Treiber von Empathie? In S. Heinemann & D. Matusiewicz (Eds.), *Digitalisierung und Ethik in Medizin und Gesundheitswesen*. MWV.

Kolpatzik, K., Mohrmann, M., & Zeeb, H. (2020). *Digitale Gesundheitskompetenz in Deutschland*. KomPart.

Wennker, P. (2020). *Künstliche Intelligenz in der Praxis*. Springer Books.

Zimmermann, J. (2021). Die Zukunft der Prävention. Zukunftsstudie des 2b AHEAD ThinkTank. Leipzig. www.thinktank-universe.com/studien/. Accessed: 19. Juli 2021.

Data-Based Innovations in the Health Sector and Strategic Preparation of Well-Known Global IT Companies

Eckhard Hempel

"When you look at the future and ask retrospectively: What was Apple's greatest contribution to humanity? Then it will be about health."—Tim Cook, Apple CEO

1 Introduction

Google, Amazon, Microsoft, Apple and other technology giants have changed the way billions of us communicate, shop, socialise and work. Now, as both consumers and medical centres and insurers increasingly use health-tracking apps, technology companies want a bigger share of the trillions of dollars spent on health care each year.

2 Status Quo

Health is certainly not unfamiliar to the tech industry. IBM, Intel and Microsoft have long provided enterprise services to the health care industry. But increasingly, they and other companies are focusing more visibly on developing or investing in new types of technologies for doctors, patients and consumers.

In recent years, Amazon was one of the investors in a funding round for Grail, a cancer detection startup that raised more than $ 900 million. Apple acquired

E. Hempel (✉)
Invandus E&R GbR, Höchstadt, Germany

I. Knappertsbusch and K. Gondlach (eds.), *Work and AI 2030*,
https://doi.org/10.1007/978-3-658-40232-7_37

Beddit, a maker of sleep-tracking technology, for an undisclosed amount. And Google, perhaps the most active American consumer tech giant in health and biotechnology, acquired Senosis Health, a developer of apps that use smartphone sensors to monitor certain health signals, also for an undisclosed amount. Alphabet, Google's parent company, also has its own research unit, Verily Life Sciences, dedicated to developing new tools for collecting and analysing health data.

"There is no shortage of hype," says Dr. Eric Topol, an expert in digital medicine who leads the Scripps Translational Science Institute in San Diego. "We are in the early stage of learning these tools: Who do they help? Who do they not help? For whom do they only provide anxiety and false alarms?" (Singer, 2017).

3 Challenges and Solutions: Strategies and Innovations for Market Entry into the Health Care Sector

"The main reason many of these technology companies are now entering the health care sector is that the market is too big, too important and too personal for their users to ignore," said John Prendergass, associate health care director at Ben Franklin Technology Partners, a non-profit organisation in Philadelphia (Singer, 2017).

Each technology company pursues its own approach and relies on its core business strengths to ultimately improve people's health or make health care more efficient.

At **Microsoft**, more than 100,000 health care companies worldwide are paying customers today. Microsoft's focus in transforming health care is on "smart health through AI and the cloud" (Lee, 2018). The overarching strategic goal is to provide cloud and AI-based tools to unlock the enormous potential of health data. The company's Healthcare NExT initiative focuses on:

- developing the foundations for precision health care,
- enabling the health care industry's transition to the cloud, and
- supporting the people who work in health care.

Healthcare NExT's efforts are forward-looking "and a bet that cloud and AI services will really matter for improving health care—making it more efficient and giving people better access," says Peter Lee, Microsoft Research executive and corporate vice president for AI. "We're really focused on what people in health care are doing today and how we can improve it," Lee continues, and so Microsoft has developed a platform, for example, that allows medical centres to create virtual assistants for patients. Aurora Health Care, a non-profit organisation based in Milwaukee, has developed a health bot that helps patients decide which specialists to visit and schedule appointments.

Amazon could use its expertise to influence everything from the pharmaceutical supply chain to health insurance management. The e-commerce giant is serious about entering the health care sector and brings with it a non-traditional business model, infrastructure in logistics and computing and data centres, and extraordinary customer focus.

Amazon made its first major move in the pharmacy arena in 2018 with the acquisition of the online pharmacy PillPack for around $ 750 million. Unlike acquiring drugs in pharmacies, PillPack delivers users' medications directly to their homes. In particular, the company ships medications in pre-sorted bags that are taken at specific times of the day. In addition to shipping medications, PillPack includes information sheets that list when each medication should be taken, when the current batch of prescriptions will run out, when the next delivery is expected, and more. PillPack's user experience is reportedly much simpler and more intuitive than obtaining in traditional pharmacies.

Amazon's service "Amazon Comprehend Medical" uses machine learning to extract medical data from patient records. Amazon's algorithms analyse text for specific data types and return the results in an organised format. "We are able to search the medical data completely and automatically and identify patient details, including diagnoses, treatments, dosage and strengths, with incredible accuracy," says Matt Wood, managing director at Amazon Web Services. In compliance with data protection regulations, this will help identify patients who qualify for experimental studies in the drug approval process. By automatically scanning and retrieving key data from the records, the number of staff normally required for this task will be significantly reduced.

Apple positions itself to improve health care as the data backbone for patient care and pursues three strategic core elements:

- software for monitoring and sharing data via apps,
- hardware (e.g., Apple Watch and iPhone) as ubiquitous device platform, and
- expanding relationships for monitoring and sharing data.

For example, Apple introduced HealthKit in 2014 to provide Apple users with a central repository and track health and fitness data on their Apple devices.

In 2015, Apple released ResearchKit, which aims to help doctors and scientists collect more frequent and accurate data from participants using iPhone apps (Apple, 2021).

In January 2018, Apple announced that its Health app would allow users to access their electronic health records directly on their iPhones. When Apple launched the feature, the company started with more than 200 participating American hospital networks.

Google is well equipped as a champion in web search in terms of hardware and software for handling huge amounts of data. Google's main strategy is currently the transition from a "mobile first" world to an "AI first" world (D'Onfro, 2016). Google sees health care as one of the biggest areas where the benefits of AI will play out in the next 10 to 20 years. Therefore, Google relies on AI in almost all areas of the company and no other company in Silicon Valley invests as heavily in health-related companies as Google Ventures (more than 75 investments since 2013) (Meskó, 2018).

Regarding the scaling of new products, Google also has a decisive advantage: it is available almost everywhere and everyone knows it. Google currently has seven products with more than one billion users each and thus has excellent sales and customer relationships. Furthermore, Google is able and also willing to take on unpredictable and risky so-called moonshot projects, e.g. diagnostic and therapeutic wearables, nanoparticle phoresis, Google Glass, Google Lens, etc.

Google already covers a broad and diverse spectrum of adjacent technologies and capabilities that have a high innovation potential in the health sector and will create new fields of work in the combination of medicine and technology, including e.g. robotics, artificial intelligence, humanoid robots, robotic arms and AR/VR.

Google's patented and published AI-based predictive modelling with data from electronic health records is expected to improve personalised health care by creating accurate and scalable predictions for a variety of clinical scenarios (Mossin, 2019). The AI models developed by Google achieved an unprecedented accuracy for predictions on diagnoses, hospital mortality, unplanned readmission within 30 days, prolonged length of stay and final discharge times of a patient and outperformed in all cases conventional, clinically used prediction models (c.f. Mossin, 2019).

4 Outlook on AI in the Medical and Pharmaceutical Work Environment in 2030

4.1 Data-Based Business Models as the Basis for New Fields of Work in Medicine

Data-based business models will reshape the landscape for medtech and healthcare players. By tapping into the information dimension, digital transformation influences the health sector far beyond the products. The digital transformation acts here as an accelerator to integrate not only doctors, insurers and clinics, but also increasingly the patients themselves and to establish new services based on the information and its processing to increase the quality of treatment, efficiency and patient satisfaction as well as cost reduction. As a result, the roles of all stakeholders in the health care system will be redefined and expanded by new actors and associated fields of work. Patients will receive diagnoses from automated tools and actively participate in treatment decisions and in exchange actively share data and information. Doctors will increasingly decide on the basis of individual real-time data from the real patient life and share and use experiences in networks. The traditional diagnostics, the laboratories and also the pharmaceutical industry will continue to provide diagnostic services, instruments and drugs as needed and introduce complementary and accompanying diagnostics as well as innovative drugs based on the interpretation of the newly acquired data. The digital transformation will also bring forth new actors. Due to the multitude and diversity of data, services will create new fields of work around the collection, preparation and merging of data using AI. Another professional group will establish itself with the interpretation of the data in the clinical routine. In addition, it will be possible with the help of AI to evaluate health-relevant and real-time parameters that are recorded directly on or in the patient and to integrate them into the treatment paths. With applications among others in the areas of imaging and diagnostics, virtual assistants, remote monitoring and hospital care, the AI health sector has brought a record rate of new companies to the market every quarter for several years (CBINSIGHTS, 2021).

4.2 Patient-Generated Health Data and Electronic Health Records

The rapid development and acceptance of so-called wearables, Internet of Things (IoT) devices and mHealth apps is also increasingly changing the health landscape (Rebhan, 2018). Data from these devices—a form of patient-generated health data (PGHD)—provide a broad insight into the health of patients across the entire healthcare spectrum. The expectations of patients and payers will soon shift the patient care model from episodic, visit-based care to continuous patient-disease and wellness management. Characteristic of PGHD is that the patient, not the provider, is responsible for the capture and recording of the data. In addition to pure activity level data, symptom-related data and treatment outcomes, as well as drug tolerability and biometric data, will contribute to the fact that today's medical environment will be supplemented by new fields of work around patient remote monitoring, prevention and well-being.

Such electronic health records will capture a wealth of new information that can be used not only for patient treatment, but also for operational efficiency. So far, most analyses of these data have focused on structured data that were captured in forms and in the field. However, the universe of unstructured data, which can be found in clinical notes, sensors in wearables, patient-reported data, as well as genomics and data on social determinants of health (e.g., socioeconomic, environmental, behavioural), is much broader. In this context, so-called natural language processing (NLP) offers a range of techniques for distilling structured data from text notes or speech. This key technology has been dominated by the major IT companies for years and tested and continuously expanded in practice in digital voice assistants from Google, Amazon, Microsoft and Apple.

Complex algorithms will possibly soon be transferred to practice here as well and help clinicians make incredibly accurate statements about our health based on largely inexplicable correlations in these data (Burt, 2018; Froomkin, 2019).

Perhaps soon, diagnoses generated by AI will have demonstrably better success rates than those of human doctors.

5 Summary and Practical Recommendations

Digitisation and the use of AI will fundamentally change the health care sector globally and redefine the way devices and humans interact and communicate with each other, providing new efficient tools to ensure that patients are better cared for, costs are reduced and treatment outcomes are significantly improved.

Born as IT and/or e-commerce companies, the DNA of these companies is digital and in today's medical landscape, the ability to manage and leverage digital resources is of utmost importance. The expertise of technology companies in data management and analysis, as well as their considerable computing power, can help support all stakeholders in the health care sector by providing new ways of accessing and delivering health.

All major IT firms are looking for ways to tap into the electronic records of medical providers to make the data more accessible for consumers and their doctors. Many are making progress in these areas by playing to their strengths: Apple's patient-oriented vision prioritises consumers, while Google continues to apply AI to almost everything, from medical devices to lifestyle management solutions, Microsoft builds health data management on its cloud platform Azure and Amazon expands its e-commerce business to the pharmacy sector.

Health is and remains our highest good and both opportunities and expectations will continue to increase. The variety of innovative technologies and the ambitions of established and new players in the medical environment that are outlined here will change existing professions and create new ones. It can be expected that the health care sector will also be the largest employer in the world in the future.

References

Apple. (2021). Institutions that support health records on iPhone and iPod touch. Apple company web page. https://support.apple.com/en-us/HT208647. Accessed: 27. Juli 2021.

Burt, A. (2018). How health care changes when algorithms start making diagnoses. Harvard Business Review. https://hbr.org/2018/05/how-health-care-changes-when-algorithms-start-making-diagnoses. Accessed: 27. Juli 2021.

CBINSIGHTS. (2021). AI 100: The Artificial Intelligence Startups Redefining Industries, CBINSIGHTS company web page. https://www.cbinsights.com/research/report/artificial-intelligence-top-startups/. Accessed: 27. Juli 2021.

D'Onfro, J. (2016). Google's CEO is looking to the next big thing beyond smartphones. BusinessInsider. https://www.businessinsider.com/sundar-pichai-ai-first-world-2016-4. Accessed: 27. Juli 2021.

Froomkin, A. (2019). When AI's outperform doctors: Confronting the challenges of a tort-induced over-reliance on machine learning. University of Miami Legal Studies Research Paper no. 18-3. https://repository.law.miami.edu/cgi/viewcontent.cgi?article=1678&context=fac_articles. Accessed: 27. Juli 2021.

Lee, P. (2018). Microsoft's focus on transforming healthcare: Intelligent health through AI and the cloud. The official Microsoft Blog. https://blogs.microsoft.com/blog/2018/02/28/microsofts-focus-transforming-healthcare-intelligent-health-ai-cloud/. Accessed: 27. Juli 2021.

Meskó, B. (2018). Google's masterplan for healthcare. The Medical Futurist. https://medicalfuturist.com/googles-masterplan-for-healthcare/. Accessed: 27. Juli 2021.

Mossin, A. (2019). System and method for predicting and summarizing medical events from electronic health records. US patent application. https://patents.google.com/patent/US20190034591A1/en?oq=US20190034591A1. Accessed: 27. Juli 2021.

Rebhan, A. (2018). Connected care series: Patient-generated health data. AdvisoryBoard. https://www.advisory.com/topics/business-intelligence-and-analytics/2018/08/connected-care-series-patient-generated-health-data. Accessed: 27. Juli 2021.

Singer, N. (2017). How big tech is going after your health care. The New York Times. https://www.nytimes.com/2017/12/26/technology/big-tech-health-care.html. Accessed: 27. Juli 2021.

AI in education and training

Introductory Qualification on Artifical Intelligence

Productive and Humane Work Design with AI in Small and Medium-Sized Enterprises

Sebastian Terstegen, Bruno Schmalen, Andreas Hinz and Maike Pricelius

"The organisational aspect is very important for AI and is usually not recognised and taken into account by the technicians."—Managing director of a medium-sized production company

S. Terstegen (✉)
ifaa – Institut für angewandte Arbeitswissenschaft e. V. (Institute of Applied Industrial Engineering and Ergonomics), Düsseldorf, Germany
e-mail: s.terstegen@ifaa-mail.de

B. Schmalen
SCHMALEN-Kommunikation und Training, Langenhagen, Germany
e-mail: schmalen@offensive-mittelstand.de

A. Hinz
RKW Rationalisierungs- und Innovationszentrum der Deutschen Wirtschaft e. V. Kompetenzzentrum, Eschborn, Germany
e-mail: hinz@rkw.de

M. Pricelius
G-IBS Gesellschaft für Innovation, Beratung und Service mbH, Berlin, Germany
e-mail: maike.pricelius@g-ibs.de

I. Knappertsbusch and K. Gondlach (eds.), *Work and AI 2030*,
https://doi.org/10.1007/978-3-658-40232-7_38

1 Introduction

AI is undisputedly a new enabling technology that will have a similar impact on the world of work as the past industrial transformations, such as the introduction of the steam engine. On the one hand, AI opens up new opportunities and possibilities for our work and our lives. On the other hand, it also entails great dependencies and dangers.

Moreover, the new enabling technology AI differs fundamentally from old enabling technologies: While people could see and feel the change associated with the steam engine 250 years ago, AI is not visible today. It performs its work invisibly packaged in technical methods and devices. This makes access and understanding of this new technology more difficult and can easily lead to uncertainties. Therefore, we should have criteria that enable us to

- recognise developments that are related to AI,
- perceive their opportunities and dangers, and
- use AI in a human-friendly and economic way in companies.

The en[AI]ble project team[1] wants to contribute to meeting these developments and the associated changes better with an AI qualification.

2 Status Quo

According to a survey in 2019, the share of companies that use AI was just under six percent (Rammert et al., 2020). The companies cite various problems as reasons for the hesitant use. They include data security and data protection, the financing of AI, the lack of tailor-made solutions and finally also a lack of competencies in the company (Weber et al., 2018; Fraunhofer IAO, 2019).

The connection between the successful use of AI in the company and the qualification of the employees was examined in an international study commissioned by Microsoft, in which interviews were conducted with around 12,000 specialists

[1] The project en[AI]ble is funded by the Federal Ministry of Labour and Social Affairs (BMAS) within the framework of the initiative New Quality of Work (INQA) under the project number: EXP.01.00008.20 and is accompanied by the Federal Institute for Occupational Safety and Health (BAuA) in terms of content. The project sponsor is the Society for Social Business Consulting mbH (gsub). Further information: www.arbeitswissenschaft.net/enaible.

and managers from 20 countries in March 2020 (see Companies weigh AI and qualification equally, 2020). Companies were able to successfully introduce and use AI in pilot projects when they simultaneously qualified their own employees and established a corresponding learning culture.

3 Challenges and Solutions

To determine the support needs for an AI qualification, the en[AI]ble project team interviewed various target groups. The interviewees were actors who either want to qualify themselves or who want to offer qualification as multipliers: entrepreneurs and managers from small and medium-sized enterprises (SMEs), works councils and representatives of associations and consulting organisations.

A central finding of the conversations is: The perception of AI as a "future technology" in the sense that it will only become relevant in the future, obscures the fact that AI is already embedded in many things that we use every day. Everyone agreed that they have to acquire AI competencies in their own field of action.

The SMEs are open to the potentials of AI, but they lack access to the systematic use of AI, and in specific terms there is usually uncertainty about the profitability of investments in AI. The competence development of managers and employees is seen as necessary, both in terms of mastering the new technology and dealing with data. The expectations of the SMEs for qualified support are broad: basic knowledge about AI, economic benefits and change processes, social skills, pedagogical skills, ability to cope with uncertainties and fears.

The interviewed works councils (WC) identify the identification of AI applications and the associated regulatory needs as essential challenges. WC fear and experience increasing monitoring, job loss fears, loss of tasks, work intensification and surveillance pressure due to the use of AI. However, they also see potentials to design working conditions in a human-friendly way through the use of AI. In order to act competently and purposefully in this complex field, they see support needs from AI experts. These can also be recruited internally with appropriate professional qualification.

AI plays a subordinate role in the consulting practice of the intermediary organisations interviewed. Almost all are aware that training, consulting and coaching will develop and change through AI. The vast majority of consulting organisations wish for a supplementary qualification that complements the existing domain knowledge of the consultants with AI competencies.

These mentioned needs and requirements - supplemented by comprehensive research and evaluation of technical literature and existing offers for AI support -

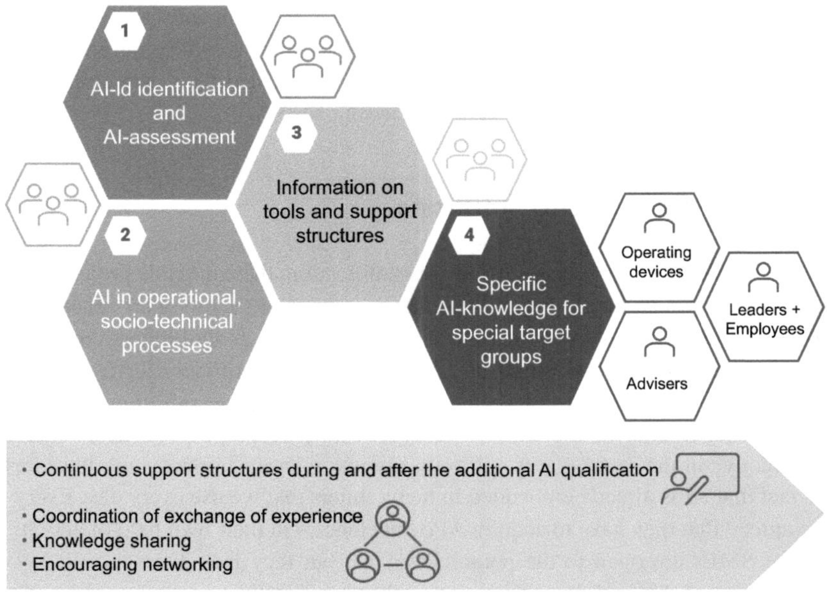

Fig. 1 Structure and components of the AI qualification developed by the en[AI]ble project team (ifaa, 2021)

form the basis for the AI qualification developed by the en[AI]ble project team. The aim is to convey basic competencies for criteria for the productive, preventive and health-friendly design of AI applications. These competencies should be integrated into the existing domain knowledge of the target groups, so that they can take into account and incorporate the topic of AI into their previous tasks (see Fig. 1).

4 Outlook on the AI Supplementary Qualification in 2030

Following the expressed problem perspectives and expectations and based on the concept of AI additional qualification, an outlook for the year 2030 will now be given.

The high development dynamics in the field of AI require continuity and adaptability in supporting small and medium-sized enterprises. The well-developed support structures provide the SMEs and the economic actors with consistent and reliable know-how that is up to date with the latest developments. This is not only about the technological potentials of AI, but also about their economic, organisational and personnel policy implications in the business landscape.

Cornerstone 1: Through experimentation arenas, companies develop competence for the planning, implementation and evaluation of AI projects.

The focus of the expanded competencies is not technological expertise, but the mediation of AI knowledge in socio-technical contexts, the exchange of experiences on process and work design and project competence. This requires new formats with experimentation arenas, in which companies and actors realise their projects directly. In close cooperation with universities, universities of applied sciences and vocational schools, products, services and complex solutions are created. The participants are not customers, but partners of a consortium.

Institutions, organisations and associations will formulate offers that, in addition to the technological tasks, take into account the preventive aspects of working with new AI-based solutions. Three different application scenarios characterise the AI activities in the companies:

- Scenario 1: Solutions and things that are already used in the company and contain AI.
- Scenario 2: Solutions and things that are to be purchased, and where it has to be checked whether AI is included.
- Scenario 3: AI applications that are to be implemented and used in the company.

In the three scenarios, a systemic and holistic approach involving the stakeholders is required. Data sovereignty, transparency and explainability are guiding criteria for data handling and process design. Project competencies will serve these criteria.

Here, "data sovereignty" means that a user of AI knows what happens to his or her captured data and agrees to its use.

"Transparency" refers to the property that the actions and functions of the system are comprehensible. The demand for maximum transparency is often not fully met, as many models are so complex that users of AI cannot see through the processes.

This is exactly where the third criterion comes in, the "explainability". The company needs the competence to explain the laws, processes and their backgrounds and thus to get all actors "on board". Companies and individuals develop an awareness of how AI works and how data is used. No in-depth IT knowledge is required for this. Mastering this ability will be an important task in the "learning field AI" (Offensive Mittelstand, 2019a, p. 43).

Cornerstone 2: Consultants accompany companies and their employees in a productive and preventive use of AI.

Dealing with AI projects and accompanying AI actors has become an important part of the consulting services. The domain knowledge of the consultants is combined with clear criteria for AI solutions and their implementation.

Consultants see their task in integrating AI into their existing competencies. An occupational safety expert does not become a change manager and a personal coach does not become a technology consultant. Rather, they see themselves challenged to expand their own domain knowledge with the knowledge about the effects of AI on their core competence.

Consultants sharpen the attention for where the use of AI makes sense. They have criteria with which they can assess what is happening in the company. And they know the corresponding networks and know who can help with an expertise. What is needed are design criteria that focus on both the benefit for the value creation process of the company and good working conditions and the motivation of the employees (Offensive Mittelstand, 2019a, p. 46 f.).

Cornerstone 3: All company actors are trained to recognise basic modes of operation of AI.

How AI is used depends to a large extent on the knowledge that managers and employees have about AI and how smartly and foresightedly they use and shape the new possibilities. The challenge is to maintain the balance between the requirements of technology, profitability and people. With regard to the structure and functioning of AI, its operators and users should know,

- which sensors are in the "object" that they want to purchase or use, and which data are collected by these sensors,
- where data are stored and who has access to the data and
- according to which rules AI processes their data, controls the processes and develops further.

Here it also becomes clear how important, in addition to technological competencies of the project service providers (often external), the work design competencies of the company actors are. This is what further training in the companies should focus on (Offensive Mittelstand, 2019a, p. 44 f.).

Cornerstone 4: The fields of action "leadership and culture", "organisation", "safety" and "health" form central elements of the AI design concepts. They are also important elements of an AI education.

With the introduction of AI, new requirements arise for managers: They have to integrate the strengths of technical AI systems effectively into their processes and at the same time preserve the specific abilities of people and social relationships in the company in order to maintain innovation capacity. The roles of the moderator, trainer, coach, organiser will become even more important (Offensive Mittelstand, 2019a, p. 61 ff.).

A "preventive organisation" is network- and team-oriented in its essence. This implies self-organised structures. Clear goal specifications and objectives are combined with the greatest possible leeway for the implementation of work tasks. Often, learning AI is mentioned; much more important are learning organisations and self-determined learning employees (Offensive Mittelstand, 2019a, p. 169 ff.).

Data protection is about the legal issues, under which conditions personal data may be collected, processed or used. Here, actors need professional competence, but also an understanding of the uncertainties and fears of the participants. Data security is about the question, which measures have to be taken to ensure the protection of operational data from the outside. Data security is thus a state that is to be achieved by appropriate and effective measures.

The health-promoting aspects of AI should already be considered in the planning and procurement of AI solutions and integrated into the processes, in order to design a productive and health-appropriate work 4.0. At the same time, there is a risk of increasing efficiency and productivity at the expense of health. From this field of action, tasks arise for further training of the actors. They have to learn to assess the consequences, opportunities and dangers in newly designed work processes and forms of work (Offensive Mittelstand, 2019a, p. 431 ff.).

Against this background, stable and efficient consulting infrastructures are emerging in the field of AI. The organisations set different thematic priorities and have different expertise. An exchange about the service portfolio takes place constantly. In this way, consulting and qualification offers are coordinated and interlinked. The operational actors can thus access a multifaceted range of topics, covering technological, economic, work science and human resource management fields of action (Offensive Mittelstand, 2019b).

5 Summary and Practical Recommendations

AI changes competence profiles and established task assignments. Dealing with AI requires new competencies from managers, employees, consultants and advisors and means additional qualification requirements.

The management should clarify for themselves the criteria according to which AI is used in the company. They should communicate how AI is integrated, in order to make the effects on monitoring, scope of action or workplace design transparent and thus reduce fears.

Managers should pay attention that the integration of AI succeeds better if they involve the ideas of the employees. They should accordingly reflect and adapt their leadership behaviour. They should involve the employees in the discussions about the development of new forms of leadership.

Works councils should actively communicate with the management throughout the process of AI planning and implementation and jointly find feasible design options. They should consider which measures are necessary for a safe and health-appropriate work design when integrating AI into operational functional areas.

Consultants and advisors should have the competence to integrate the topic of AI into their consulting offer. In this way, they can show the customers where the use of AI makes sense, what potentials are created by AI and how processes for the introduction of AI can be designed. And they know the corresponding networks and know who can help with specific expertise.

References

Fraunhofer-Institut für Arbeitswirtschaft und Organisation (IAO). (2019.). Studie zum Einsatz Künstlicher Intelligenz in Unternehmen. Präsentation der Gesamtergebnisse. März 2019. Stuttgart.

ifaa – Institut für angewandte Arbeitswissenschaft e. V. (Eds.). (2021). KI-Zusatzqualifizierung. Für eine produktive und menschengerechte Arbeitsgestaltung. ifaa, Düsseldorf. www.arbeitswissenschaft.net/enaible_broschuere. Accessed: 15. Juni 2021.

Offensive Mittelstand. (Eds.). (2019a). Umsetzungshilfe Arbeit 4.0. Künstliche Intelligenz für die produktive und präventive Arbeitsgestaltung nutzen: Hintergrundwissen und Gestaltungsempfehlungen zur Einführung der 4.0-Technologien. Offensive Mittelstand, Heidelberg. www.offensive-mittelstand.de/fileadmin/user_upload/pdf/uh40_2019/umsetzungshilfen_paperback_3103_web.pdf. Accessed: 15. Juni 2021.

Offensive Mittelstand. (Eds.). (2019b). Gemeinsames Beratungsverständnis der Partnerinstitutionen der Offensive Mittelstand. Offensive Mittelstand, Heidelberg. https://www.offensive-mittelstand.de/fileadmin/user_upload/pdf/gemeinsames_beratungsverstaendnis_factsheet_2019_1101_1012.pdf. Accessed: 17. Juni 2021.

Rammer, C., Bertschek, I., Schuck B., (ZEW), Demary, V., Goecke, H., & BMWi (Eds.). (2020). Einsatz von Künstlicher Intelligenz in der Deutschen Wirtschaft – Stand der KI-Nutzung im Jahr 2019. https://www.bmwi.de/Redaktion/DE/Publikationen/Wirtschaft/einsatz-von-ki-deutsche-wirtschaft.pdf?__blob=publicationFile&v=8. Accessed: 21. Juni 2021.

Unternehmen gewichten KI und Qualifizierung gleich (2020), IT&Production 5 (Juni) 2020.

Weber, T., Bertschek, I., Ohnemus, J., Ebert, M., & BMWi. (Eds.). (2018). Monitoring-Report Wirtschaft DIGITAL 2018. BMWi, Berlin. www.bmwi.de/Redaktion/DE/Publikationen/Digitale-Welt/monitoring-report-wirtschaft-digital-2018-langfassung.pdf?__blob=publicationFile&v=14. Accessed: 15. Juni 2021.

AI in Education: Educational Technology and AI

Challenges and Requirements for the Educational Technologies of the Future

André Renz

1 Introduction

The relevance of AI in education has increased significantly in recent years, raised high expectations and offers enormous innovation potential for the entire education market (Holmes et al., 2019). More and more educational technology (EdTech)[1] providers are entering the market with the idea of developing intelligent teaching and learning solutions through AI-driven approaches. A goal is to create adaptive, flexible, individualised and effective learning environments that complement traditional education and training formats. In addition, the use of AI in EdTech is supposed to enable deeper insights into the learning behaviour, response times or emotions of learners (e.g. Holmes et al., 2019; Luckin et al., 2016). In particular, flexibility and individualisation are used as central narratives by EdTech providers to highlight the transformative and innovative character of their applications and thus formulate a new promise for the education of the future.

[1] EdTech can be summarized as all measures that aim to facilitate learning and improve learning performance by creating, using and managing appropriate technological processes and resources (Robinson et al., 2007).

A. Renz (✉)
Helmut Schmidt University, University of the Federal Armed Forces Hamburg, Hamburg, Germany
e-mail: renza@hsu-hh.de
URL: https://www.hsu-hh.des

I. Knappertsbusch and K. Gondlach (eds.), *Work and AI 2030*,
https://doi.org/10.1007/978-3-658-40232-7_39

2 Stocktaking and Case Studies: EdTech and AI

The discourse of AI in the field of EdTech is complex. While stakeholders from business, science and politics already cultivate a in-depth discussion about the technology promises of AI, the awareness in the broader society is developed only at a basic level. Since technology acceptance always goes hand in hand with participation, such a development with two speeds is certainly disadvantageous (Arnold et al., 2020). Thus, a reluctant willingness to use has long contributed to the fact that EdTech could only position itself in the niche market of education. In particular, the development and implementation of data-driven, adaptive EdTech in the European area hasonly been of limited relevance for providers and users so far (Renz & Hilbig, 2020b). The COVID-19 pandemic finally promotes a new culture of experimenting with digital applications, which positively conditions the reduction of reactance and reservations and thus dynamises the developments of the EdTech market (Renz et al., 2020a, b; Peters & Bovenschulte, 2021).

Further complexities also show up within the definitional debate and the representation of AI. The boundaries, when to speak of digitalisation and when actually of AI in EdTech, are often (still) not clear-cut (Renz et al., 2020a, b). To better understand the application relevance and potential of AI in EdTech, it should be delimited that today's available AI in EdTech exclusively correspond to the concept of weak AI, i.e. they are designed for specific, clearly defined tasks. The development of a strong AI, is considered rather unlikely in the future (Zawacki-Richter et al., 2019; De Witt et al., 2020). Another peculiarity is shown in a (often) media overemphasis of a futuristic-abstract AI understanding, in which the human being is inferior to the technology. Such a representation in turn reinforces phenomena of distortion and rejection of AI (Popenici & Kerr, 2017; Renz & Vladova, 2021).

A sub-discourse that has gained importance in recent times is the debate about ethical guidelines that should be enforced more strongly in the development of AI-supported EdTech. For example, Renz and Vladova (2021) discuss a *Human-Centered AI approach*, which, under the premise of human values and norms as well as taking into account learning paradigms, aims to increase the transparency of AI-supported EdTech and thus strengthen trust. In addition to the development of AI, approaches can also be found at the implementation level for educational institutions, e.g. the code of conduct for *Trusted Learning Analytics,* by the DIFP/Leibniz Institute for Educational Research and Educational Information, Goethe University and TU Darmstadt.

The code serves as a framework for an ethical and responsible handling of student learning data, in order to implement learning analytics applications and prospectively AI in their interest (Hansen et al., 2020; Renz & Etsiwah, 2020a).

3 Challenges and Solutions

A central challenge is the lack of evidence on the actual impact of AI in EdTech. The current scientific discourse on the use of AI in EdTech is primarily based on theoretical considerations, with argumentation and results being characterised by conjunctive formulations (Renz et al., 2020a, b; Ifenthaler & Yau, 2019; Sclater et al., 2016). The currently existing gap between basic and applied research can prospectively be reduced by a closer cooperation between EdTech companies and scientific institutions, such as the six major competence centres for AI in Germany [2] A field survey of 120 EdTech companies based in Berlin showed that a large part of the EdTech companies have already established such collaborations and will expand them further in the future.[3]

The complexity of the question about evidence, however, cannot be resolved exclusively by promoting application-oriented research. According to Hartong (2019), EdTech is per se value occupied and subject to a perception and reality modelling (see also Gorur et al., 2019). Under this condition, individualisation as a central narrative of AI-supported EdTech remains a promise, which cannot be asserted in reality as such. Rather, one can assume a standardised individualisation at the level of the institutions. The individual maturity level is only weakly pronounced and often sets in at the level of the position/role of the learner. One reason for this is the challenge of systematically capturing and translating competencies into data-based EdTech. Often, it is still unclear which data about learners and their behaviour actually need to be collected (Vladova et al., 2020a, b).

One possibility to create awareness for challenges and potentials among the users of the AI-supported EdTech and at the same time to strengthen their own

[2] BIFOLD, MCML, ML2R, DFKI, TUE.AI Center and ScaDs.Ai Dresden/Leipzig.

[3] The currently still unpublished online survey of the research group Data-Based Business Model Innovation at the Weizenbaum Institute was conducted in the first and second quarter of 2021 among 120 Berlin EdTech companies.

data sovereignty are initiatives such as the AI Campus, a learning platform for AI, or the Data Awareness Canvas, a workshop tool that aims to raise awareness for potentials and risks of data-based learning solutions. A good overview of freely available online course offerings is also provided by the platform Class Central with currently over 850 offerings on the topic of AI.

4 Outlook on AI in EdTech in 2030

EdTech applications will become more data-driven.
As a result of the COVID-19 pandemic, the EdTech market experienced a new dynamic, which also led to an increase in the available volume of learning data. This in turn created new potentials for the development of data-driven EdTech (Peters & Bovenschulte, 2021; Renz et al., 2020a, b). According to their own statements, currently 12% of Berlin-based EdTech companies use AI. Within the next five years, the share is expected to rise to around 45%. Potentials are seen especially for the areas of management and analysis of educational data, language and text understanding, feedback systems, and the improvement of learning applications. Despite these promising possibilities, the results of the Berlin EdTech surveyidentify the lack of data in sufficient quantity and quality for the development of corresponding EdTech as the biggest challenge. The existing information and distribution asymmetry of data from users, however, is not a regional phenomenon. Market observations indicate that in the future, large data-driven platforms such as Amazon, Facebook, Netflix, Google, or LinkedIn could dominate the development of adaptive technology in the education market as well (Hilbig et al., 2019).[4] Especially for (young) European EdTech companies, the expansion of the infrastructure of educational data is essential to be able to compete in the international education market in the long term (Vladova et al., 2020a, b).

Data sovereignty and competence are central requirements for a sustainable AI development in the EdTech sector.
The promotion of data and AI competencies and thus of data sovereignty will be a central requirement for users of AI-supported EdTech, educational institu-

[4] Examples: AWS EdTech Start-Up Accelerator by Amazon, Dreambox Learning by Netflix, LinkedIn Learning etc.

tions is, and EdTech companies. The way in which data-based applications can be handled in a more self-determined manner on the user side can already be found in the area of telematics tariffs for car insurance. Reflected on EdTech applications, users can use an app to decide for themselves when they want to share data with the provider, and thus also to what degree an individualisation by releasing their own learning behaviour data is desired (see also Steinackers (2020) considerations on the introduction of individual data accounts for learners). Transparency and user participation will be a major factor in the design of more and more data-based EdTech applications in the future. For educational institutions, data sovereignty and competence not only represent a teaching subject within the framework of an educational mandate. In addition to curricular implications in the sense of a data culture, new cross-sectional references are also to be sought at the structural level, e.g. in the form of interdisciplinary cooperation (see e.g. Utrecht Data School; Renz & Etsiwah, 2020a; Schäfer & van Schie, 2019).

EdTech applications will enable a new learner-centredness.
With AI, a shift from instruction-centred to project-oriented and individualised learning can be designed in the coming years (Anthony & Clark, 2011; Jonassen et al., 2003). Administrative and standardised processes can already be replaced by corresponding AI today, e.g. Gradescope, an AI system that is supposed to support professors in an objective grading of students. The media-driven myth that AI-supported EdTech replaces teachersis also to be resolved for future decades. Researchers like Kurzweil (2013) consider a technological singularity to be quite realistic. However, it should be kept in mind that the obligation to education demands more than just automatable and objectifiable processes. Thus, education and the use of EdTech will always also be a connection between humans. Self-learning systems have only limited judgement, no intuition, and no tact (Dorn & Rojas, 2018)—qualities that are essential especially in sensitive areas such as teaching and learning.

5 Summary and Practical Recommendations

Individualised learning processes will certainly be possible in small sequences by AI in a few years, but should not gain too much importance in view of a lifelong learning process. A pragmatic view of the actual potentials and limits of AI in EdTech helps to reduce resistance and reservations and to recognise AI as a complement to traditional forms of education. The social polarisation of AI in EdTech can be levelled by creating real application relevance and transparency and thus also demystify the often unjustified self-purpose of the technologies attributed to them. Regardless of this, it is especially important to put the sub-discourse on the datafication of educational processes into more balanced relations to questions of educational theories and human experiences. Questions about the halflife of data, data accounts, and thus data autonomy of users will promote participatory forms of EdTech and open up new possibilities for action. AI will lead to a shift of competencies at all levels of the education market in the EdTech sector, but will not replace a connection between humans. A sustainable design and integration of AI in EdTech can contribute to creating new spaces of possibility and experience of learning for these connections.

References

Anthony, A. B., & Clark, L. M. (2011). Examining dilemmas of practice associated with the integration of technology into mathematics classrooms serving urban students. *Urban Education,46*(6), 1300–1331. https://doi.org/10.1177%2F0042085911416015.

Arnold, N., Frieß, H.-J., Roose, J., & Werkmann, C. (2020). *Künstliche Intelligenz in Einstellungen und Nutzung bei unterschiedlichen Milieus in Deutschland*. Konrad-Adenauer-Stiftung e. V.

De Witt, C., Rampelt, F., & Pinkwart, N. (Eds.). (2020). *Künstliche Intelligenz in der Hochschulbildung*. https://doi.org/10.5281/zenodo.4063722.

Dorn, T., & Rojas, R. (2018). Die können was! Aber können Roboter auch fühlen? *Aus Politik und Zeitgeschichte,68*(6–8), 4–7.

Gorur, R., Sellar S., & Steiner-Khamsi, G. (2019). Big data and even bigger consequences. In R. Gorur, S. Sellar, & G. Steiner-Khamsi (Eds.), *Comparative methodology in the era of big data and global networks* (World Yearbook of Education 2019, pp. 1–9). Routledge.

Hansen, J., Rensing, C., Herrmann, O., & Drachsler, H. (2020). Verhaltenskodex für Trusted Learning Analytics. Version 1.0. http://www.dipfdocs.de/volltexte/2020/18903/pdf/Hansen_Rensing_Herrmann_Drachsler_2020_Verhaltenskodex_Trusted_Learning_Analytics_A.pdf. Accessed: 30. Juni 2021.

Hartong, S. (2019). Learning Analytics und Big Data in der Bildung: Zur notwendigen Entwicklung eines datenpolitischen Alternativprogramms. https://www.gew.de/index.php?eID=dumpFile&t=f&f=91791&token=702ec8d5f9770206a4aa8a1079750ec9021b90bf&sdownload=&n=Learning-analytics-2019-web-IVZ.pdf. Accessed: 30. Juni 2021.

Hilbig, R., Renz, A., & Schildhauer, T. (2019). *Data analytics: The future of innovative teaching and learning*. ISPIM Conference Italy.

Holmes, W., Bialik, M., & Fadel, C. (2019). *Artificial intelligence in education: Promises and implications for teaching and learning*. Independently published.

Ifenthaler, D., & Yau, J.Y.-K. (2019). Higher education stakeholders' views on learning analytics policy recommendations for supporting study success. *International Journal of Learning Analytics and Artificial Intelligence for Education (iJAI),1*(1), 28–42. https://doi.org/10.3991/ijai.v1i1.10978.

Jonassen, D. H., Howland, J. L., Moore, J., & Marra, R. M. (2003). *Learning to solve problems with technology: A constructivist perspective*. Merrill.

Kurzweil, R. (2013). *Menschheit 2.0. Die Singularität naht*. Lola Books.

Luckin, R., Holmes, W., Griffiths, M., & Forcier, L. B. (2016). *Intelligence unleashed. An argument for AI in education*. Pearson.

Peters, R., & Bovenschulte, M. (2021). Learning Analytics—Potenzial von KI-Systemen für Lehrende und Lernenden. Themenkurzprofil Nr. 42 Büro für Technikfolgen-Abschätzung beim Deutschen Bundestag.

Popenici, S. A. D., & Kerr, S. (2017). Exploring the impact of artificial intelligence on teaching and learning in higher education. *Research and Practice in Technology Enhanced Learning,12*, 1. https://doi.org/10.1186/s41039-017-0062-8.

Renz, A., & Etsiwah, B. (2020). Datenkultur und KI-Kompetenzen an Hochschulen. In C. de Witt, F. Rampelt, & N. Pinkwart (Eds.), *Künstliche Intelligenz in der Hochschulbildung*, (pp. 35–38). https://doi.org/10.5281/zenodo.4063722.

Renz, A., & Hilbig, R. (2020). Prerequisites for artificial intelligence in further education: Identification of drivers, barriers, and business models of educational technology companies. *International Journal of Educational technology in Higher Education,17*, 14. https://doi.org/10.1186/s41239-020-00193-3.

Renz, A., & Vladova, G. (2021). Reinvigorating the discourse on human-centered artificial intelligence in educational technology. *Technology Innovation Management Review,11*(5), 5–16. https://doi.org/10.22215/timreview/1438.

Robinson, R., Molenda, M., & Rezabek, L. (2007). Facilitating learning. In A. Januszewski, M. Molenda (Hrsg.), *Educational technology—A definition with commentary*. Routledge. https://doi.org/10.4324/9780203054000.

Renz, A., Krishnaraja, S., & Schildhauer, T. (2020a). *A new dynamic for EdTech in the age of pandemics*. Conference Paper presented at ISPIM innovation conference virtual.

Renz, A., Krishnaraja, S., & Gronau, E. (2020b). Demystification of artificial intelligence in education—How much AI is really in the educational technology? *International Journal of Learning Analytics and Artificial Intelligence for Education,2*(1), 14–31. https://doi.org/10.3991/ijai.v2i1.12675.

Sclater, N., Peasgood, A., & Mullan, J. (2016). Learning analytics in higher education: A review of UK and international practice. JISC, Bristol.

Schäfer, M. T., & van Schie, G. (2019). Utrecht Data School. Eine unternehmerische Plattform für Unterricht und angewandte Forschung im Bereich der Datafizierung. *Zeitschrift für Politikwissenschaft,29*(3), 123–140. https://doi.org/10.1007/s41358-018-0166-8.

Steinacker, K. (2020). Individuelle Datenkonten—Oder was mein Staubsauger mit digitaler Souveränität zu tun hat. Arbeitspapier der Gesellschaft für Informatik: Schlüsselaspekte digitaler Souveränität. https://gi.de/themen/beitrag/individuelle-datenkonten-oder-was-mein-staubsauger-mit-digitaler-souveraenitaet-zu-tun-hat. Accessed: 30. Juni 2021.

Vladova, G., Heuts, A., & Teichmann, M. (2020a). Dem Mitarbeiter zu Diensten. Weiterbildung und Qualifizierung als Personennahe Dienstleistung. *HMD Praxis der Wirtschaftsinformatik,57*(4), 710–721. https://doi.org/10.1365/s40702-020-00626-7.

Vladova, G., Renz, A., Gronau, N., Schildhauer, T., Köster, A., & Brandenburger, B. (2020b). New Digital Education Action Plan. Positionspapier des Weizenbaum-Instituts für die vernetzte Gesellschaft, Berlin. https://www.weizenbaum-institut.de/media/News/Statement/Weizenbaum-Institut_Positionspapier_Digital_Education.pdf. Accessed: 30. Juni 2021.

Zawacki-Richter, O., Marín, V. I., Bond, M., & Gouverneur, F. (2019). A systematic review of research on artificial intelligence applications in higher education: where are the educators? *International Journal of Education Technology in Higher Education,16*, 1. https://doi.org/10.1186/s41239-019-0171-0.

AI in Continuing Education of the Future

Clemens Jäger and Stefan Tewes

1 Introduction

The successful transformation of organisations and their long-term viability is determined by the coupling of external trends with internal adaptability.[1] A crucial factor here is the (re-)development of organisational continuing education. The limits of conventional concepts are determined in particular by three continuing education gaps in organisations:[2] The 1) *motivation gap* is based on the different development interests of the individual (career) and the company (knowledge retention). The 2) *competence gap* shows the gap between the competencies taught and those required. The 3) *transfer gap* indicates a deficit between theoretical knowledge and practical applicability.

In addition to the continuing education gaps, the speed of digital transformation leads to a new understanding of continuing education. The massive increase in continuing education needs[3] means that organisations have to link individual

[1] (Tewes & Tewes, 2020).

[2] (Moldoveanu & Narayandas, 2019).

[3] (WEF, 2018).

C. Jäger (✉)
FOM Hochschule für Oekonomie & Management in Essen, Essen, Germany
e-mail: clemens.jaeger@fom.de

S. Tewes
FOM Hochschule für Oekonomie & Management in Essen, Essen, Germany
e-mail: stefan.tewes@fom.de

I. Knappertsbusch and K. Gondlach (eds.), *Work and AI 2030*,
https://doi.org/10.1007/978-3-658-40232-7_40

and organisational learning.[4] Consequently, organisations have to solve real problems and issues of daily work in the course of continuing education. Moreover, new methods of continuing education have to be established and implemented in order to keep pace with the constant requirement of adaptation and the temporal demand.

An important approach for the future of continuing education is the integration of artificial intelligence (AI). Although there is no uniform definition, AI is basically "the property of an IT system to show human-like, intelligent behaviour."[5] The basic idea is to create an approximation of important functions of the human brain by machines—learning, judging and problem-solving.[6] The Dartmouth Conference marks the starting point of AI. Here, a machine behaved for the first time as if it had intelligence.[7] The second milestone is the development of machine learning (ML). Here, programmes solve complex problems by algorithms and automation. As a further development, deep learning (DL) can be given as a third milestone. DL enables autonomous learning based on artificial neurons in multi-layered networks.[8] Prognostically, AI will offer great value creation potentials especially in the areas of work productivity, personalisation and quality (Fig. 1).[9]

2 Status Quo

In the context of continuing education, AI can be attributed to various applications.[10] The following will look at the analytical support *(Learning Analytics),* the optimisation of individual and adaptive learning *(Personalised Learning)* as well as the automation of tasks *(Task Automation)* and the generation of tailor-made content *(Smart Content).*[11]

[4] (Argyris & Schön, 1996; Senge, 2006).

[5] (Bitkom, 2017, p. 14).

[6] https://news.sap.com/germany/2018/03/was-ist-kuenstliche-intelligenz/

[7] (McCarthy et al., 1955).

[8] (Sarvepalli, 2015).

[9] (PWC, 2017).

[10] (Bughin et al., 2017).

[11] https://elearningindustry.com/5-main-roles-artificial-intelligence-in-education

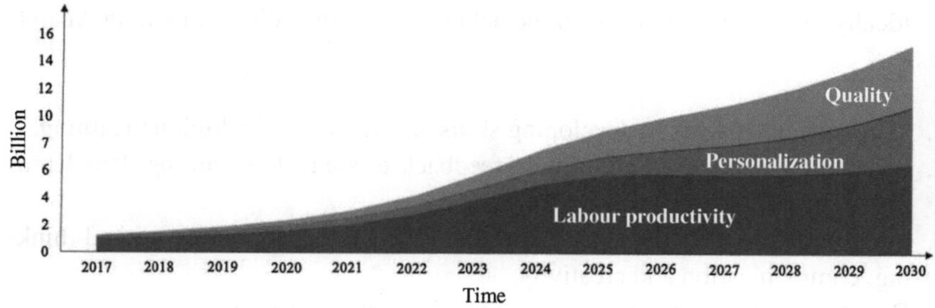

Fig. 1 Value creation through artificial intelligence. (Based on PWC, 2017)

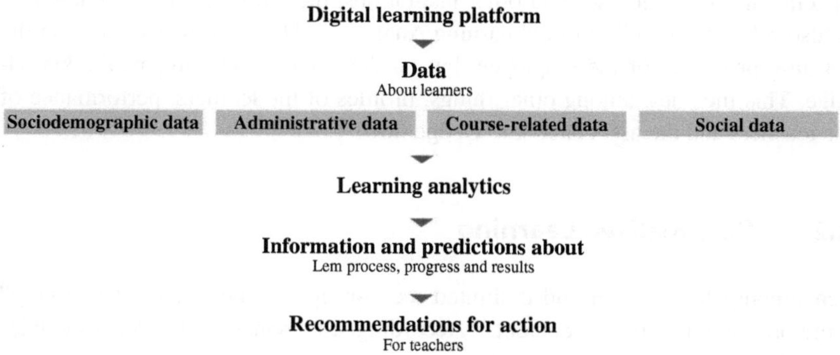

Fig. 2 Ecosystem—Learning Analytics. (Based on Gašević et al., 2016)

2.1 Learning Analytics

Learning Analytics describes "the measurement, collection, analysis and reporting of data about learners and their contexts for the purpose of understanding and optimising learning and the environments in which it takes place"[12]. The following figure illustrates the corresponding ecosystem (Fig. 2):

[12] "Learning analytics is the measurement, collection, analysis and reporting of data about learners and their contexts, for purposes of understanding and optimising learning and the environments in which it occurs", https://tekri.athabascau.ca/analytics/.

Ideally, the following goals can be achieved with the help of Learning Analytics:

1. "Supporting students in developing skills and strategies for lifelong learning,
2. Providing personalised and timely feedback to students regarding their learning,
3. Supporting the development of key skills such as collaboration, critical thinking, communication and creativity,
4. Developing student awareness by supporting self-reflection,
5. Supporting the quality of learning and teaching by providing empirical evidence for the success of pedagogical innovations"[13]

Specifically for businesses and other institutions, the term "Academic Analytics" is also relevant in addition to "Learning Analytics". Here, the focus is not on the learning processes of the employee, but on the teachers who impart the knowledge. This includes, among other things, profiles of the learners, performance of the teachers and the associated knowledge flow.[14]

2.2 Personalised Learning

Schaumburg has defined and delimited the concept of "Personalised Learning" more precisely in terms of content.[15] Accordingly, Personalised Learning includes the following design areas:

1. "Didactic decision fields (personalisation of learning objectives, learning content, learning methods, learning paths, learning time, learning location),
2. Characteristics of the students (personalisation with regard to prior knowledge, performance, interest, motivation, learning style) and
3. Locus of Control (control and selection of the personalised learning material externally, i.e. by the teacher or by a computer programme, or internally, i.e. by the employees)."[16]

---

[13] https://www.solaresearch.org/about/what-is-learning-analytics/

[14] (Long & Siemens, 2011, p. 34).

[15] (Schaumburg, 2021a, p. 137 f.).

[16] (Schaumburg, 2021b, p. 382–399).

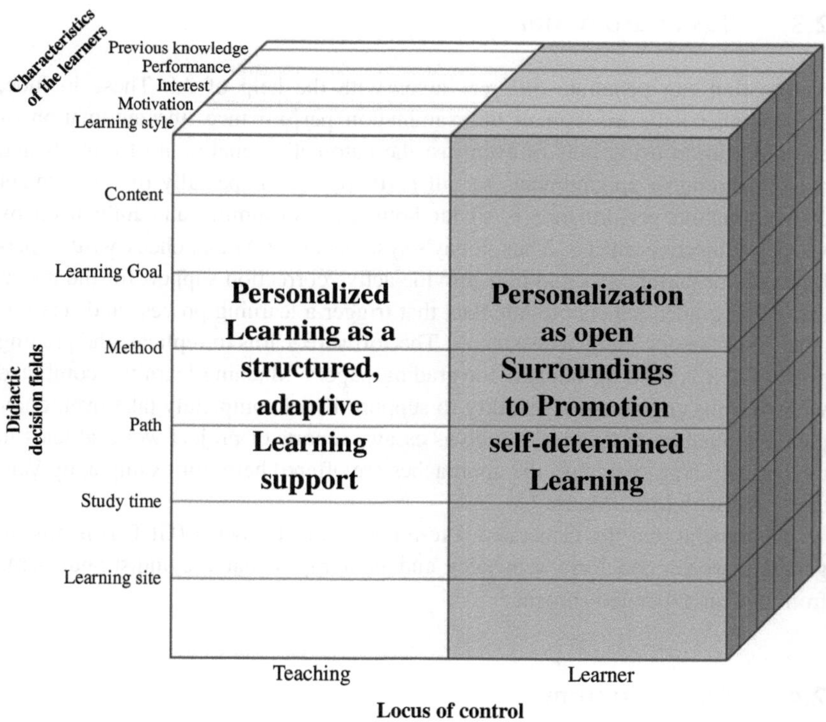

Fig. 3 Dimensions and basic concepts of personalized learning. (Based on Schaumburg, 2021a)

In the course of personalisation, two perspectives meet.[17] On the one hand, the data-based diagnosis, structuring and improvement of processes in the context of learning and on the other hand, the independence and self-responsibility of the learner within the given challenge (Fig. 3).[18]

[17] (Schaumburg, 2021a, p. 137 f.).

[18] (Kallick & Zmuda, 2017; Schratz & Westfall-Greiter, 2010, p. 18–31; Schaumburg, 2021b, p. 382–399).

2.3 Task Automation

Automation can penetrate different areas with the help of AI. These include, among others, the assessment of examination performance, the automation of communication using bots or even just the automatic generation of e-mails and the scheduling of appointments for all participants.[19] Especially the assessment of examination performance is so far both time-consuming and influenced by various subjective aspects. Thus, today's systems using AI can check written performance of participants and thus provide active correction support for the teachers.[20] The graded papers provide data that trigger a learning process and serve as a basis for the upcoming assessments. The software learns to replicate the grading process that is used by humans for grading papers. Machine learning, combined with AI, thus creates the possibility to support, if not completely take over, complex examination performance such as essays, reports or project work, at least in the correction process. Specific approaches are offered here, for example, by Vantage Labs with IntelliMetric®.[21]

It is crucial that the Generative Pre-trained Transformer 3 (GPT-3) is able to generate words and form sentences and paragraphs that are indistinguishable from human-generated content.[22]

2.4 Smart Content

"Smart Content" includes digital, adaptive learning interfaces, digital textbooks, study guides and learning units that can be individually adapted and used by learners in terms of scope.[23] This can significantly increase the acceptance and success rate of users. Furthermore, with the help of AI, the information perception can be improved in terms of visualisation. In addition, AI enables the compilation, continuous updating and adaptation of the learning content to the learning curve of the users.

---

[19] (Adamopoulou & Moussiades, 2020, p. 15; Vittorini et al., 2020).

[20] (Vittorini et al., 2020).

[21] https://www.intellimetric.com/direct/

[22] (Brown et al., 2020, p. 5); https://medium.com/analytics-vidhya/what-is-gpt-3-and-why-it-is-revolutionizing-artificial-intelligence-44d8e17c7edf.

[23] cf. in the following https://elearningindustry.com/5-main-roles-artificial-intelligence-in-education.

"Smart Content" ideally is:

1. "**TARGETED** at exactly what customers want and need, when they need it,
2. **OPTIMISED,** so that the content is more visible and findable,
3. **AVAILABLE AT ANY TIME** and technically up to date,
4. **INTEGRATED,** cross-device and activated throughout the marketing stack for maximum impact,
5. **PROFITABLE** for marketers, because the content excites and engages readers on topics they intend, and prepares them for a conversion"[24]

3 Challenges and Solutions: Human Bias

"Bias is colloquially used to describe many things, from prejudices to distortions in data-driven decision situations to the promotion or neglect of certain social groups."[25]

With the help of AI, ideally the distortions in the evaluation of employee performance can be avoided. Through this possible objectification of performance, AI contributes to the reduction of discrimination and manipulation and thus also protects employers from possible civil claims by employees. The AI should meet three requirements: "Transparency", "Explainability" and "Proveability".[26]

Balkow and Eckardt have shown that this is by no means a matter of course and that there are significant risks for the integration of one or more biases in the development of corresponding systems from the conception phase to the construction phase, the data phase, the deployment phase and the evaluation phase.[27] Kugel has shown that the use of algorithms does not always lead to ethically justifiable decisions and that it requires a multidisciplinary interplay especially in the context of the conception phase. For example, speech recognition systems are more functional for men than for women and face recognition systems are more powerful for light-skinned than for dark-skinned people.[28] Companies should critically question this in their decision-making process when acquiring such

[24] https://www.brightedge.com/glossary/what-is-smart-content

[25] (Balkow and Eckardt, 2019, p. 1).

[26] (Baccala et al., 2018, p. 16).

[27] (Balkow and Eckardt, 2019, p. 4).

[28] (Kugel, 2021, p. 67).

systems. This includes the design of the legal consequences. Due to the current developments in many countries in Europe, it can be expected that especially the aspect of equal treatment in connection with AI will gain importance in the coming years. In this context, AI can make a positive contribution to society if it is developed and applied correctly.

4 AI in Further Education 2030

In general, different settings are relevant for the further education of the future.[29] With the help of AI, it will be possible to develop even more needs-based and tailor-made offers. On the technical level, the following AI-supported approaches with exemplary fields of application are conceivable:[30]

- *Big data analytics and neural networks:* Determination of personal performance indicators, competence development and content selection
- *Deep learning:* Self-learning software for personalised recommendations
- *Immersive and virtual learning:* 3D and mixed reality for virtual learning arrangements
- *Machine learning:* Taxonomies and pattern recognition for learning advice
- *Multimodal learning offers:* Optimal formats (e.g. wearables) and design for the learner
- *Paralingual communication:* Mimetic-gestural components for human interaction and learning processes
- *Language and dialogue:* ASR/NLP as an extended human-machine interface

For a detailed examination of AI in further education, the benefit must be considered in the context of the further education gaps.[31] AI does not act on one of the further education gaps in particular, but supports the closing of all gaps.

In the context of the *motivation gap*, the combination of learning analytics and point of control is particularly noteworthy. By controlling the personalised learning material, both the company and the individual can integrate their interests into

[29] (OECD, 2018).

[30] (Schmid, 2018).

[31] (Moldoveanu & Narayandas, 2019).

the learning process.[32] The provision of personalised and timely feedback is also supportive. In addition, task automation enables a more "objective" result. Within the *competence gap*, the gap between imparted and necessary competencies is closed by AI. Especially personalised learning (individual learning objectives, content, methods, etc.) is conducive here. Possibilities for self-reflection, individual development paths and strategies for skill development offer further potentials for closing the gap. Finally, the *transfer gap* is considered. This aims at the deficit between theoretical knowledge and practical application. Smart content enables an individual adaptation of learning units, which can also adjust prior knowledge and practical relevance through learning analytics.

5 Summary and Practical Recommendations

AI will provide a supportive contribution to further education in the future. In addition to the analytical level and optimised, individual learning, task automation and tailor-made content also offer great development potentials. Likewise, the further education of the future will be characterised by the unification of individual and organisational learning. The solution of existing problems will come into focus. More than ever, further education must provide a direct benefit—for the human being, but also for the organisation. Only in this way can the increased costs be offset by a return.

AI offers considerable opportunities for new business models in further education in these contexts, which are often still in development today. Technologisation as an end in itself is the wrong approach in selecting the right technology partners. It will be interesting to see how organisations and providers jointly implement effective AI concepts.

Finally, the success of AI in further education will be determined by the bias debate. The potential for abuse must not be underestimated. For example, GPT-3 and future standards can independently create and revise text. Further education and goal achievement will thus no longer be verifiable, because in the "worst case" the AI of the learners will be checked by the AI of the teachers. The achievement of educational goals will certainly not be achieved in this way. Also, the apparent objectification of performance is at least questionable. Consequently,

[32] cf. Fig. 3: Dimensions and basic assumptions of personalized learning.

despite supportive technology in further education of the future, the human being must be in focus—with all his or her individuality, values and personality[33].

It is critical to consider that already today the Generative Pre-trained

References

Adamopoulou, E., & Moussiades, L. (2020). Chatbots: History, technology, and applications. *Machine Learning with Applications, 2*, 1–18.

Argyris, C., & Schön, D. A. (1996). *Organizational learning II. Theory, method and practice*. Addison-Wesley.

Baccala, M., Curran, C., Garrett, D. et al. (2018). 2018 AI predictions—8 insights to shape. https://www.pwc.es/es/publicaciones/tecnologia/assets/ai-predictions-2018.pdf. Accessed: 18. Juni 2021.

Balkow, C., & Eckardt, I. (2019). DENKIMPULS DIGITALE ETHIK: Bias in algorithmischen Systemen—Erläuterungen, Beispiele und Thesen. *Initiative D21, Unterarbeitsgruppe Algorithmen-Monitoring,* pp. 1.

Bitkom. (2017). Künstliche Intelligenz—Wirtschaftliche Bedeutung, gesellschaftliche Herausforderungen, menschliche Verantwortung. https://www.dfki.de/fileadmin/user_upload/import/9744_171012-KI-Gipfelpapieronline. Accessed: 10. Mai 2021.

Bughin, J., Hazan, E., Ramaswamy, S., Chui, M., Allas, T., Dahlström, P., Henke, N., & Trench, M. (2017). Artificial Intelligence—The next digital frontier? Discussion paper. McKinsey Global Institute. Business strategy. *PwC—pwc.com/us/AI2018,* pp. 1–25.

Brown, T. B., Mann, B., & Ryder, N., et al. (2020). *Language models are few-shot learners*. Cornell University.

Gašević, D., Dawson, S., Rogers, T., & Gasevic, D. (2016). Learning analytics should not promote one size fits all: The effects of instructional conditions in predicting academic success. *The Internet and Higher Education, 28*, 68–84.

Kallick, B., & Zmuda, A. (2017). *Students at the Center: Personalized Learning with Habits of Mind*. ASCD.

Kugel, J. (2021). Der Code der Diskriminierung. *Manager Magazin,* Nr. 6/ 2001, pp. 67.

Long, P., & Siemens, G. (2011). Penetrating the fog: Analytics in learning and education. *Educause Review, 46*(5), 31–40.

McCarthy, J., Minsky, M. L., Rochester, N., & Shannon, C. (1955). A proposal for the dartmouth summer research project on Artificial Intelligence. http://jmc.stanford.edu/articles/dartmouth/dartmouth.pdf. Accessed: 15. Mai 2021.

Moldoveanu, M., & Narayandas, D. (2019). Die Zukunft des Lernens. *Harvard Business Manager,* pp. 30–40.

[33] "Learning analytics is the measurement, collection, analysis and reporting of data about learners and their contexts, for purposes of understanding and optimising learning and the environments in which it occurs", https://tekri.athabascau.ca/analytics/.

Organisation for Economic Co-operation and Development. (2018). Future of education and skills 2030: Conceptual learning framework. 8th Informal Working Group (IWG) meeting.

PWC. (2017). Sizing the prize. What's the real value of AI for your business and how can you capitalise? https://www.pwc.com/gx/en/issues/data-and-analytics/publications/artificial-intelligence-study.html. Accessed: 31. Juli .2021.

Sarvepalli, S. K. (2015). *Deep learning in neural networks: The science behind an Artificial brain*. Liverpool Hope University.

Schaumburg, H. (2021a). Personalisiertes Lernen mit digitalen Medien als Herausforderung für die Schulentwicklung—Ein systematischer Forschungsüberblick. *MedienPädagogik Zeitschrift für Theorie und Praxis der Medienbildung, 41*, 137–138.

Schaumburg, H. (2021b). Personalisierung mit digitalen Medien. In H. G. Rolff & A. Brägger (Eds.), *Kompetenzorientiert Unterrichten und Lernen mit digitalen Medien* (pp. 382–399). Beltz.

Schmid, U. (2018). KI@Education: Wann kommt der LehrBot? https://schule21.blog/2018/11/06/kieducation-wann-kommt-der-lehrbot. Accessed: 11. Mai 2021.

Schratz, M., & Westfall-Greiter, T. (2010). Das Dilemma der Individualisierungsdidaktik. Plädoyer für personalisiertes Lernen in der Schule. *Journal für Schulentwicklung, 1*(2010), 18–31.

Senge, P. M. (2006). *Die fünfte Disziplin—Kunst und Praxis der lernenden Organisation*. Klett-Cotta.

Tewes, C., & Tewes, S. (2020). Megatrends und digitaler Einfluss. In S. Tewes, B. Niestroj, & C. Tewes (Eds.), *Geschäftsmodelle in die Zukunft denken—Erfolgsfaktoren für Branchen, Unternehmen und Veränderer* (pp. 21–31). Springer Gabler.

Vittorini, P., Menini, S., & Tonelli, S. (2020). An AI-based system for formative and summative assessment in data science courses. *International Journal of Artificial Intelligence in Education, 31*(2), 159.

World Economic Forum (WEF). (2018). The future of jobs report. http://www3.weforum.org/docs/WEF_Future_of_Jobs_2018.pdf. Accessed: 10. Mai 2021.

Links

https://www.brightedge.com/glossary/what-is-smart-content. Accessed: 11. Mai 2021.

https://elearningindustry.com/5-main-roles-artificial-intelligence-in-education. Accessed: 11. Mai 2021.

https://www.intellimetric.com/direct/. Accessed: 14. Mai 2021.

https://medium.com/analytics-vidhya/what-is-gpt-3-and-why-it-is-revolutionizing-artificial-intelligence-44d8e17c7edf. Accessed: 13. Mai 2021.

https://news.sap.com/germany/2018/03/was-ist-kuenstliche-intelligenz/. Accessed: 11. Mai 2021.

https://tekri.athabascau.ca/analytics/. Accessed: 20. Mai 2021.

https://www.netzwerk-digitale-bildung.de/learning-analytics-die-digitale-zukunft-des-lernens/. Accessed: 17. Mai 2021.

https://www.solaresearch.org/about/what-is-learning-analytics. Accessed: 20. Mai 2021.

AI in Vocational Rehabilitation—Intelligent Assistance for People with Disabilities

Berit Blanc, Rolf Feichtenbeiner, Susan Beudt and Niels Pinkwart

1 Introduction

In the working world, the use of artificial intelligence (AI) is often associated with the competitiveness of companies (cf. Federal Ministry of Economics and Energy, 2020) and related to automation processes and efficiency gains. Also in the AI strategies of the European Union and Germany, the innovation power of the economic locations and international competition for AI are important pillars (Federal Government, 2020; European Commission, 2020). The importance of AI systems for assistance at work, especially for the inclusion of persons with disabilities, and possible effects of increasing digitisation and automation processes on this heterogeneous target group, are much less in focus.

B. Blanc (✉) · R. Feichtenbeiner · S. Beudt · N. Pinkwart
Educational Technology Lab, Deutsches Forschungszentrum für Künstliche Intelligenz (DFKI), Berlin, Germany
e-mail: berit.blanc@dfki.de

R. Feichtenbeiner
e-mail: rolf.feichtenbeiner@dfki.de

S. Beudt
e-mail: susan.beudt@dfki.de

N. Pinkwart
e-mail: Niels.Pinkwart@dfki.de

I. Knappertsbusch and K. Gondlach (eds.), *Work and AI 2030*,
https://doi.org/10.1007/978-3-658-40232-7_41

However, AI systems can offer great potential for implementing international conventions (e.g., the UN Convention on the Rights of Persons with Disabilities) and compliance with legal regulations on inclusion and vocational participation. Vocational inclusion is also relevant for companies, as they face the challenge of finding suitable personnel. Some of the reasons for this are the demographic change and related skills shortages (cf. Federal Employment Agency, 2019), the increase of mental impairments of employees (cf. AOK Federal Association, 2020) as well as changing tasks and occupations (among others due to digitisation and automation; Nedelkoska & Quintini, 2018). Vocational inclusion is to be promoted by the system of vocational rehabilitation. If a person has or acquires a disability and cannot participate in the general labour market, they can take advantage of retraining or training offers at rehabilitation facilities, request personal aids from rehabilitation providers, and prepare for re-entry into a company in the general labour market.

2 Status Quo—AI in Vocational Rehabilitation

A systematic overview of existing and already used AI-supported assistive systems[1] is neither available for vocational rehabilitation nor for training and work in companies. In some cases, such systems have been developed specifically for persons with disabilities to compensate for disadvantages (e.g., by intelligent prostheses, smart glasses for environmental recognition and speech-to-text translation). In addition, there are AI systems not explicitly developed for persons with disabilities, but offer potential for vocational rehabilitation by supporting them at work or learning (e.g., intelligent workbenches that indicate errors in the work process).

Based on such individual examples, opportunities and risks of AI-supported assistive systems are discussed, but without being able to draw on systematic studies of the effects of AI technologies on the vocational inclusion of persons with disabilities (cf. Dziobek et al., 2017; Marzin, 2018).

---

[1] Assistance systems are distinguished in the project KI.ASSIST into *AI-based* (whose functions are based on AI methods and are enabled by them) and *AI-supported* (digital assistive technologies that are supported by AI components in the sense of a functional extension). In this article, the term *AI-supported* is used, which also includes purely AI-based systems.

3 Challenges and Solutions

The systematic development, introduction, and dissemination of AI-supported assistive systems in vocational rehabilitation face various challenges. The current technological state and the potentials and risks of AI-supported assistive systems for persons with disabilities are not sufficiently investigated. Moreover, the prerequisites, success factors, and approaches for their introduction and use in the practice of rehabilitation facilities and companies are not known.

This is where the project KI.ASSIST[2] comes in. In the joint project, the necessary technical, organisational, legal and political aspects are comprehensively, systematically and evidence-based illuminated. Also, models and approaches for the introduction and long-term use of AI-supported assistive systems in vocational rehabilitation are developed. In a systematic monitoring, on the one hand, the state of technological developments in the field of AI-supported assistive systems, which are used or developed for the work and training context, is recorded. On the other hand, inclusion and AI experts evaluate the researched technologies with regard to their potential suitability for persons with disabilities at work or training.

4 Results: Existing Technologies and Individual, Organisational and Technical Success Factors for Their Introduction, and Long-Term Use

In a comprehensive web and literature search based on a search, description, and evaluation system (on the methodology see Beudt & Pinkwart, in press), 131 **AI-supported assistive technologies** were identified. Almost half of the technologies are products available on the market (45%). More than every second technology (55%) comes from research and development projects (ongoing or completed), whose prototypes only represent applications that are practically usable to a minimal extent. Three out of four technologies address types of disabilities. Gaps are evident here, especially with regard to addressing persons with learning disabilities and mental disorders (see Fig. 1).

[2] KI.ASSIST—Assistance services and artificial intelligence for people with severe disabilities in vocational rehabilitation, funded by the BMAS from funds of the compensation fund.

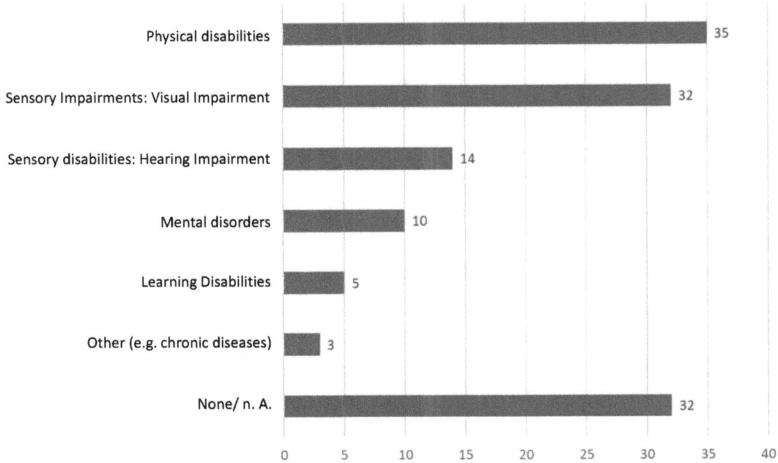

Fig. 1 Number of different types of disabilities addressed by 131 researched AI-supported assistive technologies

The qualitative content analysis for further structuring and evaluation of all recorded technologies resulted in a systematisation of the applications according to the main tasks performed i.e,. the type of support, in seven task groups (see Fig. 2).

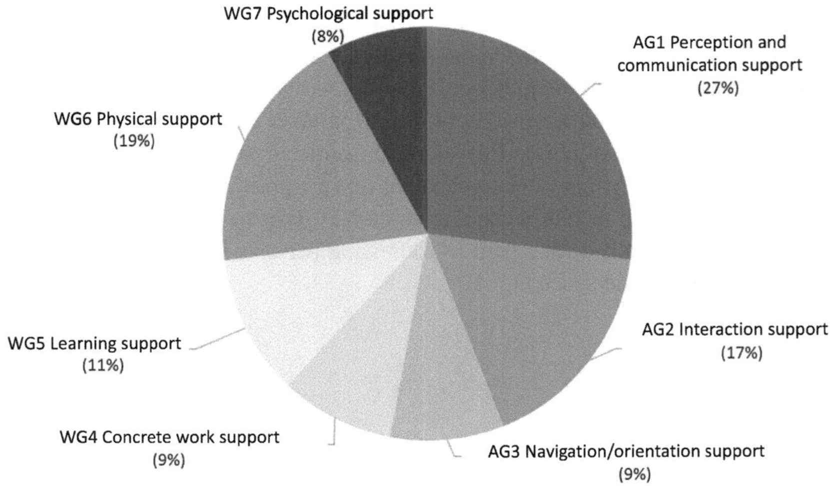

Fig. 2 AI-supported assistive technologies (n = 131) researched in the KI.ASSIST project and their distribution across identified task groups

The assistance is mainly based on recorded sensory environmental and movement data and their algorithm-based analysis. For example, a camera records movements during an assembly process, compares them with the correct sequence, and automatically gives feedback to workers about incorrect executions. Many researched technologies are often adaptive and work with image, speech, or text recognition and output.[3]

Technical, individual and organizational success factors could be identified for the introduction and long-term use of AI-based assistive systems in vocational rehabilitation.[4] On the **technical level**, the availability of suitable, functional, and legally compliant AI-supported assistive systems is a central prerequisite. The best-rated technologies are characterised by the fact that they can be used and developed in the long term, can be used to a great extent independently of others , and are easy to learn or acquire. Characteristics of the five lowest-ranked technologies are that they have higher usage requirements, lower technical adaptability to individual needs, and cannot be used independently enough. From an **individual user perspective**, AI-supported assistive systems should cover personal needs and thus generate clear added value (e.g., maintaining, facilitating, and improving work or training). In addition, the usability of the systems for persons with disabilities was considered important. Besides classic aspects such as comfort and weight, it is especially crucial for persons with disabilities that the systems are as discreet and inconspicuous as possible. Moreover, the suitability of AI-supported assistive systems to users' media usage preferences, competencies and work environment is important. On the **organisational level**, a good cost-benefit ratio is a critical success factor for the successfully introducing and using AI-supported assistive systems. Thus financing such systems, which have little immediate benefit for companies but support persons with disabilities in their work, plays a vital role in government funding systems. Another important prerequisite is that

[3] The exploratory analysis of AI components used in KI.ASSIST, based on the periodic table of AI of the platform Learning Systems (https://periodensystem-ki.de/), showed that the most frequently used concrete AI components are image recognition (26x) as well as language understanding (15x) and language generation (16x).

[4] Sample: 83 inclusion and AI experts, methods: guideline-based interviews, online survey; evaluation of the potential suitability of the technologies for people with disabilities based on profiles of 14 typical technology examples (two per task group); influencing factors and recommendations for action for the introduction of AI-supported assistive technologies into vocational rehabilitation.

institutions and companies and their internal stakeholders can practically get to know and test the systems to assess their potential for people and organisation on a sound basis.

5 Outlook on AI in Vocational Rehabilitation in 2030

In light of the presented findings, the question arises of how AI-supported assistive technologies for persons with disabilities at the workplace or vocational training could look like in 2030. Based on the results from research and validation, AI experts described the following scenario as desirable in a foresight workshop:

> The use of AI technologies is the norm in all areas in 2030. All involved actors were and are routinely involved in developing and introducing the systems. Data protection regulations are clear and understandable, and easy to implement. For this, for approval procedures, and for financing, advisory and supportive bodies and resources are available within the framework of a governmental AI and inclusion strategy.
>
> The technologies are integrated systems that promote learning, are constantly optimised, and arehighly interoperable. They are easily and independently usable by persons with disabilities and automatically adapt to the users' dynamic individual needs and characteristics. In this way, they enable all employees with disabilities to participate in working life and generally facilitate work.
>
> For the development of respective AI-supported technologies optimal conditions have also been created: For the data availability and diversity required for AI applications, an ecosystem for data exchange with large and diverse data sets is available (keyword "data lake"). The consideration of persons with disabilities in developing and using AI is thereby ensured.

How AI-supported assistive systems in vocational rehabilitation will develop in the future depends largely on the design of technology development processes in research and among AI providers, on their introduction and use in rehabilitation facilities and companies, and on the creation of conducive framework conditions. Based on the previous descriptions, two approaches of transformation movements can be distinguished, which ideally interlock.

The first approach focuses on the identified gaps between the needs of persons with disabilities and the current state of available AI-supported assistive systems.

If cross-facility or cross-company needs of persons with disabilities form the starting point for development processes and the technologies are also relevant for persons without disabilities, not only the technology acceptance can be supported. Also, for AI developers, companies, and rehabilitation facilities, a high market potential and a good cost-benefit ratio can be achieved. In addition, participatory development processes involving persons with disabilities, AI developers or AI researchers, companies, rehabilitation facilities, and rehabilitation providers are recommended. For the development of the systems, data availability and accessibility and high data diversity must be ensured (especially regarding persons with disabilities).

The second approach is based on existing AI-supported assistive systems for persons with disabilities . It aims at their widespread diffusion in vocational rehabilitation and the labour market enabling improved vocational participation opportunities for as many people as possible. Following Hastall et al. (2017), five steps for a successful technology adaptation can be identified and used as a starting point for the future diffusion of such systems.

1. Persons with disabilities, rehabilitation facilities, and companies must know and understand AI-supported assistive systems.
2. A good information base is also a prerequisite for persons with disabilities to form an opinion about these AI systems and for facilities and companies to assess their potential.
3. If persons with disabilities want to use AI-supported assistive systems at work, their financing by rehabilitation providers and their usability or use in organisational contexts such as facilities or companies is central.
4. Before and during the systems' first use(s) developing the required competencies of persons with disabilities (as users) and the accompanying professionals is essential.
5. For the continuous use of the systems, further development, as well as technical and organisational support and advice, play an important role.

6 Summary and Practical Recommendations

A systematic investigation of the status quo and the impacts of AI technologies for the vocational participation of persons with disabilities was carried out for the first time within the framework of the KI.ASSIST project. In this process, gaps in the supply of AI-supported assistive technologies and success factors for their introduction and long-term use, were identified. For AI in vocational rehabilitation in 2030, the following recommendations can be derived: With regard to the development of AI-supported assistive systems for the needs of persons with disabilities, two approaches should be pursued. Approaches of inclusive design can lead to discrete technologies for persons with and without disabilities given comparable needs, and to marketable products that can be used in companies. Approaches of accessible design can lead to targeted and tailored support services for the specific needs of persons with disabilities. For this purpose, establishing a data ecosystem is recommended, which provides developers access to diverse training data sets. For the diffusion of existing AI-supported assistive systems into the practice of vocational rehabilitation and the world of work, persons with disabilities, stakeholders of vocational rehabilitation, and companies should be informed and educated about the assistive potentials of AI-supported technologies. An effective starting point can be the networking and exchange between AI developers and AI researchers on the one hand and stakeholders of vocational rehabilitation and companies on the other hand. For rehabilitation facilities and companies, pilot projects are advisable to build competencies for AI-supported assistive systems in their own organisation and important partner networks with external competence carriers. In this context, temporal leeway for internal stakeholders should be enabled so that they can participate by bringing in their needs and opinions. For successful introduction and long-term use, AI-supported assistive systems should be integrated into the financing structures of vocational rehabilitation as personal aids or technical work aids.

References

AOK-Bundesverband. (2020). *Zahlen und Fakten 2020*. https://aok-bv.de/imperia/md/aokbv/aok/zahlen/zuf_2020_web.pdf. Accessed: 29. Juli 2021.

Beudt, S., & Pinkwart, N. (im Druck). KI-Anwendungen in der beruflichen Rehabilitation—Inklusionspotenziale und Herausforderungen. In S. Seufert, J. Guggemos, D. Ifenthaler, J. Seifried, & H. Ertl (Eds.), *Künstliche Intelligenz in der beruflichen Bildung: Zukunft der Arbeit und Bildung mit intelligenten Maschinen. Zeitschrift für Berufs- und Wirtschaftspädagogik, Beiheft 31* (pp. 293–317). Steiner.

Bundesagentur für Arbeit. (2019). *Fachkräfteengpassanalyse* (Berichte: Blickpunkt Arbeitsmarkt). Nürnberg. https://statistik.arbeitsagentur.de/Statistikdaten/Detail/201906/arbeitsmarktberichte/fk-engpassanalyse/fk-engpassanalyse-d-0-201906-pdf. Accessed: 29. Juli 2021.

Bundesministerium für Wirtschaft und Energie. (2020). *Einsatz von Künstlicher Intelligenz in der Deutschen Wirtschaft. Stand der KI-Nutzung im Jahr 2019*. Berlin. https://www.bmwi.de/Redaktion/DE/Publikationen/Wirtschaft/einsatz-von-ki-deutsche-wirtschaft.pdf. Accessed: 29. Juli 2021.

Die Bundesregierung. (2020). *Strategie Künstliche Intelligenz der Bundesregierung. Fortschreibung 2020.* https://www.ki-strategie-deutschland.de/files/downloads/201201_Fortschreibung_KI-Strategie.pdf. Accessed: 29. Juli 2021.

Dziobek, I., Lucke, U., & Manzeschke, A. (2017). Emotions-sensitive Trainingssysteme für Menschen mit Autismus. In M. Eibl & M. Gaedke (Eds.), *INFORMATIK 2017. Gesellschaft für Informatik.* https://doi.org/10.18420/in2017_30.

Europäische Kommission. (2020). *White paper on artificial intelligence—A European approach to excellence and trust.* https://ec.europa.eu/info/sites/default/files/commission-white-paper-artificial-intelligence-feb2020_en.pdf. Accessed: 29. Juli 2021.

Hastall, M. R., Dockweiler, C., & Mühlhaus, J. (2017). Achieving end user acceptance: Building blocks for an evidence-based user-centered framework for health technology development and assessment. In M. Antona & C. Stephanidis (Eds.), *Universal access in human-computer interaction. Human and technological environments* (pp. 13–25). Springer.

Marzin, C. (2018). *Plug and pray? A disability perspective on artificial intelligence, automated decision-making and emerging technologies*. European Disability Forum. https://www.edf-feph.org/content/uploads/2020/12/edf-emerging-tech-report-accessible.pdf. Zugegriffen: 29. Juli 2021.

Nedelkoska, J., & Quintini, G. (2018). *Automation, skills use and training* (OECD Social, Employment and Migration Working Papers Nr. 202). https://doi.org/10.1787/2e2f4eea-en.

MIX
Papier aus verantwortungsvollen Quellen
Paper from responsible sources
FSC® C105338

If you have any concerns about our products,
you can contact us on
ProductSafety@springernature.com

In case Publisher is established outside the EU,
the EU authorized representative is:
Springer Nature Customer Service Center GmbH
Europaplatz 3, 69115 Heidelberg, Germany

Printed by Libri Plureos GmbH
in Hamburg, Germany